The To-day and To-morrow Reader

The To-day and To-morrow book series (1923–1931) was a unique publishing phenomenon – over 100 short, often brilliant, books choosing a particular subject, outlining its present state, and then speculating about its future. This Reader brings together some of the best work in the series, including eleven complete volumes and substantial extracts from ten more.

To-day and To-morrow is one of the key documents of modernity. It contains some of the best writing of the twentieth century, and some of the most visionary predictions. The contributors were creative writers, scientists, inventors, philosophers, lawyers, doctors, and teachers. Included here are Bertrand Russell, Vera Brittain, Sarvepalli Radhakrishnan, Robert Graves, and the scientists J. B. S. Haldane, J. D. Bernal, and Sir James Jeans. The topics range from emerging technologies such as the talkies, television, robotics, and drones, to speculations about future technologies like test-tube babies, artificial wombs, cyborgs, genetic modification, hormone replacement therapy, space exploration, the internet, and the possibility of hive minds. The books consider how societies will respond to such developments; how the transformations will impact on lives, relationships, beliefs, politics.

To-day and To-morrow brings new perspectives to the literature and culture of modernism and modernity for general readers, students, and scholars. It sheds new light on twentieth-century literature, culture, and society. It offers resources for teachers and students of creative writing – and everyone – facing the challenge of thinking about our future.

Max Saunders is Interdisciplinary Professor of Modern Literature and Culture at the University of Birmingham. He directed the Arts and Humanities Research Institute at King's College London. He works on modernism, future thinking, life writing, and visual culture. His books include *Imagined Futures: Writing, Science, and Modernity in the To-Day and To-Morrow Book Series, 1923–31* (2019).

The To-day and To-morrow Reader

Future Speculations from the 1920s and Early 1930s

Edited by Max Saunders

Routledge
Taylor & Francis Group
LONDON AND NEW YORK

First published 2025
by Routledge
4 Park Square, Milton Park, Abingdon, Oxon OX14 4RN

and by Routledge
605 Third Avenue, New York, NY 10158

Routledge is an imprint of the Taylor & Francis Group, an informa business

British Library Cataloguing-in-Publication Data
A catalogue record for this book is available from the British Library

ISBN: 978-1-032-62548-5 (hbk)
ISBN: 978-1-032-62551-5 (pbk)
ISBN: 978-1-032-62552-2 (ebk)

DOI: 10.4324/9781032625522

Typeset in Times New Roman
by codeMantra

Contents

Acknowledgments viii

Introduction 1

PART I
Science and Technology 17

Complete Texts 18

Daedalus; or, Science and the Future 19
J. B. S. HALDANE (1923)

Icarus; or, the Future of Science 46
BERTRAND RUSSELL (1924)

Wireless Possibilities 60
A. M. LOW (1924)

Archimedes; or, the Future of Physics 79
L. L. WHYTE (1927)

The World, the Flesh and the Devil: An Enquiry into the Future of the Three Enemies of the Rational Soul 106
J. D. BERNAL (1929)

Excerpts 138

From: *Prometheus; or, Biology and the Advancement of Man* 139
H. S. JENNINGS (1925)

From: ***Automaton; or, the Future of the Mechanical Man*** 149
H. STAFFORD HATFIELD (1928)

From: ***Eos; or, the Wider Aspects of Cosmogony*** 173
J. H. JEANS (1929)

PART II
Society, Politics, and Gender 183

Complete Texts 184

Hypatia; or, Woman and Knowledge 185
DORA RUSSELL (1925)

What I Believe 215
BERTRAND RUSSELL (1925)

Halcyon; or, the Future of Monogamy 237
VERA BRITTAIN (1929)

Excerpts 276

From: ***Quo Vadimus? Some Glimpses of the Future*** 277
E. E. FOURNIER D'ALBE (1925)

From: ***Kalki; or, the Future of Civilization*** 282
SARVEPALLI RADHAKRISHNAN (1929)

From: ***Chiron; or, the Education of a Citizen of the World*** 296
M. CHANING PEARCE (1931)

From: ***Aphrodite; or, the Future of Sexual Relationships*** 307
RALPH DE POMERAI (1931)

PART III
Culture, Media, and the Arts 331

Complete Texts 332

Proteus; or, the Future of Intelligence 333
VERNON LEE (1925)

Timotheus; or, the Future of the Theatre 353
BONAMY DOBRÉE (1925)

The Future of Futurism 371
JOHN RODKER (1926)

Excerpts 391

From: ***Euterpe; or, the Future of Art*** 392
LIONEL R. McCOLVIN (1926)

From: ***Scheherazade; or, the Future of the English Novel*** 399
'JOHN CARRUTHERS' (PSEUD. OF J. Y. T. GREIG) (1927)

From: ***Lars Porsena; or, the Future of Swearing and Improper Language*** 409
ROBERT GRAVES (1927)

Index *417*

Acknowledgments

This Reader has benefitted from discussions with many friends and colleagues. I should particularly like to thank:

George Myerson, for inspiring and sustained discussions about the series (and everything else).

Lisa Gee, for her energy and enthusiasm in helping to start the Future Thinking Network at the University of Birmingham, and for getting launched our series of FUTURES books, re-running the To-Day and To-Morrow experiment for the twenty-first century.

Elise Schraner, for her help in annotating some of the texts.

Frances Parkes and Karen Raith from Routledge for their valuable advice.

Finn Fordham, for Joycean detective services.

The Editor is grateful to the following for permission to reprint. The text of *Halcyon; or the Future of Monogamy* is © The Literary Executors of the Vera Brittain Will Trust (1970). M. Chaning Pearce's *Chiron; or, the Education of a Citizen of the World* (1931) is extracted by permission of the author's family. I am grateful to Dr Jane Bernal for permission to reproduce J. D. Bernal's *The World, the Flesh and the Devil*; to David de Pomerai for permission to quote from Ralph de Pomerai's *Aphrodite*; to David Higham for permission to reproduce Bonamy Dobrée's *Timotheus: The Future of the Theatre*; to the executors of the estate of J. B. S. Haldane for permission to reproduce *Daedalus; or, Science and the Future*; to Tom Low for permission to reproduce A. M. Low's *Wireless Possibilities*; to Taylor and Francis for permission to quote from S. Radhakrishnan's *Kalki; or, the Future of Civilization*; and to the Estate of John Rodker for permission to reproduce *The Future of Futurism*. Extracts from Robert Graves's *Lars Porsena; or, The Future of Swearing* are reprinted with the permission of Carcanet Press. *Icarus; or, the Future of Science* and *What I Believe* are © The Bertrand Russell Peace Foundation Ltd. Dora Russell's *Hypatia; or, Woman and Knowledge* is reprinted with the kind permission of Harriet Ward.

Introduction

To-day and To-morrow was an inspired thought-experiment.[1] Just over a hundred writers, scientists, thinkers, and experts chose a subject, discussed its present state, then imagined its future. It was this speculative element that set the series apart from all the popularizing competition, like the Home University Library or the Routledge Introductions to Modern Knowledge. To-day and To-morrow wasn't just modern or up-to-date. It was ahead of its time.

It was one of several book series edited for the publisher Kegan Paul by the eccentric polymath C. K. Ogden; a man known now, if at all, for his collaboration with leading Cambridge literary critic I. A. Richards—they wrote a book together called *The Meaning of Meaning* (1923)—and for his later campaign for 'BASIC,' a radically simplified and regularized version of English designed for use as an international auxiliary language. But Odgen had an immense impact on the intellectual history of the English-speaking world through his textbook series, the International Library of Psychology, Philosophy and Scientific Method, which introduced readers to many of the major contemporary thinkers, especially continental ones. It began in 1922 with Wittgenstein's *Tractatus*, which Ogden helped translate, and included work by leading figures such as Jung, Adler, Piaget, Carnap, Malinowski, and Bertrand Russell. There were over 150 volumes in Ogden's lifetime and the series lived on for another 40 years after him.

To-day and To-morrow was in a lighter vein: often humorous, satirical, whimsical; but always intelligent and often visionary. It began provocatively, with a brilliant volume called *Daedalus, or; Science and the Future*, by the Cambridge biochemist who would become a leading population geneticist, J. B. S. Haldane. His preface says acerbically that the book 'will be criticized for its undue and unpleasant emphasis on certain topics'—in other words, sex and reproduction: 'This is necessary if people are to be induced to think about them,' he continues, 'and it is the whole business of a university teacher to induce people to think' (vii).

Daedalus was a short book. It had started as a debating society paper which Haldane had been developing since before the war. In 1923 he read it to the Cambridge

1 For the first full account of the series, see Saunders, *Imagined Futures* (Oxford: Oxford University Press, 2019).

DOI: 10.4324/9781032625522-1

society the 'Heretics,' of which Ogden was the President. Even written up for publication it was only 12,000 words. It could have been published as a sixpenny pamphlet (which is how Routledge, Kegan Paul's sister company, produced those 'Introductions to Modern Knowledge'). But Ogden knew he had something more substantial and durable. It was produced as a small-format hardback pocketbook, and sold for the high price of two shillings and sixpence. Ogden asked Bertrand Russell to respond to it, which he did in a volume called *Icarus; or, The Future of Science*.

Haldane's book was a breathtaking roller-coaster ride of prediction and speculation, from wind power to artificial foods, and from hormone replacement therapy to genetic modification. He had chosen Daedalus as his iconic scientist not so much for his invention of the art of flying, but as the 'the first modern man' (32–33), the first scientist concerned with 'biological problems': 'it is safe to say that posterity has never equalled his only recorded success in experimental genetics' (33) —by which he meant the invention which enabled King Minos' wife to mate with a bull to produce the Minotaur. But the concept which attracted most attention was what he called 'ectogenesis': the growing of babies outside the body, in artificial wombs. The idea was made famous by Haldane's close friend Aldous Huxley, who used it a decade later in *Brave New World* (1932), in which children are mass-produced in 'hatcheries' by a form of cloning (it was Haldane again who later coined the word 'clone'). It must have seemed like science fiction at the time. Nearly a hundred years later, it has been achieved for sheep, and is being developed to save extremely premature human babies. *Daedalus* was not just the inspiration for the idea of the series, but also for many of the other contributions. It is the volume most discussed by others. Often an idea Haldane throws out in a sentence gets developed by the later volumes.

Haldane wasn't naïve about the dangers of technology. Like many of To-day and To-morrow's writers, he had served in the war. He had been badly wounded in an explosion. But he remained committed to the idea of the progress of human knowledge as being humanity's best hope in future.

Russell, by contrast, was more skeptical; cynical even, arguing that science increased human power, and that history gave us no reason to trust that future leaders would use that power with anything other than malevolence. The series quickly excited reviewers, who often used the word 'brilliant' to describe it. But *Icarus* was marketed with what must be one of the best puff-quotes in publishing history: simply '"Utter pessimism."—*Observer*.'

The pair of books made a publishing sensation; both were reprinted several times in their first year. Ogden and the publisher had realized they could build a series of such books about the future. To-day and To-morrow would run from 1924 to 1931, with most of the books using a Classical or mythological title, and an explanatory subtitle. It was also published in the US by E. P. Dutton. Though the high standard was kept up till the last volumes, the series ended in the dark days of the Depression, the rise of totalitarianism, and the approach of another world war—all of which it anticipated.

Vera Brittain, in her contribution, *Halcyon; or, the Future of Monogamy* (1929), quotes Dora Russell's volume, called *Hypatia* (1925), and describes it as 'one of a famous series of twentieth-century political and sociological essays then called To-day and To-morrow but now popularly known as *The Fireworks Pamphlets*' (242). 'Famous' is only a slight exaggeration for comic effect. The series got excellent reviews—quoted in the series catalog which started to be added to the end of each volume. They sold well, many being reprinted. It's a nice joke to stretch that fame into later centuries. But the nickname she gives the series is telling. That's what many of the volumes must have felt like: a dazzling and entertaining display of pyrotechnics that aroused wonder and excitement.

These are, by any standard, some of the most interesting books of the twentieth century. It is extraordinary the series is not better-known a century on.[2] A number of its *writers* are: Brittain herself, Robert Graves, Winifred Holtby, Sylvia Pankhurst, Bertrand Russell, André Maurois, Vernon Lee, Hugh MacDiarmid, Lewis Grassic Gibbon. Its scientists were very high-profile too: Haldane himself, Sir James Jeans, J. D. Bernal. But To-day and To-morrow is not among the things they are known for. Ogden had a knack for spotting talent and commissioning writers before they were famous. The volumes are frequently their authors' first books, and sometimes very different from the work they were later associated with. Critics who know the harrowing war memoirs by Graves and Brittain, say, must struggle to get their light-hearted and witty volumes on the future of humor or of monogamy to fit with the rest of their work.

It may be that prophecy has a 'best before' date. Why read *Nineteen Eighty-Four* in 1985? Or watch *2001* in 2031? When we read past prophecies of now, or of our past, it is hard not to be mesmerized by what they got right and what they got wrong, and to attend to little else. These questions can be revealing (and I say more about them in the note below suggesting uses of this material in teaching). But to remain confined to them is to miss two of the most valuable aspects of To-day and To-morrow.

First, that one pleasure of reading past visions of the future is what they tell us about their own time of writing. Orwell chose the year 1984 because he was in 1948. We don't read H. G. Wells' *The War of the Worlds* because we think he's right about Martian life forms, but for its allegory of turn-of-the-century imperialism and militarism, and its gripping account of the terror of being on the losing side against an implacable enemy. That was T. S. Eliot's point when he reviewed some of the To-day and To-morrow books: 'We are able to peer into the future by means of that brilliant series [which] will constitute a precious document upon the present time.'[3] He was actually being less flattering than that sounds. Written in 1927, the year he declared his allegiance to Royalism, Toryism and Anglo-Catholicism, he disapproved of the socially progressive politics of most of the To-day and To-morrow network. He thought their confidence that the future could be better-engineered,

2 Some of the books, especially those by Haldane, Bertrand Russell, James Jeans and Robert Graves, were often reprinted.

3 'Charleston, Hey! Hey!' *Nation and Athenaeum*, Vol. 40 (29 January 1927), 595. In *The Complete Prose of T. S. Eliot: The Critical Edition: Literature, Politics, Belief, 1927–1929,* ed. Frances Dickey, Jennifer Formichelli, and Ronald Schuchard (Baltimore: Johns Hopkins University Press, 2015), 25–29.

and could be forecast using scientific principles, was a symptom of decadence; an escapist fantasy from the realities of original sin, and a hubristic denial of subservience to religious and political authority. But he's right in another way. The books are indispensable for the light they shed on the inter-war period. There's scarcely a topic of any significance on which they don't have something striking and illuminating to say, from Aesthetics to Zoology. Whatever you are working on in the period, you are sure to find some of the best commentary on it in a paragraph in the series, if not in a whole book. The aim here is to bring together some of the best writing in the series. Eleven important volumes are included in their entirety, with substantial extracts from ten more, so as to give readers a sense of which other volumes they might want read to explore a particular subject.

Once the series had grown to sixty volumes, Kegan Paul started including a 'Classified Index.' It helped readers navigate the titles, and presumably gave an indication of the broad subject areas Ogden felt needed coverage. The Index is reproduced below, showing how the volumes were grouped. This *Reader* organizes the material differently. The material is grouped into three broad (and inevitably overlapping) sections: Science and Technology; Society, Politics, and Gender; and Culture, Media and the Arts.

Some of the original headings have dated. 'Marriage and Morals' sounds coy now about sex, though *Hypatia* and Ralph de Pomerai's *Aphrodite* are anything but. 'Great Britain, the Empire, and America' sounds various post-colonial alarm bells to our ears. Also, the headings don't capture what was distinctive and significant about the series: that—as was clear from having Haldane and Russell as its avant-garde—its pioneering, ultra-modern speculations were informed by scientific thinking, whatever the subject in hand. It isn't just that some of the best writing is on contemporary topics such as technology, communication, media, psychology, evolution, and eugenics, which barely figure in the Classified Index (Figure 0.1). It's that even the writing on apparently distant topics—religion, morality, politics, art, and so on—is either written from, or engaging directly with, what Russell called 'The Scientific Outlook.'[4]

To-day and To-morrow reflects the immense prestige of the physical sciences at the time, as intellectuals were struggling to assimilate the new theories in Physics of Relativity and Quantum Mechanics, the discoveries of X-rays, radiation, and the photo-electric effect. That is why this selection, like the series, begins with ideas about science in general, and the philosophy of science; then moves on to look at the application of scientific ideas to various aspects of human life, from technological innovation, to ideas about evolution (and pseudo-scientific ideas about eugenics), to ideas about such topics as time, psychology, the nature of mind and consciousness, questions of sexuality, education, culture, and many more besides.

About a third of the volumes were devoted to scientific and technological subjects. But the *idea* of science, the new ideas and technologies it had introduced, and the changes these were bringing about, were much more pervasive across the

4 Russell, *The Scientific Outlook* (1931).

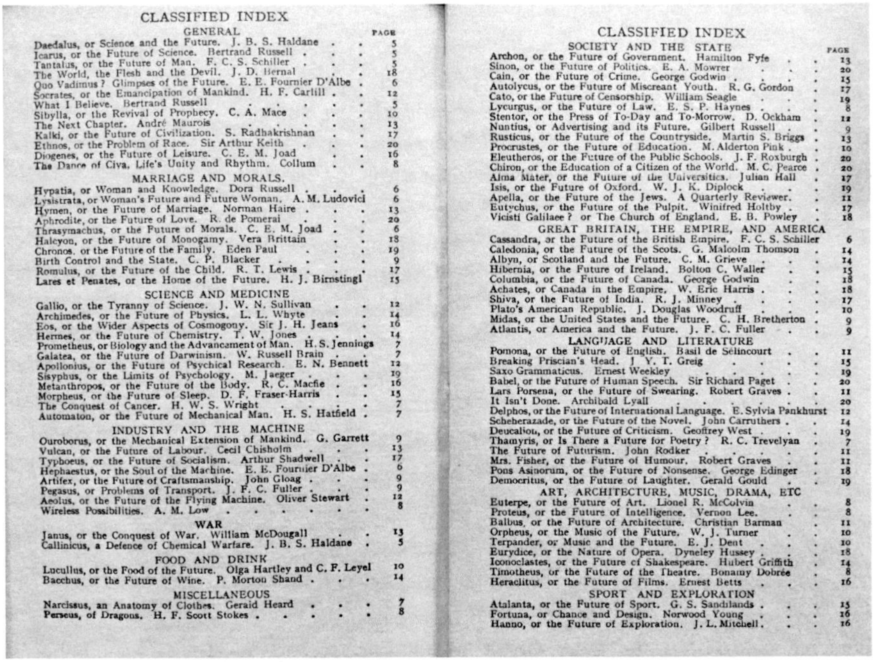

CLASSIFIED INDEX

GENERAL

	PAGE
Daedalus, or Science and the Future. J. B. S. Haldane	5
Icarus, or the Future of Science. Bertrand Russell	5
Tantalus, or the Future of Man. F. C. S. Schiller	5
The World, the Flesh and the Devil. J. D. Bernal	18
Quo Vadimus? Glimpses of the Future. E. E. Fournier D'Albe	6
Socrates, or the Emancipation of Mankind. H. F. Carlill	12
What I Believe. Bertrand Russell	5
Sibylla, or the Revival of Prophecy. C. A. Mace	10
The Next Chapter. André Maurois	13
Kalki, or the Future of Civilization. S. Radhakrishnan	17
Ethnos, or the Problem of Race. Sir Arthur Keith	20
Diogenes, or the Future of Leisure. C. E. M. Joad	16
The Dance of Civa, Life's Unity and Rhythm. Collum	8

MARRIAGE AND MORALS.

Hypatia, or Woman and Knowledge. Dora Russell	6
Lysistrata, or Woman's Future and Future Woman. A. M. Ludovici	6
Hymen, or the Future of Marriage. Norman Haire	13
Aphrodite, or the Future of Love. R. de Pomerai	20
Thrasymachus, or the Future of Morals. C. E. M. Joad	6
Halcyon, or the Future of Monogamy. Vera Brittain	18
Chronos, or the Future of the Family. Eden Paul	19
Birth Control and the State. C. P. Blacker	9
Romulus, or the Future of the Child. R. T. Lewis	17
Lares et Penates, or the Home of the Future. H. J. Birnstingl	15

SCIENCE AND MEDICINE

Gallio, or the Tyranny of Science. J. W. N. Sullivan	12
Archimedes, or the Future of Physics. L. L. Whyte	14
Eos, or the Wider Aspects of Cosmogony. Sir J. H. Jeans	16
Hermes, or the Future of Chemistry. T. W. Jones	14
Prometheus, or Biology and the Advancement of Man. H. S. Jennings	7
Galatea, or the Future of Darwinism. W. Russell Brain	7
Apollonius, or the Future of Psychical Research. E. N. Bennett	12
Sisyphus, or the Limits of Psychology. M. Jaeger	19
Metanthropos, or the Future of the Body. R. C. Macfie	16
Morpheus, or the Future of Sleep. D. F. Fraser-Harris	15
The Conquest of Cancer. H. W. S. Wright	7
Automaton, or the Future of Mechanical Man. H. S. Hatfield	7

INDUSTRY AND THE MACHINE

Ouroborus, or the Mechanical Extension of Mankind. G. Garrett	9
Vulcan, or the Future of Labour. Cecil Chisholm	13
Typhoeus, or the Future of Socialism. Arthur Shadwell	17
Hephaestus, or the Soul of the Machine. E. E. Fournier D'Albe	6
Artifex, or the Future of Craftsmanship. John Gloag	9
Pegasus, or Problems of Transport. J. F. C. Fuller	9
Aeolus, or the Future of the Flying Machine. Oliver Stewart	12
Wireless Possibilities. A. M. Low	8

WAR

Janus, or the Conquest of War. William McDougall	13
Callinicus, a Defence of Chemical Warfare. J. B. S. Haldane	5

FOOD AND DRINK

Lucullus, or the Food of the Future. Olga Hartley and C. F. Leyel	10
Bacchus, or the Future of Wine. P. Morton Shand	14

MISCELLANEOUS

Narcissus, an Anatomy of Clothes. Geraid Heard	7
Perseus, of Dragons. H. F. Scott Stokes	8

CLASSIFIED INDEX

SOCIETY AND THE STATE

	PAGE
Archon, or the Future of Government. Hamilton Fyfe	13
Sinon, or the Future of Politics. E. A. Mowrer	20
Cain, or the Future of Crime. George Godwin	15
Autolycus, or the Future of Miscreant Youth. R. G. Gordon	17
Cato, or the Future of Censorship. William Seagle	19
Lycurgus, or the Future of Law. E. S. P. Haynes	8
Stentor, or the Press of To-Day and To-Morrow. D. Ockham	12
Nuntius, or Advertising and its Future. Gilbert Russell	9
Rusticus, or the Future of the Countryside. Martin S. Briggs	13
Procrustes, or the Future of Education. M. Alderton Pink	10
Eleutheros, or the Future of the Public Schools. J. F. Roxburgh	20
Chiron, or the Education of a Citizen of the World. M. C. Pearce	20
Alma Mater, or the Future of the Universities. Julian Hall	17
Isis, or the Future of Oxford. W. J. K. Diplock	19
Apella, or the Future of the Jews. A Quarterly Reviewer.	11
Eutychus, or the Future of the Pulpit. Winifred Holtby	17
Vicisti Galilaee? or The Church of England. E. B. Powley	18

GREAT BRITAIN, THE EMPIRE, AND AMERICA

Cassandra, or the Future of the British Empire. F. C. S. Schiller	6
Caledonia, or the Future of the Scots. G. Malcolm Thomson	14
Albyn, or Scotland and the Future. C. M. Grieve	14
Hibernia, or the Future of Ireland. Bolton C. Waller	15
Columbia, or the Future of Canada. George Godwin	18
Achates, or Canada in the Empire. W. Eric Harris	18
Shiva, or the Future of India. R. J. Minney	17
Plato's American Republic. J. Douglas Woodruff	10
Midas, or the United States and the Future. C. H. Bretherton	9
Atlantis, or America and the Future. J. F. C. Fuller	9

LANGUAGE AND LITERATURE

Pomona, or the Future of English. Basil de Sélincourt	11
Breaking Priscian's Head. J. Y. T. Greig	15
Saxo Grammaticus. Ernest Weekley	19
Babel, or the Future of Human Speech. Sir Richard Paget	20
Lars Porsena, or the Future of Swearing. Robert Graves	11
It Isn't Done. Archibald Lyall	20
Delphos, or the Future of International Language. E. Sylvia Pankhurst	12
Scheherazade, or the Future of the Novel. John Carruthers	14
Deucalion, or the Future of Criticism. Geoffrey West	19
Thamyris, or Is There a Future for Poetry? R. C. Trevelyan	7
The Future of Futurism. John Rodker	11
Mrs. Fisher, or the Future of Humour. Robert Graves	11
Pons Asinorum, or the Future of Nonsense. George Edinger	18
Democritus, or the Future of Laughter. Gerald Gould	19

ART, ARCHITECTURE, MUSIC, DRAMA, ETC

Euterpe, or the Future of Art. Lionel R. McColvin	8
Proteus, or the Future of Intelligence. Vernon Lee.	8
Balbus, or the Future of Architecture. Christian Barman	11
Orpheus, or the Music of the Future. W. J. Turner	10
Terpander, or Music and the Future. E. J. Dent	10
Eurydice, or the Nature of Opera. Dyneley Hussey	18
Iconoclastes, or the Future of Shakespeare. Hubert Griffith	14
Timotheus, or the Future of the Theatre. Bonamy Dobrée	8
Heraclitus, or the Future of Films. Ernest Betts	16

SPORT AND EXPLORATION

Atalanta, or the Future of Sport. G. S. Sandilands	15
Fortuna, or Chance and Design. Norwood Young	16
Hanno, or the Future of Exploration. J. L. Mitchell	16

Figure 0.1 The Classified Index from *Aphrodite* (1931).

whole series. In 1959, C. P. Snow lamented that intellectual life had been split into 'The Two Cultures' of the humanities and science. To-day and To-morrow represents the previous generation's hope for a new synthesis. Its scientists included several of the best popularizers of science of their day. Most of the other volumes discuss the effects of science and technology on their subjects; many advocate a scientific outlook in approaching the human realm. Though not academic textbooks on sociology, psychology, education, or economics, they are inflected by the social sciences, and touch on all these disciplines (which, along with theories of science, were very much the terrain covered by Ogden's more academic projects).

Part Two thus concentrates on the topics which fall under the human or social sciences. This includes writing on society, religion and morality, gender, education, labor, and a number of everyday life topics. The whole concept of civilization had been severely challenged by the cataclysm of the war. The series frequently stands back from commenting on specific social changes to consider the nature of civilization itself, and how that was changing.

The third and final part is devoted to the products of the human mind: Culture, Media, and the Arts. These were vital concerns to many of the contributors, whether writers, scientists, or other professionals. As modernism was reinventing the styles and forms of particular arts, To-day and To-morrow was reimagining the function of the aesthetic. Though some express concerns about mass culture turning art into the consumption of a commodity, the writers generally use art as a way of thinking about society, about humanity, about science, and about the future. Here too, ideas of science and technology are crucial. And here too, they link back to social science topics—as, for example, broadcasting grows out of a new scientific and technological understanding of electromagnetism, and begins to change communication, the media, and politics. Lionel McColvin's excellent *Euterpe; or, the Future of Art* (1926), for example, considers the impact of mechanical reproduction on the reception of different art-forms—anticipating Walter Benjamin's canonical essay on the topic by a decade.

Not all the writers were so persuaded by the claims of science, and the attempts to extend the scientific outlook to the sphere of the human, whether in psychology, sociology, or anthropology. To take two examples not included in the present selection: in *Gallio; or, the Tyranny of Science* (1927), J. W. N. Sullivan mounts a powerful idealist case against precisely such claims for the validity of a scientific approach to topics outside the physical and biological sciences. But even he is addressing science, and urging scientists to think more like poets—something the series had itself exemplified from the start. Others take a balanced view. Geoffrey West, considering 'the future of criticism' in *Deucalion* (1930), contrasts the intuitive humanism of John Middleton Murry with the Cambridge criticism of I. A. Richards, inspired by psychology and social science methods, arguing that the two need to be combined. In general it is that belief in the possibilities opened up by science that distinguishes To-day and To-morrow; and in particular, a belief that science offers a new basis for thinking about the future. Science was in the *Zeitgeist*. The year the series launched, Sir Richard Gregory persuaded Ramsay Macdonald's first ever Labour government to hold a conference on science and labor. The government proved short-lived; the conference's outcome limited. But it was indicative of a phase of optimism both about the future of science, and its role in shaping the future.

The logic to this science-focused structure should not obscure the multiple interconnections between the topics. Indeed, part of the fascination of the series is the way a book on one topic will include discussion of a surprisingly different domain. A passage from *Aphrodite; or, the Future of Sexual Relationships* (1931) will suddenly introduce a discussion of quantum physics or of the relation between matter and consciousness.[5] A passage from *Archimedes; or the Future of Physics* (1927) asks how scientists might create artificial life.[6] Such moments illuminate how the plethora of exciting new ideas in the period was transforming many different fields. To-day and To-morrow was the 1920s version of interdisciplinarity, both within individual volumes and across the series as a whole. J. D. Bernal's powerfully visionary volume, *The World, the Flesh and the Devil* (1929), moves from space colonies to the re-engineering of the human body, to aesthetics. In the process, his imagining of interplanetary cyborgs connected by wireless to form a collective mind anticipates both the internet and AI, all before the advent of the

5 *Aphrodite* (London: Kegan Paul, 1931), 74, 87.
6 See 92.

computer. The disciplinary separation implied here of Sciences, Social Sciences, and Arts and Humanities is not intended to confine readers to their own specialist area. On the contrary, much of the interest of the series comes from the fact that it was popularizing its ideas to general readers. It tells us what major modern writers like Joyce, Huxley, Eliot, Evelyn Waugh or Wyndham Lewis were reading about science or society. We know, for example, that both Joyce and Winston Churchill read *Daedalus*: Joyce starting in the middle and reading backwards and forwards, taking extensive notes; Churchill writing an essay immediately afterwards called "Shall We All Commit Suicide?" Einstein too had a copy, as did Virginia Woolf's husband Leonard, who reviewed at least nine volumes. To-day and To-morrow tells us what politicians, intellectuals, journalists were reading about the arts; what scientists were reading about society; and so on. The main rationale for organizing the material around the idea of science is to convey how liberating it seemed, in the jazz age, that new ways of thinking deriving from and inspired by science seemed to be opening up in every field.

Such a large series inevitably has its unevennesses. The handful of duller volumes are not represented here. Some intelligent volumes such as Arthur Shadwell's *Typhoeus, or The Future of Socialism* (1929) are valuable for their detailed, extended arguments, and suffer in excerpts. In general, material liable to make contemporary readers wince has been excluded; especially the more eugenicist or elitist passages. Ogden wanted to set up debates between opposing views, as he had with *Daedalus* and *Icarus*. Sometimes the views are too toxic to reprint. Anthony Ludovici's neo-Nietzschean and proto-fascist eugenicist desire in *Lysistrata; or, Woman's Future and Future Woman* to suppress 'wrecks, cripples, and incurables' (115) will appal today's readers. His anti-Feminist, anti-democratic views are certainly not typical of the series, but he states them with a clarity and eloquent elaboration one rarely encounters. For those studying such debates in the inter-war period, it would be valuable to see the kinds of argument the progressive writers in the series were countering (like Dora Russell, who was explicitly rebutting him in *Hypatia*). Other troubling instances have been included where views which seem to us prejudicial are inseparable from otherwise original and important ideas—as when Haldane's trenchant account of biological inventions and their effect on religion gets tangled up with racist anthropological views of the period.

Needless to say, inclusion does not indicate editorial endorsement of all the writers' views; rather, that the material is thought revealing and potentially useful. Some of the books were clearly meant to be provocative—Haldane's *Callinicus; a Defence of Chemical Warfare* (1925), for example; or de Pomerai's advocacy in *Aphrodite* of multiple sexual partners. As Haldane put it ironically in *Callinicus*:

> The views which I have expressed do not coexist in the mind of any party leader or newspaper proprietor, and must therefore be those of a crank. But until some stronger argument can be waged against them than that they are unusual and unpleasant, there remains the possibility that they are true.
>
> (83–84)

Others are just eccentric, such as Gerald Heard's lengthy Carlylean *Narcissus; an Anatomy of Clothes* (1924), with its argument that evolution has now moved outside the body and into fashion, excerpts from which would have made even less sense than the whole. Some are knowingly nonsensical, notably *Pons Asinorum; or, the Future of Nonsense* (1929), which manages both to anatomize a particular kind of humor, and to satirize modish contemporary preoccupations as inane.

It is to be hoped that this selection will illustrate how, in Eliot's words, the series constitutes 'a precious document' upon its own time, less because it enhances what we already know of the period, and more because it shows the period in a different light. To-day and To-morrow includes some of the best writing of the 1920s and early 1930s on topics such as feminism, science, war, radio, time, cosmology, and others, because it reveals thinking about these and many other topics that has largely disappeared off the radar of literary and cultural history. In particular, it shows a new face of modernism: a movement which Eliot's work did so much to shape for Anglo-American readers, and which was duly established as classicist, backward looking, with a distaste for modern life and conditions. The New Modernist Studies of the twenty-first century has been much more inclusive than that in terms of time, space, gender, and genre. But To-day and To-morrow puts the idea of the future back into modernism in a way that alters the whole field.

One thing this means is that the series challenges the predominant narrative about World War I and its aftermath: the narrative of disillusion, despair, and trauma. The selection brings out how pervasive the experience of the conflict was for these generations. Whatever subject they are writing about—be it science, technology, sex, politics, music, or humor—war's shadow is a felt presence. Nevertheless, the series provides the best evidence that the experience of the war did not foreclose hope, enthusiasm, optimism, progressive ideals, and utopian politics. On the contrary, in the case of many of these writers, it appears to have given them greater impetus; greater determination to work for improvements; to make sure such a cataclysm would not be repeated. So reading To-day and To-morrow can alter our view of the canonical war narratives of the period. After all, it is the argument of Brittain's *Testament of Youth* that the devastation of the war was what determined her subsequent pacifism; and Graves needed to say *Goodbye to All That* so he could go on to write about other subjects.

There is anxiety about the likelihood of future war, to be sure. Several of the volumes—such as Oliver Stewart's otherwise delightful *Aeolus; or, the Future of the Flying Machine* (1927)—give vivid accounts of air attacks on cities. London had experienced air raids in the World War I from Zeppelins and Gotha bombers. H. G. Wells, whose visions of the future often inspire To-day and To-morrow's writers, had already foreseen such tactics in *The War in the Air* (1908). But by the 20s and 30s these writers had a clearer sense of what the technologies could do and how they would be deployed and countered. Yet for all that, their prevailing mood was not to turn against science, technology, and the Enlightenment—it would take World War II to effect that—but to hope that such disasters might be averted if mankind could use its science and its intelligence to plan the future better.

The second reason why past speculations can be worth our attention even once history has caught up with them, is to do with the processes of speculation rather than the events they predicted. J. Leslie Mitchell, who would later become known as the novelist Lewis Grassic Gibbon, was one of the writers who produced his first book for To-day and To-morrow: *Hanno, or The Future of Exploration* (1928). He had been in the RAF since 1920, and imagined the possibilities which the new technology of the helicopter might open up. He was also surprisingly accurate in predicting that man would reach the Moon within half a century. After Apollo 11 landed there in 1969, the passage reads differently. It no longer makes us wonder whether such a future were really possible, and what the world might be like if it were. Instead, we start assessing its accuracy. Hmmm. Correct to within nine years. Presumably because his aeronautic training gave him a sharper sense of the technology that might be necessary. Yet still limited by the propulsion technologies available at the time. The jet had not yet been invented, so like Jules Verne and H. G. Wells, Mitchell assumes the capsule will be shot at the Moon like a bullet. His speculation that 'the main explosive force behind the projectile may be Signor Mussolini in pursuit of an Italian Empire' (84) is both amusing and disturbing. Mussolini's forceful personality may not have got Italians to the Moon, but it did contribute to the World War II. He's the wrong dictator though. Hitler hadn't yet come to power. But it was German rocket science which underpinned the American space program. And, once we remember Neil Armstrong and Buzz Aldrin planting and saluting the American flag on the lunar surface, perhaps that vision of space exploration as imperial mission isn't so wrong.

In other words, after the predicted event has occurred (or not), the interest shifts, from whether it will happen, and whether it will happen as predicted; to why the predictor thought it might happen then and thus. It becomes a question of what it was possible to see, and to imagine, at the time; and of what the blind spots were. And if we can understand that about futurists from the 1920s or 1930s, perhaps it can tell us something about what it is possible for us to see and to imagine now; about where our predictions are likely to succeed and where they might fail. If we can understand why the inter-war period elicited so much inspired writing about the future, it may help us understand why it's much rarer today, when we need it more than ever. In short, looking at past predictions can enhance our understanding of how we think about the future.

That is why To-day and To-morrow still matters. Many of its individual predictions were extraordinarily good: some because they were surprisingly accurate; others because, though what they predicted may not yet have come to pass, their visions can still inspire current thinking. Bonamy Dobrée's imagining in *Timotheus: The Future of the Theatre* of a professor who invents a machine which 'recorded the feelings' (362) must have appeared wild satirical fantasy to its first readers in 1925. Less so now, with the advent of emotion recognition algorithms, brain-computer interfaces, and neural implants. But regardless of its individual predictions, the series is illuminating about the nature of possibility; how we think about it; and how, in thinking about it with intelligence and imagination, we can expand the potential for human development. The best of the writers understood this perfectly. As Bertrand Russell put it in *Icarus*, as the series was still being formulated: 'I do not wish to be supposed to be making prophecies: I am only suggesting possibilities which it may be instructive to consider' (55). He didn't want to sound prophetic because that might imply the whole world-view of divine foreknowledge that his atheism rejected. 'God and immortality [. . .] find no support in science,' he wrote (217). But it is interesting he doesn't use the word 'prediction' either. True to his belief in the mathematical and logical bases of reality, his imagined futures are hypotheses, to be explored with scientific rigor. To-day and To-morrow, in short, is significant not so much for its predictions as its speculative experiments—its exploration of new ways of thinking about human lives and technological or cultural developments.

In our era of climate emergency, global pandemic, the rise of populism, and fears about the existential and economic threats posed by AI and machine learning, we are in danger of losing that imagination of other possibilities; the ability to think constructively about the future. J. D. Bernal argues: 'Perhaps after all it is hope that really determines whether an age is or is not creative. But the existence of hope in a society at any time itself, depends on many unexplored psychological, economic and political causes' (126). The inter-war period was certainly a highly creative age. The label 'the Jazz Age' conveys dance and parties and high spirits. But it also evokes improvisation, the exhilaration of creating ideas and inventions which hadn't been seen before. A century on, we are in danger of abandoning hope. True, there are signs of the imagining of radical alternatives, whether in social activist movements like Extinction Rebellion, or speculation about new technologies for carbon capture, gene editing, and sustainable power. But, as Marc Augé argues in his book *The Future*, the idea of the future has been co-opted by technological capitalism to such an extent that it hinders us from imagining anything beyond it.

> The real problems with democratic life today stem from the fact that technological innovations exploited by financial capitalism have replaced yesterday's myths in the definition of happiness for all, and are promoting an ideology of the present, an ideology of the future *now*, which in turn paralyses all thought about the future.[7]

In his famous essay 'The End of History?,' Francis Fukuyama argued that neoliberal capitalism having won the ideological battle of the Cold War, there was no need for further political and economic change. As Augé's comment suggests, that is really the end of the future as much as the end of history. But History, as T. S. Eliot said in 'Gerontion,' has many 'cunning passages.' After the financial crash of 2008, the economic rise of China, the fires of 2019, the existential crises of COVID-19 and the war in Ukraine, we can no longer assume business as usual. Rebecca Solnit has written about the way such crises can usher in radical change:

7 Augé, *The Future*, translated by John Howe (London: Verso, 2014), 3.

> The outcome of disasters is not foreordained. It's a conflict, one that takes place while things that were frozen, solid and locked up have become open and fluid—full of both the best and worst possibilities. We are both becalmed and in a state of profound change.[8]

It will take more than reading To-day and To-morrow to change the world. But the wonderful writing from this unparalleled series—how its provocative and imaginative arguments open up the debates in creative ways—may inspire us to get ready. The new scientific ideas its writers were responding to made them feel they were at the beginning of something—as indeed they were, poised on the threshold of the atomic age, the invention of computers and the discovery of the structure of DNA. It is another reason their moment speaks to ours, as we step over the threshold of gene editing, nuclear fusion reactors, the multiverse, and AI.

A Note on Selection, Further Reading, Uses in Teaching, and Presentation

Selection

In 2008 Routledge published a library edition reproducing 104 of the 107 titles published in the UK (a further three were published only in the US). The three volumes then omitted—the two by Bertrand Russell, and Dora Russell's *Hypatia*—are all included here in full. The *Reader* is intended to supplement the library edition, to contextualize it, and to provide annotations for some of its key volumes. The arrangement with Dutton for US publication broke down in the late 1920s. So a number of the volumes included here—*Automaton, Halcyon, The World, the Flesh and the Devil, Chiron*, and *Aphrodite*, did not have original US editions, and consequently received less attention than they deserve.

The main criterion for inclusion is excellence. All the works and extracts reprinted here are outstanding examples of the speculative imagination. Each gives a compelling vision of a possible future world. I have selected those books which seem the best; the most interesting; the most influential; the most engaging and entertaining. But there are many more strong and fascinating volumes, and cases could be made for different selections. (None of the volumes is without interest. It would even be good to be able to read some which were rejected, like Evelyn Waugh's *Noah; or the Future of Intoxication*; or which were advertised but never appeared, like Rebecca West's *The Future of Sex*.) One aim of the present collection is to encourage readers to explore other volumes.

8 Rebecca Solnit, '"The Impossible Has Already Happened": What Coronavirus Can Teach Us About Hope', *Guardian* (7 April 2020): https://www.theguardian.com/world/2020/apr/07/what-coronavirus-can-teach-us-about-hope-rebecca-solnit; accessed 3 May 2020.

Suggestions for Further To-day and To-morrow Reading

Betts, Ernest (1896–1975), *Heraclitus; or, the Future of Films,* 1928
Collum, *Dance of Çiva, the; or Life's Unity and Rhythm,* 1927
Fournier d'Albe, E. E., *Hephaestus; or the Soul of the Machine,* 1925
Garrett, Garet, *Ouroboros; or, the Mechanical Extension of Mankind,* 1926
Grieve, C. M. [i.e. Hugh MacDiarmid], *Albyn; or, Scotland and the Future,* 1927
Hartley, Olga and C. F. Leyel, *Lucullus; or, the Food of the Future,* 1926
Holtby, Winifred, *Eutychus; or, the Future of the Pulpit,* 1928
Joad, C. E. M., *Thrasymachus; the Future of Morals,* 1925
Joad, C. E. M., *Diogenes; or, the Future of Leisure,* 1928
Keyser, Cassius Jackson, *Thinking about Thinking,* 1926 (published only in the US by E. P. Dutton)
Lyall, Archibald, *It isn't Done; or, the Future of Taboo among the British Islanders,* 1930
McDougall, William, *Janus: the Conquest of War: A Psychological Inquiry,* 1927
Pankhurst, E. Sylvia, *Delphos: the Future of International Language,* 1927
Sandilands, G. S., *Atalanta; or, the Future of Sport,* 1928
Stewart, Oliver, *Aeolus; or, the Future of the Flying Machine,* 1927
Sullivan, J. W. N., *Gallio; or, the Tyranny of Science,* 1927
West, Geoffrey, *Deucalion; or, the Future of Criticism,* 1930
Woodruff, J. Douglas, *Plato's American Republic,* 1926
For a full list, a discussion of the series as a whole, and its book history and literary and critical reception, see my book *Imagined Futures.*

Teaching Possibilities

The books were written for thoughtful general readers, and remain of interest to such readers today. They offer a different kind of interest to scholars and students of the 1920s and 1930s, and one hope in making them available in this form is that teachers and students may find fresh resources here to draw upon. Annotations have been provided to aid all such readers.

Given the literariness of many of the volumes, they continually suggest ways they might be read productively alongside familiar landmarks in modern literature. Here are some thought-prompts by way of starting that process.

Daedalus and *Icarus* could be juxtaposed with novels concerned with science, technology, and sex, like *Lady Chatterley's Lover* or *Brave New World.*

The imagining of future culture in *Timotheus* could be contrasted with *The Time Machine*'s emphases on biology and brutality.

Quo Vadimus could be taught alongside E. M. Forster's story 'The Machine Stops' as works offering diametrically opposite attitudes to technological connectivity.

Hypatia and *Aphrodite* could be taught in conjunction with texts about sexuality, such as *Women in Love, Ulysses*, or *The Well of Loneliness.*

Scheherazade could be read with Woolf's essays on modern fiction—or alongside her novels—leading to a discussion of the formal choices novelists make.

Kalki could join a conversation between *The Waste Land* and *A Passage to India.*

Eos could be compared with Wyndham Lewis' *Enemy of the Stars* or *The Childermass.*

The Future of Futurism could offer new ways into the poetry of T. S. Eliot and Ezra Pound, or the prose of Gertrude Stein.

Lars Porsena could be read with *Ulysses* to discuss obscenity and censorship.

The books could equally lend themselves to comparative work with other literatures and other media. One could compare *Timotheus* to Strindberg's *A Dream Play*; or *Automaton* to Fritz Lang's *Metropolis* and Chaplin's *Modern Times*; or *The World, the Flesh and the Devil* to *2001: A Space Odyssey*. And so on. . . Or one could make more provocative leaps. *Proteus; or, the Future of Intelligence* and Oscar Wilde—or even Barry Humphries, who owned a copy. Or, *Proteus* and AI.

To-day and To-morrow is concerned with much more besides literature and the arts, though; and its reach across science, society, culture, politics, and everyday life means that the books have much to offer disciplines working across these areas too, including the Social Sciences, History, and Philosophy. Students of Future Studies—of course!—but also Media, Cultural Studies, Gender, Social History, Social Psychology, Cultural Geography, or Anthropology, for example, will find stimulating and surprising ideas in these books.

Here one could work via direct comparisons between today's conceptions of a discipline and their inter-war precursors; introducing Low on wireless or McColvin on recordings into a course on media; or Dora Russell, Vera Brittain, and Ralph de Pomerai into one on feminism or sexuality.

But the very nature of the series—as itself one vast interdisciplinary experiment in future thinking—might prompt us to think more creatively across disciplines. Contrasting the predictions which have been most closely fulfilled with those which have not, or not yet, opens up other forms of pedagogical potential. To give some examples:

Predictions fulfilled the most (so far)

Daedalus on *in vitro* fertilization, HRT, genetic modification (as opposed to *Prometheus*, which thinks it unfeasible).

The World, the Flesh, and the Devil on cyborgs, brain-electronics interfaces, the internet.

Automaton on telephone answering machines, automatic dictation, programmable central heating.

Wireless Possibilities on drone warfare.

Hypatia and *Halcyon* on sex equality, sex education, maternity leave.

Quo Vadimus?, *Timotheus*, and *Euterpe* on the entertainment industry, special effects, and how recording technologies affect our sense of life and art.

Electricity was already well-established; but these writers understood how an electric lifestyle would soon be the norm. In general, they get the everydayness of the future; how, when it comes, it will be as ordinary, as familiar, as the present is to us. As such, they represent a challenge to science fiction's virtual monopoly on future thinking, and its tendency to try to arouse awe about the future via the technological sublime.

Predictions (so far) unfulfilled

Daedalus (and others) on artificial foods.
The World, the Flesh, and the Devil on interplanetary travel and artificial bodies.
Wireless Possibilities on mobile phones in aeroplanes.
Kalki on religious and cultural tolerance.
Chiron on the idea of international federalist (rather than nationalist or imperial) education.

That leaves many areas still up for discussion. For instances: Vera Brittain's speculation that home entertainment will reduce adultery. The discussions in *Archimedes* and *The World, the Flesh, and the Devil* on artificial life. *Icarus* on leaders using technology to manipulate populations. *Daedalus* predicting a return to philosophical idealism. *Archimedes*, *Aphrodite* (and *Daedalus* again) on the inseparability of matter and consciousness. *Hypatia* and *Aphrodite* on sexual relationships as a form of knowledge and creativity.

We might ask how far such texts, especially in their debates between progressive ideas and reactionary resistances, predict the culture wars which have become so entrenched in our time. Or whether their anti-nationalism and anti-imperialism anticipates the globalization and decolonization which followed; or if their advocacy of a world state was effectively a new form of Western hegemony. Hegel notoriously said that Africa had no history. There are plenty of statements in the series against slavery and oppression in the Empire, and advocacy of independence for its colonies; but less of a sense that Africa or India would have their own futures.

Then there are the things they thought unfeasible but which have been achieved, such as atomic energy and gene editing; things they barely glimpsed, such as the computer, or the dangers of Eugenics; and things they didn't foresee at all, like GPS, or AI, or the way the main uses of electronic interconnectivity would not be to promote science and rationality but commerce and social media.

Thinking about the books in this way raises questions about *how* their writers could see the things they did; and how they could not see the things they didn't, or didn't quite. It raises questions about the role of such writing in putting such ideas into people's imaginations, and thereby creating the desires that drive future innovation.

These are questions of future thinking; and they could also be addressed more directly by focusing on way the texts are self-conscious about their futurological imaginations. This is especially noticeable in *Proteus* and *The World, the Flesh, and the Devil*, say, though it runs through most of the volumes here—and others which aren't here, like C. A. Mace's *Sibylla; or, the Revival of Prophecy* (1926). One future thinking mode which proves especially fruitful in the series is future history; the kind of futurology-in-reverse which occurs when Haldane, Graves, and Brittain imagine how the future might think of *them*.

One of the most important lessons of To-day and To-morrow is that effective future-thinking texts like these are a unique form of creative writing—one that is rarely taught, but should be if the next generation of future thinkers are to be up to the challenges they will undoubtedly face. The texts collected here could provide a starting point for students learning how to write the future. Whether it is done by considering what works like *Timotheus*, *The Future of Futurism*, *Scheherazade,* and *Euterpe* have to say on the future(s) of art and writing; or through juxtaposition with students' own assumptions about either the present or the future; these volumes are sure to prompt thoughts and provide inspirations.

Presentation

Uniformity across the series was minimal; confined to a classical title (though not in all cases); contents which would fit into 96 pages (though some were shorter, some longer; the font sometimes reduced to get the longer volumes to fit). Some had chapters and chapter titles, some did not. Some had sub-headings, and when they did, presented them in different ways. Some had footnotes or endnotes, some did not. Standardization has been limited to essential aspects of house style, so as to preserve as much as possible the individual ways the writers conceptualized their tasks. Citations from the volumes not included here are all from the Kegan Paul original editions, and normally omit the publisher's name for brevity.

Part I
Science and Technology

Complete Texts

Daedalus[1]; *or, Science and the Future*[2]

J. B. S. Haldane (1923)

Headnote

Daedalus was J. B. S. Haldane's first book. It was originally published in November 1923. Its title-page gave the date as 1924 (presumably to stop the book looking out of date after a few weeks), and calls it 'a paper read to the Heretics, Cambridge on February 4th, 1923.' Haldane (1892–1964) had recently arrived in Cambridge as a Reader in Biochemistry. The Heretics was the discussion group for advanced ideas presided over by C. K. Ogden. Ogden encouraged Haldane to publish it with Kegan Paul, the firm he acted for as an advisor and editor. It was an instant sensation, and was reprinted eight times over the following seven years. Its success prompted Ogden to commission Bertrand Russell to write a response—*Icarus* (the next volume in this collection)—and to launch a new series, To-day and To-morrow, including them.

Daedalus was thus in the series, but had preceded—and inspired—it. One reason for its impact was that it crystallized ideas Haldane had been mulling over for more than a decade. In a new Preface, appearing only in the fifth impression (1925), he said it had begun before the war, in 'a paper read to the now defunct Shaftesbury Society at Oxford in June, 1914. Subsequent versions were read to various other Societies in Oxford in 1919–1922.' The version given to the 'Heretics' was also called 'Daedalus.' But its first publication was in the New York *Century* magazine in, August 1923 under the title 'If You Were Alive in 2123 A. D.' That shorter version excludes the war memories, any mention of Daedalus, and also the coda describing the 'ghastly mission' of the 'scientific worker of the future.'

1 Daedalus, whose name means 'cunning worker,' is the Greek mythological inventor known for designing the Labyrinth and crafting wings so that he and his son Icarus could escape from Crete. Haldane has other ideas.

2 In the Preface to the fifth impression, Haldane wrote: 'Of the ideas expressed in the book very few are my own. A large number of the more startling are to be found in Renan's *Dialogues Philosophiques.* Unfortunately that great writer is best known in England by his worst work, the *Vie de Jesus*.' Ernest Renan (1823–1892) was a widely read French scholar of civilization, religion and philosophy. In addition to the *Dialogues et fragments philosophiques* (1876), he had published *L'Avenir de la science, pensées de 1848* (1890). Its title (*The Future of Science*) doubtless contributed to those of Haldane's *Daedalus*, and Bertrand Russell's *Icarus; or, the Future of Science*.

DOI: 10.4324/9781032625522-4

Whether because of its long gestation, or because of Haldane's multi-disciplinary intelligence, *Daedalus* has an unusual range and compression, as it considers not just the futures of all the sciences, but their impact on society, psychology, morality, and religion. Numerous passages—sometimes individual sentences—throw out ideas and arguments which later To-day and To-morrow volumes would explore.

Haldane made important theoretical contributions to evolutionary biology, genetics, and population genetics.

> His greatest achievement was in uniting Darwinian evolution theory with Mendelian genetics. In a series of papers starting in 1924 he used a Mendelian system of heredity to work out, mathematically, the consequences of natural selection. He assigned selective values to genes, and deduced the number of generations required to bring about changes in gene frequency under many alternative assumptions. By developing a quantitative theory of evolution Haldane's work, together with that of R. A. Fisher and Sewall Wright, re-established Darwinian natural selection as the accepted mechanism of evolutionary change.
>
> (Oxford DNB)

He had become a socialist during the First World War. He wrote a second provocative volume for To-day and To-morrow, called *Callinicus; a defence of chemical warfare* (1925). At the time of the Spanish Civil War he supported the Communist Party, and joined it in 1942. His reputation was tarnished during the 'Lysenko Affair', due to his unwillingness to speak out publicly against Lysenko's Soviet-sanctioned pseudo-scientific claims.

****

Introduction

I have slightly expanded certain parts of this paper since reading it. It has therefore probably lost any unity which it may once have possessed. It will be criticized for its undue emphasis on certain unpleasant topics. This is necessary if people are to be induced to think about them, and it is the whole business of a university teacher to induce people to think.

Daedalus; or, Science and the Future

As I sit down to write these pages I can see before me two scenes from my experience of the late war.[3] The first is a glimpse of a forgotten battle of 1915.[4] It has a

3 When war was declared between Britain and the central powers Haldane immediately enlisted as an officer in a crack Scottish regiment, the Black Watch.

4 The Battle of Aubers Ridge, on 9 May 1915 (continued as the Battle of Festubert from 15 to 25 May), which proved disastrous for the British Army. Haldane was wounded and knocked unconscious by an exploding shell.

curious suggestion of a rather bad cinema film. Through a blur of dust and fumes there appear, quite suddenly, great black and yellow masses of smoke which seem to be tearing up the surface of the earth and disintegrating the works of man with an almost visible hatred. These form the chief part of the picture, but somewhere in the middle distance one can see a few irrelevant looking human figures, and soon there are fewer. It is hard to believe that these are the protagonists in the battle. One would rather choose those huge substantive oily black masses which are so much more conspicuous, and suppose that the men are in reality their servants, and playing an inglorious, subordinate, and fatal part in the combat. It is possible, after all, that this view is correct.

Had I been privileged to watch a battle three years later,[5] the general aspect would have been very similar, but there would have been fewer men and more shell-bursts. There would probably, however, have been one very significant addition. Then men would have been running, with mad terror in their eyes, from gigantic steel slugs, which were deliberately, relentlessly and successfully pursuing them.[6]

The other picture is of three Europeans in India looking at a great new star in the milky way. These were apparently all of the guests at a large dance who were interested in such matters. Amongst those who were at all competent to form views as to the origin of this cosmoclastic[7] explosion, the most popular theory attributed it to a collision between two stars, or a star and a nebula. There seem, however, to be at least two possible alternatives to this hypothesis. Perhaps it was the last judgment of some inhabited world, perhaps a too successful experiment in induced radioactivity the part of some of the dwellers there. And perhaps also these two hypotheses are identical, and what we were watching that evening was the detonation of a world on which too many men came out to look at the stars when they should have been dancing.

These two scenes suggest, very briefly, a part of the case against science. Has mankind released from the womb of matter a Demogorgon[8] which is already beginning to turn against him, and may at any moment hurl him into the bottomless void? Or is Samuel Butler's even more horrible vision correct, in which man becomes a mere parasite of machinery, an appendage of the reproductive system of huge and complicated engines which will successively usurp his activities, and end by ousting him from the mastery of this planet?[9] Is the machine-minder engaged on

5 In France he was wounded by artillery fire, and sent back to Scotland, where he served as instructor of grenades for Black Watch recruits. In 1916, he was posted to Mesopotamia (now Iraq), where he was wounded by a British shell and sent to India for recuperation. He was released from active duty upon his return to Britain in 1917.

6 The first use of tanks in the First World War was during the Battle of the Somme, when British Mark 1 tanks were deployed on 15 September 1916.

7 World-destroying.

8 A terrible ancient god or demon, associated with hell or the underworld.

9 Samuel Butler (1835–1902), writer and artist. His satirical utopia *Erewhon* (approximately 'No-where' reversed) was first published anonymously in 1872. *'May Not Man Himself Become a Sort of Parasite Upon the Machines? An Affectionate Machine-Tickling Aphid?'* Chapter xxiv (London: Trübner, 1872), 199.

repetition-work the goal and ideal to which humanity is tending? Perhaps a survey of the present trend of science may throw some light on these questions.

But first we may consider for a moment, the question of whether there is any hope of stopping the progress of scientific research. It is after all a very recent form of human activity, and a sufficiently universal protest of mankind would be able to arrest it even now. In the middle ages public opinion made it so dangerous as to be practically impossible, and I am inclined to suspect that Mr. Chesterton, for example, would not be averse to a repetition of this state of things.[10] The late M. Joseph Reinach, an able and not wholly illiberal thinker, publicly advocated it.[11]

I think, however, that so long as our present economic and national systems continue, scientific research has little to fear. Capitalism, though it may not always give the scientific worker a living wage, will always protect him, as being one of the geese which produce golden eggs for its table. And competitive nationalism, even if war is wholly or largely prevented, will hardly forego the national advantages accruing from scientific research.

If we look at the other most probable alternative the prospect is little more hopeful. In this country the labour party alone among political organizations includes the fostering of research in its official programme. Indeed as far as biological research is concerned labour may prove a better master than capitalism, and there can be little doubt that it would be equally friendly to physical and chemical research if these came to lead immediately to shortened hours rather than to unemployment. In particular there is perhaps reason to think that that form of sentimentalism which hampers medical research in this country by legislation would be less likely to flourish in a robust and selfish labour party of the Australian type than in parties whose members enjoy the leisure which seems necessary to the development of such emotional luxuries.

It is of course possible that civilisation may collapse throughout the world as it has done in parts of Russia, and science with it, but such an event would, in all probability, only postpone the problem for a few thousand years. And even in Russia we must not forget that first-rate scientific research is still being carried on.

The possibility has been suggested—I do not know how seriously—that the progress of science may cease through lack of new problems for investigation. Mr. Chesterton in *The Napoleon of Notting Hill*,[12] a book written fifteen years or so ago, prophesied that hansom-cabs would still be in existence a hundred years hence owing to a cessation of invention. Within six years there was a hansom-cab in a museum, and now that romantic but tardy vehicle is a memory like the trireme, the velocipede, and the 1907 Voisin biplane. I do not suggest that Mr. Chesterton be dragged—a heavier Hector[13]—behind the last hansom cab, but I do contend

10 The writer G[ilbert] K[eith] Chesterton (1874–1936) had converted to Catholicism in 1922.

11 Joseph Reinach (1856–1921) French author and politician, is best known for his support of Alfred Dreyfus during the Affair which divided France. He also wrote a four-volume history of the case.

12 Chesterton's 1904 futuristic satire of a London in 1984 in which little has changed.

13 Hector, the son of King Priam of Troy was killed by Achilles in single combat in the Trojan wars. Achilles then dragged his body around the city tied to his chariot.

that, in so far as he claims to be a prophet rather than the voice of one crying in the wilderness, he may be regarded as negligible for the purposes of our discussion. I shall try shortly to show how far from complete are any branches of science at the present time.

But first a word on Mr. H. G. Wells might not be out of place. The very mention of the future suggests him. There are two points I wish to make about Mr. Wells. In the first place, considered as a serious prophet, as opposed to a fantastic romancer, he is singularly modest. In 1902, for example, in a book called "Anticipations," he gave it as his personal opinion that by 1950 there would be heavier than air flying machines capable of practical use in war.[14] That, he said, was his own view, though he was well aware that it would excite considerable ridicule. *I propose in this paper to make no prophecies rasher than the above.*

The second and more important point is that he is a generation behind the time. When his scientific ideas were formed, flying and radio-telegraphy, for example, were scientific problems, and the centre of scientific interest still lay in physics and chemistry. Now these are commercial problems, and I believe that the centre of scientific interest lies in biology. A generation hence it may be elsewhere, and the views expressed in this paper will appear as modest, conservative, and unimaginative as do many of those of Mr. Wells to-day.

I will only touch very briefly on the future of physics, as the subject is inevitably technical. At present physical theory is in a state of profound suspense. This is primarily due to Einstein—the greatest Jew since Jesus. I have no doubt that Einstein's name will still be remembered and revered when Lloyd George, Foch, and William Hohenzollern[15] share with Charlie Chaplin that ineluctable oblivion which awaits the uncreative mind. I trust that I may be excused if I trespass from the strict subject of my theme to add my quota to the rather numerous misstatements of Einstein's views which have appeared during the last few years.

Ever since the time of Berkeley it has been customary for the majority of metaphysicians to proclaim the ideality of Time, of Space, or of both.[16] But they soon made it clear that in spite of this, time would continue to wait for no man, and space to separate lovers. The only practical consequences that they generally drew was that their own ethical and political views were somehow inherent in the structure of the universe. The experimental proof or disproof of such deductions is difficult,

14 Wells, *Anticipations of the Reactions of Mechanical and Scientific Progress upon Human Life and Thought* (London: Chapman and Hall, 1901), 191, wrote that people familiar with contemporary experiments in flying 'will be inclined to believe that long before the year A.D. 2000, and very probably before 1950, a successful aeroplane will have soared and come home safe and sound. Directly that is accomplished the new invention will be most assuredly applied to war.'

15 David Lloyd George (1863–1945) was Prime Minister of the United Kingdom during the First World War and after, from 1916 to 1922. Ferdinand Foch (1851–1929) was the supreme Allied military commander. 'William Hohenzollern' was the German Emperor, Kaiser Wilhelm II (1859–1941).

16 George Berkeley 1685–1753), Anglo-Irish Bishop and philosopher, who developed a theory 'immaterialism' (later termed 'subjective idealism') arguing that objects only exist as they are perceived by the mind, not due to having any material substance.

and—if the late war may be regarded as an experimental disproof of certain of Hegel's political tenets—costly and unsatisfactory.[17]

Einstein, so far from deducing a new Decalogue,[18] has contented himself with deducing the consequences to space and time themselves of their ideality. These are mostly too small to be measurable, but some, such as the deflection of light by the sun's gravitational field, are susceptible of verification, and have been verified. The majority of scientific men are now being constrained by the evidence of these experiments to adopt a very extreme form of Kantian idealism. The Kantian *Ding-an-sich*[19] is an eternal four-dimensional manifold, which we perceive as space and time, but what we regard as space and what as time is more or less fortuitous.

It is perhaps interesting to speculate on the practical consequences of Einstein's discovery. I do not doubt that he will be believed. A prophet who can give signs in the heavens is always believed. No one ever seriously questioned Newton's theory after the return of Halley's comet.[20] Einstein has told us that space, time and matter are shadows of the fifth dimension, and the heavens have declared his glory. In consequence Kantian idealism will become the basal working hypothesis of the physicist and finally of all educated men, just as materialism did after Newton's day. We may not call ourselves materialists, but we do interpret the activities of the moon, the Thames, influenza, and aeroplanes in terms of matter. Our ancestors did not, nor, in all probability, will our descendants. The materialism (whether conscious or sub-conscious does not matter very much) of the last few generations has led to various results of practical importance, such as sanitation, Marxian socialism, and the right of an accused person to give evidence on his or her own behalf. The reign of Kantian idealism as the basal working hypothesis, first of physics, and then of every-day life, will in all probability last for some centuries. At the end of that time a similar step in advance will be taken. Einstein showed that experience cannot be interpreted in terms of space and time. This was a well-known fact, but so long as space and time did not break down in their own special sphere, that of explaining the facts of motion, physicists continued to believe in them, or at any rate, what was much more important, to think in terms of them for practical purposes.

A time will however come (as I believe) when physiology will invade and destroy mathematical physics, as the latter have destroyed geometry.[21] The basic metaphysical working hypothesis of science and practical life will then, I think, be

17 The influential German idealist philosopher Georg Wilhelm Friedrich Hegel (1770–1831). In his *Lectures on the Philosophy of World History* (published in 1837; ed. Duncan Forbes, Hugh Barr Nisbet, Johannes Hoffmeister; Cambridge: Cambridge University Press, 1980), 54, he argued that what he called the 'Germanic nations' provided the first realization of human freedom.

18 The Ten Commandments.

19 The German philosopher Immanuel Kant (1724–1804) developed a theory of transcendental idealism, according to which space and time are 'forms of intuition.' The *Ding-an-sich* (thing in itself) means an object as it is in itself independent of observation or perception—and as such inaccessible to human understanding.

20 Seen every 75–76 years. It is the only short-period comet to be seen visibly with the naked eye, and the only comet to return within a human lifetime.

21 This prediction anticipates the argument of L. L. Whyte's *Archimedes* (also included here).

something like Bergsonian activism.[22] I do not for one moment suggest that this or any other metaphysical system has any claims whatever to finality.

Meanwhile we are in for a few centuries during which many practical activities will probably be conducted on a basis, not of materialism, but of Kantian idealism. How will this affect our manners, morals and politics? Frankly I do not know, though I think the effect will be as great as that of Newton's work, which created most of the intellectual forces of the eighteenth century. The Condorcets, Benthams, and Marxs of the future will I think be as ruthlessly critical of the metaphysics and ethics of their times as were their predecessors, but not quite so sure of their own;[23] they will lack a certain heaviness of touch which we may note in Utilitarianism and Socialism. They will recognise that perhaps in ethics as in physics, there are so to speak fourth and fifth dimensions that show themselves by effects which, like the perturbations of the planet Mercury, are hard to detect even in one generation, but yet perhaps in the course of ages are quite as important as the three-dimensional phenomena.

If the quantum hypothesis[24] is generally adopted even more radical alterations in our thinking will be necessary. But I feel it premature even to suggest their direction in the present unsatisfactory state of quantum mechanics. It may be that as Poincaré (the other Poincaré)[25] suggested we shall be forced to conceive of all changes as occurring in a series of clicks, and all space as consisting of discrete points. However this may be it is safe to say that a better knowledge of radiation will permit us to produce it in a more satisfactory manner than is at present possible. Almost all our present sources of light are hot bodies, 95% of whose radiation is invisible. To light a lamp as a source of light is about as wasteful of energy as to burn down one's house to roast one's pork. It is a fairly safe prophecy that in 50 years light will cost about a fiftieth of its present price, and there will be no more night in our cities. The alternation of day and night is a check on the freedom of human activity which must go the way of other spatial and temporal checks. In the long run I think that all that applied physics can do for us is abolish these checks. It enables us to possess more, travel more, and communicate more. I shall not attempt to predict in detail the future developments of transport and communication. They are only limited by the velocity of light. We are working towards a condition when any two persons on earth will be able to be completely present to one another in not

22 See *Archimedes*, Chapter 2.

23 The Marquis of Condorcet (1743–1794) was a leading French Enlightenment rationalist and egalitarian. The British philosopher Jeremy Bentham (1748–1832) was the founder of Utilitarianism, and a major inspiration for C. K. Ogden's work on language.

24 First proposed by Max Planck in 1900, this stated that atoms or molecules could only emit energy in discrete packets or 'quanta.' In the later 1920s it was developed into quantum mechanics—a detailed account of the interactions of matter and energy at the atomic or sub-atomic level.

25 Jules Henri Poincaré (1854–1912). French mathematician and scientist who contributed to several branches of mathematics, including celestial mechanics, fluid mechanics, chaos theory, topology, and the special theory of relativity; and also to the philosophy of science. The other Poincaré because Raymond Poincaré (1860–1934) had been President of France from 1913 to 1920.

more than 1/1500 of a second.[26] We shall never reach it, but that is the limit which we shall approach indefinitely.

Developments in this direction are tending to bring mankind more and more together, to render life more and more complex, artificial, and rich in possibilities—to increase indefinitely man's powers for good and evil.

But there are two prerequisites for all progress of this kind, namely continuous supplies of human and mechanical power. As industries become more and more closely interwoven, so that a dislocation of any one will paralyse a dozen others (and that is the position towards which we are rapidly moving), the ideal of the leaders of industry, under no matter what economic system, will be directed less and less to the indefinite increase of production in the intervals between such dislocations, and more and more to stable and regular production, even at the cost of reduction of profits and output while the industry is proceeding normally. It is quite possible that capitalism itself may demand that the control of certain key industries be handed over completely to the workers in those industries, simply in order to reduce the number of sporadic strikes in them. And as industrial progress continues an ever larger number—perhaps the majority—of industries will become key industries. The solution may be entirely different—we may well see a return to feudalism. But the probability is that the problem will be solved. This view may seem optimistic, but it is more likely than the alternative thesis which may be briefly stated as follows: *"No human society will ever succeed in producing a stable organization in which the majority of the population is employed otherwise than in agriculture, animal-rearing, hunting or fishing."* It took some thousands of years to produce the stable agricultural society which forms the basis of European life and whose morals we are too apt to regards as eternal truths. It should take a shorter time to evolve a stable industrial society. The people that do so will inherit the earth. In sum, I believe that the progress of science will ultimately make industrial injustice as self-destructive as it is now making international injustice.

As for the supplies of mechanical power, it is axiomatic that the exhaustion of our coal and oil-fields is a matter of centuries only. As it has often been assumed that their exhaustion would lead to the collapse of industrial civilization, I may perhaps be pardoned if I give some of the reasons which lead me to doubt this proposition.

Water-power is not, I think, a probable substitute, on account of its small quantity, seasonal fluctuation, and sporadic distribution. It may perhaps, however, shift the centre of industrial gravity to well-watered mountainous tracts such as the Himalayan foothills, British Columbia, and Armenia. Ultimately we shall have to tap those intermittent but inexhaustible sources of power, the wind and the sunlight. The problem is simply one of storing their energy in a form as convenient as coal or petrol. If a windmill in one's back garden could produce a hundredweight of coal daily (and it can produce its equivalent in energy), our coalmines would be shut down to-morrow. Even to-morrow a cheap, foolproof, and durable storage battery

26 By the second impression of March 1924, Haldane had corrected the figure to 1/24 second.

may be invented, which will enable us to transform the intermittent energy of the wind into continuous electric power.

Personally, I think that four hundred years hence the power question in England may be solved somewhat as follows: The country will be covered with rows of metallic windmills working electric motors which in their turn supply current at a very high voltage to great electric mains. At suitable distances, there will be great power stations where during windy weather the surplus power will be used for the electrolytic decomposition of water into oxygen and hydrogen. These gasses will be liquefied, and stored in vast vacuum jacketed reservoirs, probably sunk in the ground. If these reservoirs are sufficiently large, the loss of liquid due to leakage inwards of heat will not be great; thus the proportion evaporating daily from a reservoir 100 yards square by 60 feet deep would not be 1/1000 of that lost from a tank measuring two feet each way. In times of calm, the gasses will be recombined in explosion motors working dynamos which produce electrical energy once more, or more probably in oxidation cells. Liquid hydrogen is weight for weight the most efficient known method of storing energy, as it gives about three times as much heat per pound as petrol. On the other hand it is very light, and bulk for bulk has only one third of the efficiency of petrol. This will not, however, detract from its use in aeroplanes, where weight is more important than bulk. These huge reservoirs of liquified gasses will enable wind energy to be stored, so that it can be expended for industry, transportation, heating and lighting, as desired. The initial costs will be very considerable, but the running expenses less than those of our present system. Among its more obvious advantages will be the fact that energy will be as cheap in one part of the country as another, so that industry will be greatly decentralized; and that no smoke or ash will be produced.

It is on some such lines as these, I think, that the problem will be solved. It is essentially a practical problem, and the exhaustion of our coal-fields will furnish the necessary stimulus for its solution. Even now perhaps Italy might achieve economic independence by the expenditure of a few million pounds upon research on the lines indicated. I may add in parenthesis that, on thermodynamical grounds which I can hardly summarize shortly, I do not much believe in the commercial possibility of induced radio-activity.

Before I turn to the principal part of my subject I should like to consider very briefly the influence on art and literature of our gradual conquest of space and time. I think that the blame for the decay of certain arts rests primarily on the defective education of the artists. An artist must understand his subject matter. At present not a single competent poet and very few painters and etchers outside the Glasgow School[27] understand industrial life, and I believe that there is only one architect of any real originality who understands the possibilities of ferro-concrete. I do not know his name, but he produced in Soissons before the war a market-place with

27 The group of artists and designers around Charles Rennie Mackintosh in Glasgow from the mid-1890s to around1910, including the Macdonald sisters and Herbert McNair, who contributed to the international Art Nouveau movement.

the dignity and daring of an ancient Egyptian temple.[28] If I knew that he had been entrusted with the rebuilding of Soissons, I could not regret its destruction.

Now if we want poets to interpret physical science as Milton and Shelley did (Shelley and Keats were the last English poets who were at all up-to-date in their chemical knowledge), we must see that our possible poets are instructed, as their masters were, in science and economics. I am absolutely convinced that science is vastly more stimulating to the imagination than are the classics, but the products of this stimulus do not normally see the light because scientific men as a class are devoid of any perception of literary form. When they can express themselves we get a Butler or a Norman Douglas.[29] Not until our poets are once more drawn from the educated classes (I speak as a scientist), will they appeal to the average man by showing him the beauty in his own life as Homer and Virgil appealed to the street urchins who scrawled their verses on the walls of Pompeii.

And if we must educate our poets and artists in science, we must educate our masters, labour and capital, in art. Personally I believe that we may have good hopes of both. The capitalist's idea of art in industry at present tends to limit itself to painting green and white stripes on the front of his factories in certain cases. This is a primitive type of decoration, but it has, I think, the root of the matter in it. Before long someone may discover that frescoes inside a factory increase the average efficiency of the worker 1.03% and art will become a commercial proposition once more. Even now it is being discovered that artistic advertising often pays. Similarly I do not doubt that labour will come to find that it cannot live by bread (or shall we say bread and beer) alone. But it can hardly be expected to make this discovery until it is assured of its supply of bread and beer.

Applied chemistry has introduced into human life no radical novelty of the importance of the heat-engine[30] or the telegraph. It has vastly increased the production of various types of substance the most important being metals. But there were explosives, dyes and drugs before chemistry was a science, and its progress along present lines will mainly alter life in a quantitative manner. Perhaps the biggest problems before it in metallurgy are the utilization of low-grade iron ores, and the production of aluminium from clay, which contains up to 24% of that metal. I do not think that even when this is accomplished aluminium will oust iron and steel as they ousted bronze and flint, but it and its alloys will certainly take the second, and possibly the first place as industrial metals. There is just a hope, though I fear little more, that a large-scale production of perfume may form the basis of a re-education of our rather rudimentary sense of smell, but the most interesting possibilities of chemical invention are very clearly in biological chemistry, and for the following reasons.

28 The covered market, built 1908–1911, was designed by architect Albert-Désiré Guilbert (1866–1949).

29 Norman Douglas (1868–1952), British writer best known for his novel *South Wind* (1917), and travel books such as *Old Calabria* (1915).

30 Any engine which converts heat into usable (and usually mechanical) energy; as for example steam or internal combustion engines.

Desirable substances fall on the whole into two classes. The first are desirable on account of their physical or chemical properties, for example iron, wood or glass, which we use as a part of systems such as fires, houses or razors, which procure us certain benefits. The second are desirable on account of their physiological properties. Such substances include foods, drinks, tobacco and drugs. Colours and scents occupy an intermediate position. The value of this second class of substances rests on a quite special relationship to the human organism which depends in the most intimate way on the constitution of the latter, and has not in general been at all fully explained in terms of physics and chemistry. For example fires can be made of coal or peat instead of wood, but no other chemical substance has the same effect as water or alcohol. So unless a chemical substance has new physiological properties its production will merely serve to improve or make possible some appliance whose use lies within the sphere of applied physics. Within historical time two and only two substances of the second class have come into universal use in Europe, namely caffeine and nicotine, which were introduced into this country in the sixteenth and seventeenth centuries. There are others of immense importance, such as chloroform and quinine, but their use is not universal. But coffee, tea, and tobacco, with alcohol, are as much a part of normal life as food and water. There is no reason to suppose that the list of such substances is exhausted. During the war Embden[31] the professor of physiology in Frankfurt University discovered that a dose of about 7 grams of acid sodium phosphate increased a man's capacity for prolonged muscular work by about 20% and probably aids in prolonged mental work. It can be taken over very lengthy periods. A group of coal-miners took it for nine months on end with very great effects on their output. It has no after-effects like those of alcohol, and one cannot take a serious overdose as it merely acts as a purgative. (They gave certain Stosstruppen[32] too much!) Thousands of people in Germany take it habitually. It is possible that it may become as normal a beverage as coffee or tea. It costs 1/9 per pound, or 1/3d. per dose.

The vast majority of chemical substances with physiological properties are unsuited for daily use like castor oil, or dangerous like morphine; probably none are without bad effects in certain cases. Those which are susceptible of daily use are of the utmost social importance. Tobacco has slight but definite effects on the character. Coffee-houses in London in the seventeenth and eighteenth centuries and cafés in modern Europe were and are civilizing influences of incalculable value. But these substances are profoundly obnoxious to a certain type of mind. It would perhaps be fantastic to suggest that Sir Walter Raleigh owed his death in part to his sovereign's objection to tobacco. But if he is not its proto-martyr it is at least probable that more have died for tobacco smoking at the hands of Sikhs, Senussis,[33] and Wahabis,[34] whose religions forbid this practice, than died under the Roman empire

31 [Author's note] Embden, Grafe, and Schmitz. *Zeitschrift für physiologische Chemie*, Vol. 113, p. 67, 1921.

32 German: Stormtroopers.

33 Islamic Sufi reformist movement in Libya and Sudan.

34 Islamic Sunni reformist movement prevalent in Saudi Arabia.

for professing Christianity. Should it ever be generally realised that temperance is a mean we may expect that mankind will ultimately have at its disposal a vast array of substances like wine, coffee and tobacco, whose intelligent use can add to the amenity of life and promote the expression of man's higher faculties.

But before that day comes chemistry will be applied to the production of a still more important group of physiologically active substances, namely foods. The facts about food are rather curious. Everyone knows that food is ultimately produced by plants, though we may get it at second or third hand if we eat animals or their products. But the average plant turns most of its sugar not into starch which is digestible, but into cellulose which is not, but forms its woody skeleton. The hoofed animals have dealt with this problem in their own way, by turning their bellies into vast hives of bacteria that attack cellulose, and on whose by-products they live. We have got to do the same, but outside our bodies. It may be done on chemical lines. Irvine has obtained a 95% yield of sugar from cellulose, but at a prohibitive cost. Or we may use micro-organisms, but in any case within the next century sugar and starch will be about as cheap as sawdust. Many of our foodstuffs, including the proteins, we shall probably build up from simpler sources such as coal and atmospheric nitrogen. I should be inclined to allow 120 years, but not much more, before a completely satisfactory diet can be produced in this way on a commercial scale.

This will mean that agriculture will become a luxury, and that mankind will be completely urbanized. Personally I do not regret the probable disappearance of the agricultural labourer in favour of the factory worker, who seems to me a higher type of person from most points of view. Human progress in historical time has been the progress of cities dragging a reluctant countryside in their wake. Synthetic food will substitute the flower garden and the factory for the dunghill and the slaughterhouse, and make the city at last self-sufficient.

There's many a strong farmer
Whose heart would break in two
If he could see the townland
That we are riding to.
Boughs have their fruit and blossom
At all times of the year,
Rivers are running over
With red beer and brown beer,
An old man plays the bagpipes
In a golden and silver wood,
Queens, their eyes blue like the ice,
Are dancing in a crowd.[35]

35 W. B. Yeats, 'The Happy Townland.' (Haldane's lineation of the poem—as heptameters with extra spaces for caesuras instead of line-breaks between the four- and three-stress lines—may indicate quoting from memory.)

I should have liked had time allowed to have added my quota to the speculations which have been made with regard to inter-planetary communication. Whether this is possible I can form no conjecture; that it will be attempted I have no doubt whatever.

With regard to the application of biology to human life, the average prophet appears to content himself with considerable if rather rudimentary progress in medicine and surgery, some improvements in domestic plants and animals, and possibly the introduction of a little eugenics. The eugenic official, a compound, it would appear, of the policeman, the priest and the procurer, is to hale us off at suitable intervals to the local temple of Venus Genetrix[36] with a partner chosen, one gathers, by something of a glorified medical board. To this prophecy I should reply that it proceeds from a type of mind as lacking in originality as in knowledge of human nature. Marriage "by numbers", so to speak, was a comparatively novel idea when proposed by Plato 2,300 years ago, but it has already actually been practised in various places, notably among the subject of the Jesuits in Paraguay. It is moreover likely, as we shall see, that the ends proposed by the eugenist will be attained in a very different manner.

But before we proceed to prophecy I should like to turn back to the past and examine very briefly the half dozen or so important biological inventions which have already been made. By a biological invention I mean the establishment of a new relationship between man and other animals or plants, or between different human beings, provided that such relationship is one which comes primarily under the domain of biology rather than physics, psychology or ethics. Of the biological inventions of the past, four were made before the dawn of history. I refer to the domestication of animals, the domestication of plants, the domestication of fungi for the production of alcohol, and to a fourth invention, which I believe was of more ultimate and far-reaching importance than any of these, since it altered the path of sexual selection, focussed the attention of man as a lover upon woman's face and breasts, and changed our idea of beauty from the steatapygous Hottentot[37] to the modern European, from the Venus of Brassempouy[38] to the Venus of Milo.[39] There are certain races which have not yet made this last invention. And in our own day two more have been made, namely bactericide and the artificial control of conception.

The first point that we may notice about these inventions is that they have all had a profound emotional and ethical effect. Of the four earlier there is not one

36 Venus in her role as the ancestress of the Roman people. She is granted the role of the goddess of motherhood and domesticity.

37 'Steatapygous' denotes an unusual accumulation of fat on the buttocks. 'Hottentot' was a name for the Khoisan tribe of hunter-gatherers who lived in the southernmost tip of Africa. In 1810 a British ship's doctor took Saarti Baartman to London, exhibiting her to a paying audience because of her pronounced buttocks and genitals. She became known as the 'Hottentot Venus.' She was later sold to a French businessman, and died in Paris in 1816.

38 An Upper Paleolithic ivory figurine of a woman of which only the head and neck survive.

39 Famous ancient Greek marble statue of Aphrodite, the arms missing, at the Louvre museum.

which has not formed the basis of a religion. I do not know what strange god[40] will have the hardihood to adopt Charles Bradlaugh and Annie Besant in the place of Triptolemus and Noah,[41] but one may remark that it is impossible to keep religion out of any discussion of the practices which they popularized.

The second point is perhaps harder to express. The chemical or physical inventor is always a Prometheus.[42] There is no great invention, from fire to flying, which has not been hailed as an insult to some god. But if every physical and chemical invention is a blasphemy, every biological invention is a perversion. There is hardly one which, on first being brought to the notice of an observer from any nation which has not previously heard of their existence, would not appear to him as indecent and unnatural.

Consider so simple and time-honored a process as the milking of a cow. The milk which should have been an intimate and almost sacramental bond between mother and child is elicited by the deft fingers of a milk-maid, and drunk, cooked, or even allowed to rot into cheese. We have only to imagine ourselves as drinking any of its other secretions, in order to realise the radical indecency of our relation to the cow.[43]

No less disgusting a priori is the process of corruption which yields our wine and beer. But in actual fact the process of milking and of the making and drinking of beer appear to us profoundly natural; they have even tended to develop a ritual of their own whose infraction nowadays has a certain air of impropriety. There is something slightly disgusting in the idea of milking a cow electrically or drinking beer out of tea-cups. And all this of course applies much more strongly to the sexual act.

I fancy that the sentimental interest attaching to Prometheus has unduly distracted our attention from the far more interesting figure of Daedalus. It is with infinite relief that amidst a welter of heroes armed with gorgon's heads or protected by Stygian baptisms[44] the student of Greek mythology comes across the first

40 *Judges* 10: 16: 'And they put away the strange gods from among them, and served the LORD: and his soul was grieved for the misery of Israel.'

41 Bradlaugh 1833–1891) was an atheist MP who founded the National Secular Society. Besant (1847–1933) was a socialist and progressive campaigner. They were prosecuted in 1877 for publishing a book about birth control. In Greek mythology Triptolemus taught humans to sow and reap crops. Noah is the first person in the bible to plant a vineyard, produce wine, and get drunk.

42 Prometheus was a Titan in Greek mythology known for his intelligence. In mythology he is credited with stealing fire from Zeus and giving it to mortals for their own use. Because of this role in mythology, he is acknowledged with playing a pivotal role in the early history of humankind.

43 [Author's note] The Hindus have recognized the special and physiological relation of man to the cow by making the latter animal holy. A good Hindu would no more kill a cow than his foster-mother. But the holiness of the cow has unfortunately extended to all its products, and the extensive use of cowdung in Indian religious ceremonies is disgusting to the average European. The latter however, is insensitive to the equally loathsome injunctions of the Catholic Church with regard to human marriage. It would perhaps be better if both marriage and milking could be secularized.

44 Perseus killed Medusa, the Gorgon with snakes for hair, and a look so terrifying she turned onlookers to stone. Achilles' mother Thetis, trying to avert a prophecy of his early death, dipped him in the river Styx to make him invulnerable. But she held him by the heel, and that was where Paris' fatal arrow struck him during the Trojan war.

modern man. Beginning as a realistic sculptor (he was the first to produce statues whose feet were separated) it was natural that he should proceed to the construction of an image of Aphrodite whose limbs were activated by quicksilver. After this his interest inevitably turned to biological problems, and it is safe to say that posterity has never equalled his only recorded success in experimental genetics.[45] Had the housing and feeding of the Minotaur been less expensive it is probable that Daedalus would have anticipated Mendel.[46] But Minos held that a labyrinth and an annual provision of 50 youths and 50 virgins were excessive as an endowment for research, and in order to escape from his ruthless economies Daedalus was forced to invent the art of flying.[47] Minos pursued him to Sicily and was slain there. Save for his valuable invention of glue, little else is known of Daedalus. But it is most significant that, although he was responsible for the death of Zeus' son Minos he was neither smitten by a thunderbolt, chained to a rock, nor pursued by furies. Still less did any of the rather numerous visitors to Hades discover him either in Elysium or Tartarus. We can hardly imagine him as a member of the throng of shades who besieged Charon's[48] ferry like sheep at a gap. He was the first to demonstrate that the scientific worker is not concerned with gods.

The unconscious mind of the early Greeks, who focussed in this amazing figure the dim traditions of Minoan science, was presumably aware of this fact. The most monstrous and unnatural action in all human legend was unpunished in this world or the next. Even the death of Icarus must have weighed lightly with a man who had already been banished from Athens for the murder of his nephew. But if he escaped the vengeance of the gods he has been exposed to the universal and agelong reprobation of a humanity to whom biological inventions are abhorrent, with one very significant exception. Socrates was proud to claim him as an ancestor.

The biological invention then tends to begin as a perversion and end as a ritual supported by unquestioned beliefs and prejudices. Even now surgical cleanliness is developing its rites and its dogmas, which, it may be remarked, are accepted most religiously by women. With the above facts in your minds I would ask you to excuse what at first sight might appear improbable or indecent in any speculations which appear below, and to dismiss from your minds the belief that biology will consist merely in physical and chemical discoveries as applied to men, animals and plants.

45 When King Minos of Crete refused to sacrifice his most beautiful bull to Poseidon, the god made Minos' queen, Pasiphaë, lust after the bull. She got Daedalus to construct an artificial cow which she hid inside, to fool the bull into mating with her. Their offspring was the Minotaur, half bull, half man.

46 Gregor Mendel (1822–1884) the pioneer of genetics.

47 Minos had forced the Athenians to send regular tributes of young men and virgins (usually seven of each) to be fed to the Minotaur. Minos' daughter Ariadne fell in love with Theseus, one of the Athenian youths, and showed him how to find his way out of the labyrinth Daedalus had designed to house the Minotaur. After Theseus killed the minotaur, Daedalus and his son Icarus were imprisoned. Daedalus designed the wings for them to fly to safety. See the notes to Bertrand Russell's *Icarus*, the next volume, for what happened next. . .

48 The ferryman of Hades in Greek mythology who ferried the dead across the river Styx or Acheron.

I say advisedly "will consist", for we are at present almost completely ignorant of biology, a fact which often escapes the notice of biologists, and renders them to presumptuous in their estimates of the present position of their science, too modest in their claims for its future. If for example we take a typical case of applied biology such as the detection and destruction of the cholera bacillus, we find a great deal of science involved, but the only purely biological principle is the very important but not very profound one that some bacteria kill some men. The really scientific parts of the process are the optical and chemical methods involved in the magnification, staining and killing of the bacilli. When on the other hand we come to immunization to typhoid we find certain purely biological principles involved which are neither simple nor at all completely understood.

Actually biological theory consists of some ancient but not very easily stated truths about organisms in general, due largely to Aristotle, Hippocrates and Harvey,[49] a few great principles such as those formulated by Darwin, Mayer, Claude Bernard,[50] and Mendel, and a vast mass of facts about individual organisms and their parts which are still awaiting adequate generalization.

Darwin's results are beginning to be appreciated, with alarming effects on certain types of religion, those of Weismann[51] and Mendel will be digested in the course of the present century, and are going to affect political and philosophical theories almost equally profoundly. I need hardly say that these latter results deal with the question of reproduction and heredity. We may expect, moreover, as time goes on, that a series of shocks of the type of Darwinism will be given to established opinions on all sorts of subjects. One cannot suggest in detail what these shocks will be, but since the opinions on which they will impinge are deep-seated and irrational, they will come upon us and our descendants with the same air of presumption and indecency with which the view that we are descended from monkeys came to our grandfathers. But owing to man's fortunate capacity for thinking in watertight (or rather idea-tight) compartments, they will probably not have immediate and disruptive effects upon society any more than Darwinism had.

Far more profound will be the effect of the practical applications of biology. I believe that the progress of medicine has had almost, if not quite, as deep an effect on society in Western Europe as the industrial revolution. Apart from the important social consequences which have flowed from the partial substitution of the doctor for the priest, its net result has been that whereas four hundred years ago most people died in childhood, they now live on an average, (apart from the late war),

49 Hippocrates was a classical Greek physician of the fifth–fourth centuries BCE, known as the 'father of medicine.' William Harvey (1578–1657) was the English physician who first described the circulation of the blood.

50 Johann Friedrich Mayer (1719–1798), German agricultural reformer who developed new methods of fertilizing and rotating crops, and wrote an influential textbook on animal husbandry. French physiologist Claude Bernard (1813–1878) established the functions of the pancreas and liver, and the dilations and contractions of blood vessels.

51 German evolutionary biologist August Weismann (1834–1914) developed 'germ plasm theory' (also known as Weismannism), stating that inheritance only operates via the gametes ('germ cells') such as egg and sperm cells.

until forty-five. Bad as our urban conditions are, there is not a slum in the country which has a third of the infantile death-rates of the royal family in the middle ages. Largely as a result of this religion has come to lay less and less stress on a good death, and more and more on a good life, and its whole outlook has gradually changed in consequence. Death has receded so far into the background of our normal thoughts that when we came into somewhat close contact with it during the war most of us failed to take it seriously.

Similarly institutions which were based on short lives have almost wholly collapsed. For example the English land system postulated that the landowner should die aged about forty, and be succeeded by his eldest son, aged about twenty. The son had spent most of his life on the estate, and had few interests outside it. He managed it at least as well as anyone else could have done. Nowadays the father dodders on till about eighty, and is generally incompetent for ten years before his death. His son succeeds him at the age of fifty or so, by which time he may be a fairly competent colonel or stockbroker, but cannot hope to learn the art of managing an estate. In consequence he either hands it over to an agent who is often corrupt, or runs it unscientifically, gets a low return, and ascribes to Bolshevism what he should really lay at the door of vaccination.

But to return, if I may use the expression, to the future, I am going to suggest a few obvious developments which seem probable in the present state of biological science, without assuming any great new generalizations of the type of Darwinism. I have the very best precedents for introducing a myth at this point, so perhaps I may be excused if I reproduce some extracts from an essay on the influence of biology on history during the 20th century which will (it is hoped) be read by a rather stupid undergraduate member of this university to his supervisor during his first term 150 years hence.

> As early as the first decade of the twentieth century we find a conscious attempt at the application of biology to politics in the so-called eugenic movement. A number of earnest persons, having discovered the existence of biology, attempted to apply it in its then very crude condition to the production of a race of super-men, and in certain countries managed to carry a good deal of legislation. They appear to have managed to prevent the transmission of a good deal of syphilis, insanity, and the like, and they certainly succeeded in producing the most violent opposition and hatred amongst the classes whom they somewhat gratuitously regarded as undesirable parents. (There was even a rebellion in Nebraska). However, they undoubtably prepared public opinion for what was to come, and so far served a useful purpose. Far more important was the progress in medicine which practically abolished infectious diseases in those countries which were prepared to tolerate the requisite amount of state interference in private life, and finally, after the league's[52] ordinance of 1958, all over the world; though owing to Hindu opposition, parts of India were still quite unhealthy up to 1980 or so.

52 The League of Nations had been established in 1920 with the aim of preventing future wars.

But from a wider point of view the most important biological work in the first third of the century was in experimental zoology and botany. When we consider that in 1912 Morgan had located several Mendelian factors in the nucleus of Drosophila, and modified its sex-ratio,[53] while Marmorek[54] had taught a harmless bacillus to kill guinea-pigs, and finally in 1913 Brachet[55] had grown rabbit embryos in serum for some days, it is remarkable how little the scientific workers of that time, and a fortiori the general public, seem to have foreseen the practical bearing of such results.

As a matter of fact it was not until 1940 that Selkovski invented the purple alga *Porphyrococcus fixator* which was to have so great an effect on the world's history. In the 50 years before this date the world's average wheat yield per hectare had been approximately doubled, partly by the application of various chemical manures, but most of all by the results of systematic crossing work with different races; there was however little prospect of further advance on any of these lines. *Porphyrococcus* is an enormously efficient nitrogen-fixer and will grow in almost any climate where there are water and traces of potash and phosphates in the soil, obtaining its nitrogen from the air. It has about the effect in four days that a crop of vetches[56] would have had in a year. It could not, of course, have been produced in the course of nature, as its immediate ancestors would only grow in artificial media and could not have survived outside a laboratory. Wherever nitrogen was the principal limiting factor to plant growth it doubled the yield of wheat, and quadrupled the value of grass land for grazing purposes. The enormous fall in food prices and the ruin of purely agricultural states was of course one of the chief causes of the disastrous events of 1943 and 1944. The food glut was also greatly accentuated when in 1942 the Q strain of *Porphyrococcus* escaped into the sea and multiplied with enormous rapidity. Indeed for two months the surface of the tropical Atlantic set to a jelly, with disastrous results to the weather of Europe. When certain of the plankton organisms developed ferments capable of digesting it the increase of the fish population of the seas was so great as to make fish the universal food that it is now, and to render even England self-supporting in respect of food. So great was the prosperity in England that in that year the coal-miner's union entered its first horse for the Derby (a horse-race which still took place annually at that time).

It was of course as a result of its invasion by *Porphyrococcus* that the sea assumed the intense purple colour which seems so natural to us, but which so distressed the more aesthetically minded of our great grand-parents who witnessed the change. It is certainly curious to us to read of the sea as having

53 Thomas Hunt Morgan (1866–1945) established the concept of linkage between pairs of genes.

54 The bacteriologist Alexander Marmorek (1865–1923) identified the toxin produced by the bacillus responsible for tuberculosis; he also discovered a vaccine against the disease.

55 Belgian physician and embryologist Albert Brachet (1869–1930).

56 The genus *Vicia*; part of the legume family, including lentils and broad beans as well as common vetch, which is used for animal fodder.

been green or blue. I need not detail the work of Ferguson and Rahmatullah who in 1957 produced the lichen which has bound the drifting sand of the world's deserts (for it was merely a continuation of that of Selkovski), nor yet the story of how the agricultural countries dealt with their unemployment by huge socialistic windpower schemes.

It was in 1951 that Dupont and Schwarz produced the first ectogenctic child.[57] As early as 1902 Heape had transferred embryo rabbits from one female to another,[58] in 1925 Haldane had grown embryonic rats in serum for ten days, but had failed to carry the process to its conclusion, and it was not till 1940 that Clark succeeded with the pig, using Kehlmann's solution as medium. Dupont and Schwarz obtained a fresh ovary from a woman who was the victim of an aeroplane accident, and kept it living in their medium for five years. They obtained several eggs from it and fertilized them successfully, but the problem of nutrition and support of the embryo was more difficult, and was only solved in the fourth year. Now that the technique is fully developed, we can take an ovary from a woman, and keep it growing in a suitable fluid for as long as twenty years, producing a fresh ovum each month, of which 90 per cent can be fertilized, and the embryos grown successfully for nine months, and then brought out into the air. Schwarz never got such good results, but the news of his first success caused an unprecedented sensation throughout the entire world, for the birthrate was already less than the death rate in most civilised countries. France was the first country to adopt ectogenesis officially, and by 1968 was producing 60,000 children annually by this method. In most countries the opposition was far stronger, and was intensified by the Papal Bull "Nunquam prius audito",[59] and by the similar fetwa of the Khalif, both of which appeared in 1960.

As we know ectogenesis is now universal, and in this country less than 30 per cent of children are now born of woman. The effect on human psychology and social life of the separation of sexual love and reproduction which was begun in the 19th century and completed in the 20th is by no means wholly satisfactory. The old family life had certainly a good deal to commend it, and although nowadays we bring on lactation in women by injection of placentin[60] as a routine, and thus conserve much of what was best in the former instinctive cycle, we must admit that in certain respects our great grandparents had the advantage of us. On the other hand it is generally admitted that the effects of selection have more than counterbalanced these evils.

57 The term 'ectogenesis' had been in use since at least 1909 to mean 'The production of structures or bodies outside the organism' (*OED*).

58 British zoologist and embryologist Walter Heape FRS (1855–1929) had successfully transferred the fertilized ova from one breed of rabbit to another in 1890.

59 Public decrees issued by the Pope are given Latin titles; here 'never previously heard.'

60 The 1920 edition of Thomas Lathrop Stedman's *A Practical Medical Dictionary* defines placentin as 'A placental extract employed in the cutireaction test for pregnancy.' (The Placenta insulin-like peptide known as placentin was not identified until 1995.)

> The small proportion of men and women who are selected as ancestors for the next generation are so undoubtedly superior to the average that the advance in each generation in any single respect, from the increased output of first-class music to the decreased convictions for theft, is very startling. Had it not been for ectogenesis there can be little doubt that civilisation would have collapsed within a measurable time owing to the greater fertility of the less desirable members of the population in almost all countries.
>
> It is perhaps fortunate that the process of becoming an ectogenetic mother of the next generation involves an operation which is somewhat unpleasant, though now no longer disfiguring or dangerous, and never physiologically injurious, and is therefore an honour but by no means a pleasure. Had this not been the case, it is perfectly possible that popular opposition would have proved too strong for the selectionist movement. As it was the opposition was very fierce, and characteristically enough this country only adopted its present rather stringent standard of selection a generation later than Germany, though it is now perhaps more advanced than any other country in this respect. The advantages of thorough-going selection, have, however, proved to be enormous. The question of the ideal sex ratio is still a matter of violent discussion, but the modern reaction towards equality is certainly strong.

Our essayist would then perhaps go on to discuss some far more radical advances made about 1990, but I have only quoted his account of the earlier applications of biology. The second appears to me to be neither impossible nor improbable, but it has those features which we saw above to be characteristic of biological inventions. If reproduction is once completely separated from sexual love mankind will be free in an altogether new sense. At present the national character is changing slowly according to quite unknown laws. The problem of politics is to find institutions suitable to it. In the future perhaps it may be possible by selective breeding to change character as quickly as institutions. I can foresee the election placards of 300 years hence, if such quaint political methods survive, which is perhaps improbable, “Vote for Smith and more musicians”, “Vote for O’Leary and more girls”, or perhaps finally “Vote for Macpherson and a prehensile tail for your great-grandchildren”. We can already alter animal species to an enormous extent, and it seems only a question of time before we shall be able to apply the same principles to our own.

I suggest then that biology will probably be applied on lines roughly resembling the above. There are perhaps equally great possibilities in the way of direct improvement of the individual, as we come to know more of the physiological obstacles to the development of different faculties. But at present we can only guess at the nature of these obstacles, and the line of attack suggested in the myth is the one which seems most obvious to a Darwinian. We already know however that many of our spiritual faculties can only be manifested if certain glands, notably the thyroid and sex-glands, are functioning properly, and that very minute changes in such glands affect the character greatly. As our knowledge of this subject increases we may be able, for example, to control our passions by some more direct method than fasting and flagellation, to stimulate our imagination by some reagent with less after-effects than alcohol, to deal with perverted instincts by physiology rather than prison.[61] Conversely there will inevitably arise possibilities of new vices similar to but even more profound than those opened up by the pharmacological discoveries of the 19th century.[62]

The recent history of medicine is as follows. Until about 1870 medicine was largely founded on physiology, or, as the Scotch called it "Institutes of medicine". Disease was looked at from the point of view of the patient, as injuries still are. Pasteur's discovery of the nature of infectious disease transformed the whole outlook, and made it possible to abolish one group of diseases. But it also diverted scientific medicine from its former path, and it is probable that, were bacteria unknown, though many more people would die of sepsis and typhoid, we should be better able to cope with kidney disease and cancer. Certain diseases such as cancer are probably not due to specific organisms, whilst others such as phthisis are due to forms which are fairly harmless to the average person, but attack others for unknown reasons. We are not likely to deal with them effectually on Pasteur's lines, we must divert our view from the micro-organism to the patient. Where the doctor cannot deal with the former he can often keep the patient alive long enough to be able to do so himself. And here he has to rely largely on a knowledge of physiology. I do not say that a physiologist will discover how to prevent cancer. Pasteur started life as a crystallographer. But whoever does so is likely at least to make use of physiological data on a large scale.

The abolition of disease will make death a physiological event like sleep. A generation that has lived together will die together. I suspect that man's desire for a future life is largely due to two causes, a feeling that most lives are incomplete, and a desire to meet friends from whom we have parted prematurely. A gentle decline into the grave at the end of a completed life's work will largely do away with the first, and our contemporaries will rarely leave us sorrowing for long.

Old age is perhaps harder on women than on men. They live longer, but their life is too often marred by the sudden change which generally overtakes them between forty and fifty, and sometimes leaves them a prey to disease, though it may improve their health. This change seems to be due to a sudden failure of a definite chemical substance produced by the ovary. When we can isolate and synthesize this body we shall be able to prolong a woman's youth, and allow her to age as gradually as the average man.[63]

Psychology is hardly a science yet. Like biology it has arrived at certain generalizations of a rather abstract and philosophic character, but these are still to some extent matters of controversy. And though a vast number of most important

61 Compare Russell's imagining in *Icarus*, 57–58, of the more sinister possibility that such knowledge might enable those in power to render their subjects more docile.

62 Morphine was discovered between 1803 and 1805; quinine in 1820. The first synthetic sleeping drug, chloral hydrate, was produced in 1832; aspirin in 1853 and paracetamol in 1877. The local anesthetic benzocaine was synthesized in 1890.

63 This idea is the basis of hormone replacement therapy for post-menopausal women, using estrogen and progesterone.

empirical facts are known, only a few great generalizations from them—such as the existence of the subconscious mind—have yet been made. But anyone who has seen even a single example of the power of hypnotism and suggestion must realise that the face of the world and the possibilities of existence will be totally altered when we can control their effects and standardize their application, as has been possible, for example, with drugs which were once regarded as equally magical. Infinitely greater, of course, would be the results of the opening up of systematic communication with spiritual beings in another world, which is claimed as a scientific possibility. Spiritualism is already Christianity's most formidable enemy, and we have no data which allow us to estimate the probable effect on man of a religion whose dogmas are matters of experiment, whose mysteries are prosaic as electric lighting, whose ethics are based on the observed results in the next world of a good or bad life in this. Yet that is the prospect before us if spiritualism obtains the scientific verification which it is now demanding, not perhaps with great success.[64]

I have only been able, in the time at my disposal, to traverse a very few of the possible fields of scientific advance. If I have convinced anyone present that science has still a good deal up her sleeve, and that of sufficiently startling character, I shall be amply repaid. If anything I have said appears to be of a gratuitously disgusting nature, I would reply that certain phenomena of normal life do seem to many to be of that nature, and that these phenomena are of the utmost scientific and practical importance.

I have tried to show why I believe that the biologist is the most romantic figure on earth at the present day. At first sight he seems to be just a poor little scrubby underpaid man, groping blindly amid the mazes of the ultra-microscopic, engaging in bitter and lifelong quarrels over the nephridia[65] of flatworms, waking perhaps one morning to find that someone whose name he has never heard has demolished by a few crucial experiments the work which he had hoped would render him immortal. There is real tragedy in his life, but he knows that he has a responsibility which he dare not disclaim, and he is urged on, apart from all utilitarian considerations, by something or someone which he feels to be higher than himself.

The conservative has but little to fear from the man whose reason is the servant of his passions, but let him beware of him in whom reason has become the greatest and most terrible of the passions. These are the wreckers of outworn empires and civilisations, doubters, disintegrators, deicides.[66] In the past they have been, in general, men like Voltaire, Bentham, Thales,[67] Marx, and very possibly the divine

64 Spiritualism—the purported ability to communicate with spirits of the dead—had become popular again during and after the First World War. The Society for Psychical Research had been founded in 1882 to conduct scientific research on paranormal phenomena such as ghosts and telepathy, and is still extant. The subject was represented in To-day and To-morrow by E. N. Bennett's *Apollonius; or, the Present and Future of Psychical Research* (1927).

65 Waste-removing organs.

66 The killers of gods.

67 Thales of Miletus (seventh to sixth centuries BCE) was a leading pre-Socratic Greek philosopher, one of the 'Seven Sages.' He is credited as the first thinker to use natural philosophy, mathematics, science, and deductive reasoning rather than mythology to explain natural phenomena.

Julius,[68] but I think that Darwin furnishes an example of the same relentlessness of reason in the field of science. I suspect that as it becomes clear that at present reason not only has a freer play in science than elsewhere, but can produce as great effects on the world through science as through politics, philosophy, or literature, there will be more Darwins. Such men are interested primarily in truth as such, but they can hardly be quite uninterested in what will happen when they throw down their dragon's teeth into the world.

I do not say that biologists as a general rule try to imagine in any detail the future applications of their science. The central problems of life for them may be the relationship between the echinoderms and the brachiopods,[69] and the attempt to live on their salaries. They do not see themselves as sinister and revolutionary figures. They have no time to dream. But I suspect that more of them dream than would care to confess it.

I have given above a very small selection from my dreams. Perhaps they are bad dreams. It is of course almost hopeless to attempt any very exact prophecies as to how in detail scientific knowledge is going to revolutionize human life, but I believe that it will continue to do so, and even more profoundly than I have suggested. And though personally I am Victorian enough in my sympathies to hope that after all family life, for example, may be spared, I can only reiterate that not one of the practical advances which I have predicted is not already foreshadowed by recent scientific work. If a chemist or physicist living at the end of the seventeenth century had been asked to predict the future application of his science he would doubtless have made many laughable errors in the best Laputan style,[70] but he would have been certain that it would somehow be applied, and his faith would have been justified.

We must regard science then from three points of view. First it is the free activity of man's divine faculties of reason and imagination. Secondly it is the answer of the few to the demands of the many for wealth, comfort and victory, for "νόσων τ' ἀπείρους και μακραίωνας βιους,"[71] gifts which it will grant only in exchange for peace, security and stagnation. Finally it is man's gradual conquest, first of space

68 After Julius Caesar's assassination, the Roman Senate declared him a god, and the Temple of Divus Iulius was built in the Forum.

69 Echinoderms are sea-creatures including starfish and sea-urchins. Brachiopods are shellfish resembling bivalve mollusks such as cockles and mussels, but are anatomically unrelated to them.

70 Laputa is a magnetically-levitating flying island in Jonathan Swift's *Gulliver's Travels* (1726) populated by mad scientists and their adulterous wives.

71 In his preface for the fifth impression Haldane gave the source for this quotation as 'Aeschylus, fragment of *Niobe.*' However, in *Aeschylus*, ed. Herbert Weir Smyth, Loeb Classical Library (Cambridge, MA: Harvard University Press, 1963), 2 vols, II: 490, it appears as no. 189 of 'Fragments of Uncertain Plays'; translation: '[He dwelt on my happiness in my children,] whose days were to be many and unacquainted with disease.' In its longer source in Plato's *Republic*, it refers to the loss of a single son rather than Niobe's many children, and has been taken to be Thetis lamenting the death of her son Achilles.

and time, then of matter as such, then of his own body and those of other living beings, and finally the subjugation of the dark and evil elements in his own soul.[72]

None of these conquests will ever be complete, but all, I believe will be progressive. The question of what he will do with these powers is essentially a question for religion and aesthetic[s]. It may be urged that they are only fit to be placed in the hands of a being who has learned to control himself, and that man armed with science is like a baby armed with a box of matches.

The answer to this contention may, I think, be found in the daily papers. For scores of centuries idealists had urged that wars must cease and all the earth be united under one rule. As long as any other alternative was possible it was persisted in. The events of the last nine years constituted a reductio ad absurdum of war, but when we ask who was responsible for this we shall find that it was not the visionaries but men like Black, Kekulé, and Langley,[73] who enlarged man's power over nature until he was forced by the inexorable logic of facts to form the nucleus of an international government.

We have already reacted against the frame of mind that engendered the league of nations, but we have not reacted at all completely. The league exists and is working, and in every country on earth there are many people, and ordinary normal people, who favour the idea in one form or another of a world state.[74] I do not suggest that a world-state will arise from the present league—or for the matter of that from the third international.[75] I merely observe that there is a widespread and organized desire for such an institution, and several possible nuclei for it. It may take another world-war or two to convert the majority. The prospect of the next world-war has at least this satisfactory element. In the late war the most rabid nationalists were to be found well behind the front line. In the next war no one will be behind the front line. It will be brought home to all whom it may concern that war is a very dirty business.

No doubt there is a fair chance that the possibility of human organization on a planetary scale may be rendered impossible by such a war. If so mankind will probably have to wait for a couple of thousand years for another opportunity. But to the student of geology such a period is negligible. It took man 250,000 years to transcend the hunting pack. It will not take him so long to transcend the nation.

I think then that the tendency of applied science is to magnify injustices until they become too intolerable to be borne, and the average man whom all the prophets and poets could not move, turns at last and extinguishes the evil at its

72 See J. D. Bernal's *The World, the Flesh and the Devil* (1929), also in this volume, which explores the tripartite division Haldane outlines here.

73 Scottish physicist and chemist Joseph Black (1728–1799) discovered carbon dioxide and magnesium, and the concepts of latent heat and specific heat. August Kekulé (1829 –1896) was a German organic chemist famous for his work on the structure of benzene. British physiologist John Newport Langley (1852–1925) pioneered the theory of chemical receptors and neurotransmitters.

74 The most prominent British advocate of a World State was H. G. Wells. His 1914 novel *The World Set Free* imagined a nuclear war making leaders realize the need for an internationalist world order.

75 Also known as the Communist International (or Comintern); a Soviet-led international organization formed in 1919 to work towards world communism.

source. Marx' theory of industrial evolution is a particular example of this tendency, though it does not in the least follow that his somewhat artificial solution of the problem will be adopted.

It is probable that biological progress will prove to be as incompatible with certain of our social evils as industrial progress has proved to be with war or certain systems of private ownership. To take a concrete example it is clear that the second biological invention considered by my future essayist would be intolerable in conjunction with our present system of relations between classes and sexes. Moral progress is so difficult that I think any developments are to be welcomed which present it as the naked alternative to destruction, no matter how horrible may be the stimulus which is necessary before man will take the moral step in question.

To sum up, then, science is as yet in its infancy, and we can foretell little of the future save that the thing that has not been is the thing that shall be; that no beliefs, no values, no institutions are safe. So far from being an isolated phenomenon the late war is only an example of the disruptive result that we may constantly expect from the progress of science. The future will be no primrose path. It will have its own problems. Some will be the secular problems of the past, giant flowers of evil blossoming at last to their own destruction. Others will be wholly new. Whether in the end man will survive his accessions of power we cannot tell. But the problem is no new one. It is the old paradox of freedom re-enacted with mankind for actor and the earth for stage. To those who believe in the divinity of that part of man which aspires after knowledge for its own sake, who are, in the words of Boethius:

> te cernere finis
> Principium, vector, dux, semita,
> terminus idem.[76]

the prospect will appear most hopeful. But it is only hopeful if mankind can adjust its morality to its powers. If we can succeed in this, then science holds in her hands one at least of the keys to the thorny and arduous path of moral progress, then:

> Per cruciamina leti
> Via panditur ardua justis,
> Et ad astra doloribus itur.[77]

76 Boethius' *De Consolatione Philosophiae* was written in 523 while he was in prison awaiting execution. III.m9.26–28: 'Thou only calm and rest / To pious men whose end is to behold Thy ray, / Who their beginning art, their guide, their bound, and way'; *The Theological Tractates / The Consolation of Philosophy*, Loeb Classical Library (London: Heinemann/ New York: Putnam's, 1918), 267.

77 Hymns of Prudentius, from the *Liber Cathemerinon*: X. Hymn for the Burial of the Dead, ll. 90–92: '[Therefore is death itself more blessed, in that] through the pains of death a way on high is opened for the righteous and by their sufferings they pass to the skies'; trans. H. J. Thomson, *Prudentius,* 2 vols (London: Heinemann/Cambridge, MA: Harvard University Press, 1949), I: 91.

That is possibly a correct large-scale view, but it is only for short periods that one can take views of history sufficiently broad to render the fate of one's own generation irrelevant. The scientific worker is brought up with the moral values of his neighbours. He is perhaps fortunate if he does not realize that it is his destiny to turn good into evil. The moral and physical (though not the intellectual) virtues are means between two extremes. They are essentially quantitative. It follows that an alteration in the scale of human power will render actions bad which were formerly good. Our increased knowledge of hygiene has transformed resignation and inaction in face of epidemic disease from a religious virtue to a justly punishable offence. We have improved our armaments, and patriotism, which was once a flame upon the altar, has become a world-devouring conflagration.

The time has gone by when a Huxley[78] could believe that while science might indeed remould traditional mythology, traditional morals were impregnable and sacrosanct to it. We must learn not to take traditional morals too seriously. And it is just because even the least dogmatic of religions tends to associate itself with some kind of unalterable moral tradition, that there can be no truce between science and religion.

There does not seem to be any particular reason why a religion should not arise with an ethic as fluid as Hindu mythology, but it has not yet arisen. Christianity has probably the most flexible morals of any religion, because Jesus left no code of law behind him like Moses or Muhammad, and his moral precepts are so different from those of ordinary life that no society has ever made any serious attempt to carry them out, such as was possible in the case of Israel and Islam. But every Christian church has tried to impose a code of morals of some kind for which it has claimed divine sanction. As these codes have always been opposed to those of the gospels a loophole has been left for moral progress such as hardly exists in other religions. This is no doubt an argument for Christianity as against other religions, but not as against none at all, or as against a religion which will frankly admit that its mythology and morals are provisional. That is the only sort of religion that would satisfy the scientific mind, and it is very doubtful whether it could properly be called a religion at all.

No doubt many people hope that such a religion may develop from Christianity. The human intellect is feeble, and there are times when it does not assert the infinity of its claims. But even then:

> Though in black jest it bows and nods
> *   *   *   *
> I know it is roaring at the gods
> Waiting for the last eclipse.[79]

78 Thomas Henry Huxley FRS (1825–1895), biologist, anthropologist and champion of Darwinian evolution; not his grandson, and Haldane's friend, the novelist Aldous Huxley.

79 G. K. Chesterton, The Ballad of The White Horse: Book III: The Harp Of Alfred.

The scientific worker of the future will more and more resemble the lonely figure of Daedalus as he becomes conscious of his ghastly mission, and proud of it.

Black is his robe from top to toe,
His flesh is white and warm below,
All through his silent veins flow free
Hunger and thirst and venery,
But in his eyes a still small flame
Like the first cell from which he came
Burns round and luminous, as he rides
Singing my song of deicides.[80]

80 Slightly misquoted from Robert Buchanan (1841–1901), *The Book of Orm* (1870); IX: The Devil's Mystics; VIII: *Homunculus; or, The Song of Deicides*, stanza 5:

Black is his raiment, top to toe,
His flesh is white and warm below.
All thro' his silent veins flow free
Hunger, and Thirst, and Venery;
But in his eye a still small flame,
Like the first Cell from which he came,
Burns round and luminous,—as he rides
To my still Song of Deicides!

The line 'Singing my song [or 'the song'] of Deicides' alternates with this version as a refrain, suggesting that Haldane was quoting from memory. In his preface to the fifth impression of *Daedalus* he wrote: 'Buchanan was a mystical Presbyterian minister, and perhaps the greatest poet produced by Scotland during the last Century. Like many mystics he was somewhat uncritical, and published a good deal of bad verse. Some of the worst of it may be found in English anthologies.'

Icarus[1]*; or, the Future of Science*

Bertrand Russell (1924)

Headnote[1]

Bertrand Arthur William Russell, 3rd Earl Russell, OM, FRS (1872–1970), was a British philosopher, historian, logician, mathematician, and advocate for social reform. He was imprisoned during the First World War for campaigning for pacifism.

Russell was an influential logician. He wrote *Principia Mathematica* (1910–1913) with his former teacher A. N. Whitehead, and was a founder of analytic philosophy with his student Ludwig Wittgenstein and his friend and colleague G. E. Moore. He was a prolific writer. His influential 1946 survey *A History of Western Philosophy* contributed to his being awarded the Nobel Prize for Literature in 1950.

His other volume for the series, *What I Believe* (1925) is included in Part II of this *Reader*, together with his wife Dora Russell's volume, *Hypatia* (1925).

****

I. Introductory

Mr. Haldane's *Daedalus* has set forth an attractive picture of the future as it may become through the use of scientific discoveries to promote human happiness. Much as I should like to agree with his forecast, a long experience of statesmen and government has made me somewhat sceptical. I am compelled to fear that science will be used to promote the power of dominant groups, rather than to make men happy. Icarus, having been taught to fly by his father Daedalus, was destroyed by his rashness.[2] I fear that the same fate may overtake the populations whom modern men of science have taught to fly. Some of the dangers inherent in the progress of

1 In Greek mythology Icarus was the son of the Cretan inventor Daedalus. They were imprisoned by King Minos—in the labyrinth Daedalus had designed to contain the Minotaur, according to one version. Daedalus made wings so they could fly to freedom. But Icarus ignored his warning not to fly too close to the sun, lest the wax holding the feathers melted.

2 Haldane had provocatively chosen Daedalus as his emblematic scientist not for the most famous part of his story, but as a proto-geneticist. See p. 33. Russell brings the myth back down to earth through the more familiar story of Icarus's crash.

DOI: 10.4324/9781032625522-5

science while we retain our present political and economic institutions are set forth in the following pages.

This subject is so vast that it is impossible, within a limited space, to do more than outline some of its aspects. The world in which we live differs profoundly from that of Queen Anne's time,[3] and this difference is mainly attributable to science. That is to say, the difference would be very much less than it is but for various scientific discoveries, but resulted from those discoveries by the operation of ordinary human nature. The changes that have been brought about have been partly good, partly bad; whether, in the end, science will prove to have been a blessing or a curse to mankind, is to my mind, still a doubtful question.

A science may affect human life in two different ways. On the one hand, without altering men's passions or their general outlook, it may increase their power of gratifying their desires. On the other hand, it may operate through an effect upon the imaginative conception of the world, the theology or philosophy which is accepted in practice by energetic men. The latter is a fascinating study, but I shall almost wholly ignore it, in order to bring my subject within a manageable compass. I shall confine myself almost wholly to the effect of science in enabling us to gratify our passions more freely, which has hitherto been far the more important of the two.

From our point of view, we may divide the sciences into three groups: physical, biological, and anthropological. In the physical group I include chemistry, and broadly speaking any science concerned with the properties of matter apart from life. In the anthropological group I include all studies specially concerned with man: human physiology and psychology (between which no sharp line can be drawn), anthropology, history, sociology, and economics. All these studies can be illuminated by considerations drawn from biology; for instance, Rivers threw a new light on parts of economics by adducing facts about landed property among birds during the breeding season.[4] But in spite of their connection with biology—a connection which is likely to grow closer as time goes on—they are broadly distinguished from biology by their methods and data, and deserve to be grouped apart, at any rate in a sociological inquiry.

The effect of the biological sciences, so far, has been very small. No doubt Darwinism and the idea of evolution affected men's imaginative outlook; arguments were derived in favour of free competition, and also of nationalism. But these effects were of the sort that I propose not to consider. It is probable that great effects will come from these sciences sooner or later. Mendelism[5] might have revolutionized agriculture, and no doubt some similar theory will do so sooner or

3 Anne (1665–1714) was Queen of England, Scotland, and Ireland from 1702, and Queen of Great Britain and Ireland following the ratification of the Acts of Union in 1707.

4 William Halse Rivers (W. H. R.) Rivers (1864–1922), anthropologist, neurologist, and psychologist, is now best-known for treating the First World War poets Wilfred Owen, Robert Graves and Siegfried Sasson at Craiglockhart Hospital in Edinburgh.

5 "The fundamental principles of inheritance (especially the laws of segregation and independent assortment and the existence of dominant and recessive characters), propounded originally by Mendel

later. Bacteriology may enable us to exterminate our enemies by disease. The study of heredity may in time make eugenics an exact science, and perhaps we shall in a later age be able to determine at will the sex of our children. This would probably lead to an excess of males, involving a complete change in family institutions. But these speculations belong to the future. I do not propose to deal with the possible future effects of biology, both because my knowledge of biology is very limited, and because the subject has been admirably treated by Mr. Haldane.[6]

The anthropological sciences are those from which, *a priori*, we might have expected the greatest social effects, but hitherto this has not proved to be the case, partly because these sciences are mostly still at an early stage of development. Even economics has not so far had much effect. Where it has seemed to have, this is because it advocated what was independently desired. Hitherto, the most effective of the anthropological sciences has been medicine, through its influence on sanitation and public health, and through the fact that it has discovered how to deal with malaria and yellow fever. Birth-control is also a very important social fact which comes into this category. But although the future effect of the anthropological sciences (to which I shall return presently) is illimitable, the effect up to the present has been confined within fairly narrow limits.

One general observation to begin with. Science has increased man's control over nature, and might therefore be supposed likely to increase his happiness and well-being. This would be the case if men were rational, but in fact they are bundles of passions and instincts. An animal species in a stable environment, if it does not die out, acquires an equilibrium between its passions and the conditions of its life. If the conditions are suddenly altered, the equilibrium is upset. Wolves in a state of nature have difficulty in getting food, and therefore need the stimulus of a very insistent hunger. The result is that their descendants, domestic dogs, over-eat if they are allowed to do so. When a certain amount of something is useful, and the difficulty of obtaining it is diminished, instinct will usually lead an animal to excess in the new circumstances. The sudden change produced by science has upset the balance between our instincts and our circumstances, but in directions not sufficiently noted. Over-eating is not a serious danger, but over-fighting is. The human instincts of power and rivalry, like the dog's wolfish appetite will need to be artificially curbed, if industrialism is to succeed.

II. Effects of the Physical Sciences

Much the greatest part of the changes which science has made in social life is due to the physical sciences, as is evident when we consider that they brought about the industrial revolution. This is a trite topic, about which I shall say as little as my subject permits. There are, however, some points which must be made.

---

and forming the basis for the science of classical genetics." Named after Gregor Mendel (1822–1884) the 'father of genetics' (*OED*).

6 [Author's note] See his *Daedalus, or Science and the Future,* Kegan Paul, 1923, 2/6 net.

First, industrialism still has great parts of the earth's surface to conquer. Russia and India are very imperfectly industrialized; China hardly at all. In South America there is room for immense development. One of the effects of industrialism is to make the world an economic unit: its ultimate consequences will be very largely due to this fact. But before the world can be effectively organized as a unit, it will probably be necessary to develop industrially all the regions capable of development that are at present backward. The effects of industrialism change as it becomes more wide-spread; this must be remembered in any attempt to argue from its past to its future.

The second point about industrialism is that it increases the productivity of labour, and thus makes more luxuries possible. At first, in England, the chief luxury achieved was a larger population with an actual lowering of the standard of life. Then came a golden age when wages increased, hours of labour diminished, and simultaneously the middle-class grew more prosperous. That was while Great Britain was still supreme. With the growth of foreign industrialism, a new epoch began. Industrial organizations have seldom succeeded in becoming world-wide, and have consequently become national. Competition, formerly between individual firms, is now mainly between nations, and is therefore conducted by methods quite different from those contemplated by the classical economists.

Modern industrialism is a struggle between nations for two things, markets and raw materials, as well as for the sheer pleasure of domination. The labour which is set free from providing the necessaries of life tends to be more and more absorbed by national rivalry. There are first the armed forces of the State; then those who provide munitions of war, from the raw minerals up to the finished product; then the diplomatic and consular services; then the teachers of patriotism in schools; then the Press. All of these perform other functions as well, but the chief purpose is to minister to international competition. As another class whose labours are devoted to the same end, we must add a considerable proportion of the men of science. These men invent continually more elaborate methods of attack and defence. The net result of their labours is to diminish the proportion of the population that can be put into the fighting line, since more are required for munitions. This might seem a boon, but in fact war is now-a-days primarily against the civilian population, and in a defeated country they are liable to suffer just as much as the soldiers.

It is science above all that has determined the importance of raw materials in international competition. Coal and iron and oil, especially, are the bases of power, and thence of wealth. The nation which possesses them, and has the industrial skill required to utilize them in war, can acquire markets by armed force, and levy tribute upon less fortunate nations. Economists have underestimated the part played by military prowess in the acquisition of wealth. The landed aristocracies of Europe were, in origin, warlike invaders. Their defeat by the bourgeoisie in the French Revolution, and the fear which this generated in the Duke of Wellington, facilitated the rise of the middle class. The wars of the eighteenth century decided that England was to be richer than France. The traditional economist's rules for the distribution of wealth hold only when men's actions are governed by law, i.e. when most people think the issue unimportant. The issues that people have considered vital have been decided by civil war or wars between nations. And for the present, owing to science, the art of war consists in possessing coal, iron, oil, and the industrial skill to work them. For the sake of simplicity, I omit other raw materials, since they do not affect the essence of our problem.

We may say, therefore, speaking very generally, that men have used the increased productivity which they owe to science for three chief purposes in succession: first, to increase the population; then, to raise the standard of comfort; and, finally, to provide more energy to war. This last result has been chiefly brought about by competition for markets, which led to competition for raw materials, especially the raw materials of munitions.

III. The Increase of Organization

The stimulation of nationalism which has taken place in modern times is, however, due very largely to another factor, namely the increase of organization, which is of the very essence of industrialism. Wherever expensive fixed capital is required, organization, on a large scale is of course necessary. In view of the economies of large scale production, organization in marketing also becomes of great importance. For some purposes, if not for all, many industries come to be organized nationally, so as to be in effect one business in each nation.

Science has not only brought about the need for large organizations, but also the technical possibility of their existence. Without railways, telegraphs and telephones, control from a centre is very difficult. In ancient empires, and in China down to modern times, provinces were governed by practically independent satraps or proconsuls,[7] who were appointed by the central government, but decided almost all questions on their own initiative. If they displeased the sovereign, they could only be controlled by civil war, of which the issue was doubtful. Until the invention of the telegraph[8], ambassadors had a great measure of independence, since it was often necessary to act without waiting for orders from home. What applied in politics applied also in business: an organization controlled from the centre had to be very loosely knit, and to allow much autonomy to subordinates. Opinion as well as action was difficult to mould from a centre, and local variations marred the uniformity of party creeds.

Now-a-days all this is changed. Telegraph, telephone, and wireless make it easy to transmit orders from a centre: railways and steamers make it easy to transport troops in case the orders are disobeyed. Modern methods of printing and advertising make it enormously cheaper to produce and distribute one newspaper with a large circulation than many with small circulations; consequently, in so far as the Press controls opinion, there is uniformity, and, in particular, there is uniformity of news. Elementary education, except in so far as religious denominations introduce variety, is conducted on a uniform pattern decided by the State, by means of

7 Persian or Roman provincial governors.
8 Invented by Samuel Morse in 1844.

teachers whom the State has trained, as far as possible, to imitate the regularity and mutual similarity of machines produced to standard. Thus the material and psychological conditions for a great intensity of organization have increased *pari passu*,[9] but the basis of the whole development is scientific invention in the purely physical realm. Increased productivity has played its part, by making it possible to set apart more labour for propaganda, under which head are to be included advertisement, the cinema, the Press, education, politics, and religion. Broadcasting is a new method likely to acquire great potency as soon as people are satisfied that it is *not* a method of propaganda.

Political controversies, as Mr. Graham Wallas[10] has pointed out, ought to be conducted in quantitative terms. If sociology were one of the sciences that had affected social institutions (which it is not), this would be the case. The dispute between anarchism and bureaucracy at present tends to take the form of one side maintaining that we want no organization, while the other maintains that we want as much as possible. A person imbued with the scientific spirit would hardly even examine these extreme positions. Some people think that we keep our rooms too hot for health, others that we keep them too cold. If this were a political question, one party would maintain that the best temperature is the absolute zero, the other that it is the melting point of iron. Those who maintained any intermediate position would be abused as timorous time-servers, concealed agents of the other side, men who ruined the enthusiasm of a sacred cause by tepid appeals to mere reason. Any man who had the courage to say that our rooms ought to be neither very hot nor very cold would be abused by both parties, and probably shot in No Man's Land. Possibly some day politics may become more rational, but so far there is not the faintest indication of a change in this direction.

To a rational mind, the question is not: Do we want organization or do we not? The question is: How much organization do we want, and where and when and of what kind? In spite of a temperamental leaning to anarchism, I am persuaded that an industrial world cannot maintain itself against internal disruptive forces without a great deal more organization than we have at present. It is not the amount of organization, buts its kind and its purpose, that causes our troubles. But before tackling this question, let us pause for a moment to ask ourselves what is the measure of the intensity of organization in a given community.

A man's acts are partly determined by spontaneous impulse, partly by the conscious or unconscious effects of the various groups to which he belongs. A man who works (say) on a railway or in a mine is, in his working hours almost entirely determined in his actions by those who direct the collective labour of which he forms part. If he decides to strike, his action is again not individual, but determined by his Union. When he votes for Parliament, party caucuses have limited his choice to one of two or three men, and party propaganda has induced him to accept *in toto*

9 Latin: 'at the same rate.'

10 An educationist and leading member of the Fabian Society, Graham Wallas (1858–1932) became a lecturer at the London School of Economics. His *Human Nature in Politics* (1908) was a founding text of the modern study of political psychology.

one of two or three blocks of opinions which form the rival party programmes. His choice between the parties may be individual, but it may also be determined by the action of some group, such as a trade union, which collectively supports one party. His newspaper-reading exposes him to great organized forces; so does the cinema, if he goes to it. His choice of a wife is probably spontaneous, except that he must choose a woman of his own class. But in the education of his children he is almost entirely powerless: they must have the education which is provided. Organization thus determines many vital things in his life. Compare him to a handicraftsman or peasant-proprietor who cannot read and does not have his children educated, and it becomes clear what is meant by saying that industrialism has increased the intensity of organization. To define this term we must, I think, exclude the unconscious effects of groups, except as causes facilitating the conscious effects. We may define the intensity of organization to which a given individual is subject as the proportion of his acts which is determined by the orders or advice of some group, expressed through democratic decisions or executive officers. The intensity of organization in a community may then be defined as the average intensity for its several members.

The intensity of organization is increased not only when a man belongs to more organizations, but also when the organizations to which he already belongs play a larger part in his life, as, for example, the State plays a larger part in war than in peace.

Another matter which needs to be treated quantitatively is the degree of democracy, oligarchy, or monarchy in an organization. No organization belongs completely to any one of the three types. There must be executive officers, who will often in practice be able to decide policy, even if in theory they cannot do so. And even if their power depends upon persuasion, they may so completely control the relevant publicity that they can always rely upon a majority. The directors of a railway company, for instance, are to all intents and purposes uncontrolled by the shareholders, who have no adequate means of organizing an opposition if they should wish to do so. In America, a railroad president is almost a monarch. In party politics, the power of leaders, although it depends upon persuasion, continually increases as printed propaganda becomes more important. For these reasons, even where formal democracy increases, the real degree of democratic control tends to diminish, except on a few questions which rouse strong popular passions.

The result of these causes is that, in consequence of scientific inventions which facilitate centralization and propaganda, groups become more organized, more disciplined, more group-conscious and more docile to leaders. The effect of leaders on followers is increased, and the control of events by a few prominent personalities becomes more marked.

In all this there would be nothing very tragic, but for the fact, with which science has nothing to do, that organization is almost wholly national. If men were actuated by the love of gain, as the older economists supposed, this would not be the case; the same causes which have led to national trusts would have led to international trusts. This has happened in a few instances, but not on a sufficiently wide scale to affect politics or economics very vitally. Rivalry is, with most well-to-do energetic people, a stronger motive than love of money. Successful rivalry requires organization of rival forces; the tendency is for a business such as oil, for example, to organize itself into two rival groups, between them covering the world. They might, of course, combine, and they would no doubt increase their wealth if they did so. But combination would take the zest out of life. The object of a football team, one might say, is to kick goals. If two rival teams combined, and kicked the ball alternately over the two goals, many more goals would be scored. Nevertheless no one suggest that this should be done, the object of a football team being not to kick goals but to win. So the object of a big business is not to make money, but to win in the contest with some other business. If there were no other business to be defeated, the whole thing would become uninteresting. This rivalry has attached itself to nationalism, and enlisted the support of the ordinary citizens of the countries concerned; they seldom know what it is that they are supporting, but, like the spectators at a football-match, they grow enthusiastic for their own side. The harm that is being done by science and industrialism is almost wholly due to the fact that, while they have proved strong enough to produce a *national* organization of economic forces, they have not proved strong enough to produce an international organization. It is clear that political internationalism such as the League of Nations was supposed to inaugurate, will never be successful until we have economic internationalism, which would require, as a minimum, an agreement between various national organizations dividing among them the raw material and markets of the world. This, however, can hardly be brought about while big business is controlled by men who are so rich as to have grown indifferent to money, and to be willing to risk enormous losses for the pleasure of rivalry.

The increase of organization in the modern world has made the ideals of liberalism wholly inapplicable. Liberalism, from Montesquieu to President Wilson,[11] was based upon the assumption of a number of more or less equal individuals or groups, with no differences so vital that they were willing to die sooner than compromise. It was supposed that there was to be free competition between individuals and between ideas. Experience has shown, however, that the existing economic system is incompatible with all forms of free competition except between States by means of armaments. I should wish, for my part, to preserve free competition between ideas, though not between individuals and groups, but this is only possible by means of what an old-fashioned liberal would regard as interferences with personal liberty. So long as the sources of economic power remain in private hands, there will be no liberty except for the few who control those sources.

Such liberal ideals as free trade, free press, unbiased education, either already belong to the past or soon will do so. One of the triumphs of early liberalism in England was the establishment of parliamentary control of the army; this was the

11 Charles-Louis de Secondat, Baron de La Brède et de Montesquieu (1689–1755). One of the foremost political philosophers of the Enlightenment, he developed the theory of the separation of powers in *The Spirit of the Laws* (1748), which influenced the US Constitution. It also argued against slavery. Woodrow Wilson (1856–1924) was the 28th president of the United States from 1913 to 1921.

casus belli[12] in the Civil War, and was decided by the Revolution of 1688.[13] It was effective so long as Parliament represented the same class from which army officers were drawn. This was still the case with the late Parliament, but may cease to be the case with the advent of a Labour Government. Russia, Hungary, Italy, Spain and Bavaria have shown in recent years how frail democracy has become; east of the Rhine it lingers only in outlying regions. Constitutional control over armaments must, therefore, be regarded as another liberal principle which is rapidly becoming obsolete.

It would seem probable that, in the next fifty years or so, we shall see a still further increase in the power of governments, and a tendency for governments to be such as are desired by the men who control armaments and raw materials. The forms of democracy may survive in western countries, since those who possess military and economic power can control education and the press, and therefore can usually secure a subservient democracy. Rival economic groups will presumably remain associated with rival nations, and will foster nationalism in order to recruit their football teams.

There is, however, a hopeful element in the problem. The planet is of finite size, but the most efficient size for an organization is continually increased by new scientific inventions. The world becomes more and more of an economic unity. Before very long the technical conditions will exist for organizing the whole world as one producing and consuming unit. If, when that time comes, two rival groups contend for mastery, the victor may be able to introduce that single world-wide organization that is needed to prevent the mutual extermination of civilized nations. The world which would result would be, at first, very different from the dreams of either liberals or socialists; but it might grow less different with the lapse of time. There would be at first economic and political tyranny of the victors, a dread of renewed upheavals, and therefore a drastic suppression of liberty. But if the first half-dozen revolts were successfully repressed, the vanquished would give up hope, and accept the subordinate place assigned to them by the victors in the great world-trust. As soon as the holders of power felt secure, they would grow less tyrannical and less energetic. The motive of rivalry being removed, they would not work so hard as they do now, and would soon cease to exact such hard work from their subordinates. Life at first might be unpleasant, but it would at least be possible, which would be enough to recommend the system after a long period of warfare. Given a stable world-organization, economic and political, even if, at first, it rested upon nothing but armed force, the evils which now threaten civilization would gradually diminish, and a more thorough democracy than that which now exists might become possible. I believe that, owing to men's folly, a world-government will only be established by force, and therefore be at first cruel and despotic. But I believe that it is necessary for the preservation of a scientific civilization, and that, if once realized, it will gradually give rise to the other conditions of a tolerable existence.

12 Latin: 'occasion for war': an act or event which provokes or is used to justify a war.

13 The 'Glorious Revolution' which deposed Catholic James II in favor of his Daughter Mary and her Protestant husband William of Orange.

IV. The Anthropological Sciences

It remains to say something about the future effects of the anthropological sciences. This is of course extremely conjectural, because we do not know what discoveries will be made. The effect is likely to be far greater than we can now imagine, because these sciences are still in their infancy. I will, however, take a few points on which to hang conjectures. I do not wish to be supposed to be making prophecies: I am only suggesting possibilities which it may be instructive to consider.

Birth-control is a matter of great importance, particularly in relation to the possibility of a world-government, which could hardly be stable if some nations increased their population much more rapidly than others. At present, birth-control is increasing in all civilized countries, though in most it is opposed by governments. This opposition is due partly to mere superstition and desire to conciliate the Catholic vote, partly to the desire for large armies and severe competition between wage-earners, so as to keep down wages. In spite of the opposition of governments, it seems probable that birth-control will lead to a stationary population in most white nations within the next fifty years. There can be no security that it will stop with a stationary population; it may go on to the point where the population diminishes.

The increase in the practice of birth-control is an example of a process contrary to that seen in industrialism: it represents a victory of individual over collective passions. Collectively, Frenchmen desire that France should be populous, in order to be able to defeat her enemies in war. Individually, they desire that their own families should be small, in order to increase the inheritance of their children and to diminish the expense of education. The individual desire has triumphed over the collective desire, and even, in many cases, over religious scruples. In this case, as in most others, the individual desire is less harmful to the world than the collective desire: the man who acts from pure selfishness does less damage than the man who is actuated by "public spirit." For, since medicine and sanitation have diminished the infant death-rate, the only checks to over-population that remain (apart from birth-control) are war and famine. So long as this continues to be the case, the world must either have a nearly stationary population, or employ war to produce famine. The latter method, which is that favoured by opponents of birth-control, has been adopted on a large scale since 1914; it is however somewhat wasteful. We require a certain number of cattle and sheep, and we take steps to secure the right number. If we were as indifferent about them as we are about human beings, we should produce far too many, and cause the surplus to die by the slow misery of under-feeding. Farmers would consider this plan extravagant, and humanitarians would consider it cruel. But where human beings are concerned, it is considered the only proper course, and works advocating any other are confiscated by the police if they are intelligible to those whom they concern.

It must be admitted, however, that there are certain dangers. Before long the population may actually diminish. This is already happening in the most intelligent sections of the most intelligent nations; government opposition to birth-control propaganda gives a biological advantage to stupidity, since it is chiefly stupid people who governments succeed in keeping in ignorance. Before long, birth-control may become nearly universal among the white races; it will then not deteriorate their quality, but only diminish their numbers, at a time when uncivilized races are still prolific and are preserved from a high death-rate by white science.

This situation will lead to a tendency—already shown by the French—to employ more prolific races as mercenaries. Governments will oppose the teaching of birth-control among Africans, for fear of losing recruits. The result will be an immense numerical inferiority of the white races, leading probably to their extermination in a mutiny of mercenaries. If, however, a world-government is established, it may see the desirability of making subject races also less prolific, and may permit mankind to solve the population question. This is another reason for desiring a world-government.

Passing from quantity to quality of population, we come to the question of eugenics. We may perhaps assume that, if people grow less superstitious, governments will acquire the right to sterilize those who are not considered desirable as parents. This power will be used, at first, to diminish imbecility, a most desirable object. But probably, in time, opposition to the government will be taken to prove imbecility, so that rebels of all kinds will be sterilized. Epileptics, consumptives, dipsomaniacs and so on will gradually be included; in the end, there will be a tendency to include all who fail to pass the usual school examinations. The result will be to increase the average intelligence; in the long run, it may be greatly increased. But probably the effect upon really exceptional intelligence will be bad. Mr. Micawber,[14] who was Dickens's father, would hardly have been regarded as a desirable parent. How many imbeciles ought to outweigh one Dickens I do not profess to know.

Eugenics has, of course, more ambitious possibilities in a more distant future. It may aim not only at eliminating undesired types, but at increasing desired types. Moral standards may alter so as to make it possible for one man to be the sire of a vast progeny by many different mothers. When men of science envisage a possibility of this kind, they are prone to a type of fallacy which is common also in other directions. They imagine that a reform inaugurated by men of science would be administered as men of science would wish, by men similar in outlook to those who have advocated it. In like manner women who advocated votes for women used to imagine that the woman voter of the future would resemble the ardent feminist who won her the vote; and socialist leaders imagine that a socialist State would be administered by idealistic reformers like themselves. These are, of course, delusions; a reform, once achieved, is handed over to the average citizen. So, if eugenics reached the point where it could increase desired types, it would not be the types desired by present-day eugenists that would be increased, but rather the type desired by the average official. Prime Ministers, Bishops, and others whom the State considers desirable might become the fathers of half the next generation. Whether

14 Wilkins Micawber is a fictional character in Dickens' novel *David Copperfield* (1850), based on Dickens' father, John Dickens, who ended up in the King's Bench Prison for debt.

this would be an improvement it is not for me to say, as I have no hope of ever becoming either a Bishop or a Prime Minister.[15]

If we knew enough about heredity to determine, within limits, what sort of population we would have, the matter would of course be in the hands of State officials, presumably elderly medical men. Whether they would really be preferable to Nature I do not feel sure. I suspect that they would breed a subservient population, convenient to rulers but incapable of initiative. However, it may be that I am too sceptical of the wisdom of officials.

The effects of psychology on practical life may in time become very great. Already advertisers in America employ eminent psychologists to instruct them in the technique of producing irrational belief; such men may, when they have grown more proficient, be very useful in persuading the democracy that governments are wise and good. Then, again, there are the psychological tests of intelligence, as applied to recruits for the American army during the war. I am very sceptical of the possibility of testing anything except average intelligence by such methods, and I think that, if they were widely adopted, they would probably lead to many persons of great artistic capacity being classified as morons. The same thing would have happened to some first-rate mathematicians. Specialized ability not infrequently goes with general disability, but this would not be shown by the kind of tests which psychologists recommend to the American government.

More sensational than tests of intelligence is the possibility of controlling the emotional life through the secretions of the ductless glands.[16] It will be possible to make people choleric or timid, strongly or weakly sexed, and so on, as may be desired. Differences of emotional disposition seem to be chiefly due to secretions of the ductless glands, and therefore controllable by injections or by increasing or diminishing the secretions. Assuming an oligarchic organization of society, the State could give to the children of holders of power the disposition required for command, and to the children of the proletariat the disposition required for obedience. Against the injections of the State physicians the most eloquent Socialist oratory would be powerless. The only difficulty would be to combine this submissiveness with the necessary ferocity against external enemies; but I do not doubt that official science would be equal to the task.

It is not necessary, when we are considering political consequences, to pin our faith to the particular theories of the ductless glands, which may blow over, like other theories. All that is essential in our hypothesis is the belief that physiology will in time find ways of controlling emotion, which it is scarcely possible to doubt. When that day comes we shall have the emotions desired by our rulers, and the chief business of elementary education will be to produce the desired disposition, no longer by punishment or moral precept, but by the far surer method of injection or diet. The men who will administer this system will have a power beyond the

15 Ray Monk writes that Russell 'received from private tutors an education fit for the future prime minister his grandmother hoped and expected him to become' (*Oxford DNB*), his grandfather having been Prime Minister twice under Queen Victoria.

16 The endocrine glands which secrete hormones into the bloodstream.

dreams of the Jesuits, but there is no reason to suppose that they will have more sense than the men who control education to-day. Technical scientific knowledge does not make men sensible in their aims, and administrators in the future, will be presumably no less stupid and no less prejudiced than they are at present.

Conclusion

It may seem as though I had been at once gloomy and frivolous in some of my prognostications. I will end, however, with the serious lesson which seems to me to result. Men sometimes speak as though the progress of science must necessarily be a boon to mankind, but that, I fear, is one of the comfortable nineteenth-century delusions which our more disillusioned age must discard. Science enables the holders of power to realize their purposes more fully than they could otherwise do. If their purposes are good, this is a gain; if they are evil, it is a loss. In the present age, it seems that the purposes of the holders of power are in the main evil, in the sense that they involve a diminution, in the world at large, of the things men are agreed in thinking good. Therefore, at present, science does harm by increasing the power of rulers. Science is no substitute for virtue; the heart is as necessary for a good life as the head.

If men were rational in their conduct, that is to say, if they acted in the way most likely to bring about the ends that they deliberately desire, intelligence would be enough to make the world almost a paradise. In the main, what is in the long run advantageous to one man is also advantageous to another. But men are actuated by passions which distort their view; feeling an impulse to injure others, they persuade themselves that it is to their interest to do so. They will not, therefore, act in the way that is in fact to their own interest unless they are actuated by generous impulses which make them indifferent to their own interest. This is why the heart is as important as the head. By the "heart" I mean, for the moment, the sum-total of kindly impulses. Where they exist, science helps them to be effective; where they are absent, science only makes men more cleverly diabolic.

It may be laid down as a general principle to which there are few exceptions that, when people are mistaken as to what is to their own interest, the course they believe to be wise is more harmful to others than the course that really is wise. There are innumerable examples of men making fortunes because, on moral grounds, they did something which they believed to be contrary to their own interests. For instance, among early Quakers there were a number of shopkeepers, who adopted the practice of asking no more for their goods than they were willing to accept, instead of bargaining with each customer, as everybody else did. They adopted this practice because they held it to be a lie to ask more than they would take. But the convenience to customers was so great that everybody came to their shops and they grew rich. (I forget where I read this, but if my memory serves me it was in some reliable source). The same policy *might* have been adopted from shrewdness, but in fact no one was sufficiently shrewd. Our unconscious is more malevolent than it pays us to be; therefore the people who do most completely what is in fact to their interest are those who, on moral grounds, do what they believe to be against their interest.

For this reason, it is of the greatest importance to inquire whether any method of strengthening kindly impulses exists. I have no doubt that their strength or weakness depends upon discoverable physiological causes; let us assume that it depends upon the glands. If so, an international secret society of physiologists could bring about the millennium by kidnapping, on a given day, all the rulers of the world, and injecting into their blood some substance which would fill them with benevolence towards their fellow-creatures. Suddenly M. Poincaré would wish well to Ruhr miners, Lord Curzon to Indian nationalists, Mr. Smuts to the natives of what was German South West Africa, the American government to its political prisoners and its victims in Ellis Island.[17] But alas, the physiologists would first have to administer the love-philtre to themselves before they would undertake such a task. Otherwise, they would prefer to win titles and fortunes by injecting military ferocity into recruits. And so we come back to the old dilemma: only kindliness can save the world, and even if we knew how to produce kindliness we should not do so unless we were already kindly. Failing that, it seems that the solution which the Houy[h]nhnms adopted towards the Yahoos,[18] namely extermination, is the only one; apparently the Yahoos are bent on applying it to each other.

We may sum up this discussion in a few words. Science has not given men more self-control, more kindliness, or more power of discounting their passions in deciding upon a course of action. It has given communities more power to indulge their collective passions, but, by making society more organic, it has diminished the part played by private passions. Men's collective passions are mainly evil; far the strongest of them are hatred and rivalry directed towards other groups. Therefore at present all that gives men power to indulge their collective passions is bad. That is why science threatens to cause the destruction of our civilization. The only solid hope seems to lie in the possibility of world-wide domination by one group, say the United States, leading to the gradual formation of an orderly economic and political world-government. But perhaps, in view of the sterility of the Roman Empire, the collapse of our civilization would in the end be preferable to this alternative.

17 At the time Raymond Poincaré was serving the second of his three terms as French Prime Minister. Lord Curzon (1859–1925) had been Viceroy of India and Foreign Secretary. Field Marshal Jan Smuts, OM, FRS (1870–1950) was prime minister of the Union of South Africa from 1919 to 1924. Ellis Island was the busiest immigration processing station in the US; from 1924 it was used to detain migrants.

18 In the last part of Jonathan Swift's *Gulliver's Travels*, the Houyhnhnms are horse-like, super-rational beings; the Yahoos, filthy and brutal versions of humanity.

Wireless Possibilities

A. M. Low (1924)

Headnote

Archibald Montgomery Low has been described as the father of drones and radio guidance systems. He is credited on the title page of *Wireless Possibilities* as 'Late Hon. Ass.t Professor of Physics/Royal Artillery College/Author of The "Two Stroke Engine" and "The Oscillographic Marograph"/"Cinematographic Study of Engineering Design," etc.' He was a television pioneer before the war. While in the Royal Flying Corps he worked on radio-guided rocket science, testing a guided missile on Salisbury Plain in 1917. In the 1920s he specialised in acoustics, but returned to research on rockets and radio, working for the Air Ministry during the Second World War.

Low was a prolific writer on technology and futurology, also publishing *The Future* (London: G. Routledge & Sons, 1925), *Tendencies of Modern Science; or, Science and Modern Life* (London: E. Mathews & Marrot, 1930), *Our Wonderful World of Tomorrow: A Scientific Forecast of the Men, Women, and the World of the Future* (London: Ward, Lock & Co., Limited, 1934), and the wonderfully titled *Science in Wonderland* (London: L. Dickson & Thomson, 1935).

******

To
John Low[1]

Preface

The effects of history upon the advance of science are often noted, but the result of the march of progress is more often entirely neglected.

It would seem desirable that the future should be studied with reasonable accuracy if we are to protect ourselves from the ill-effects and obtain the benefit from the good fortunes of invention.

A. M. L.

1 A. M. Low's father.

DOI: 10.4324/9781032625522-6

Invention

Considering the very evident fact that we owe every detail of our lives, every little comfort which separates us from the cave-man, to the science of invention, it seems strange that so long should have elapsed before this remarkable faculty received proper recognition.

Invention, in many ways, is the science and art of continuity of thought. The inventor is often referred to as a strange person; very true, very necessarily true when we realise that his doings must be strange or new, to be of value. To train oneself to forget the smell of the beefsteak when hungry and to continue the natural sequence of ideas which may be passing through the mind, is to train the brain to improve. If we can but sweep a crossing[2] a very little cleaner than that next to our own, perhaps we have surely accomplished one of the greatest duties of all.

If not one day is spent without something learnt, surely we have achieved the greatest object of work and enabled ourselves to realise that there are no such things as basic facts.

Invention is not labour, for the latter is doing something we do not wish to do in some one else's time, and invention like all good things is a work of love. Possibly that is why it is never paid!

We are too apt, I think, all of us, to rejoice in our greatness as her devotees rejoiced in the greatness of Diana of the Ephesians:[3] we should realise every time we undress that we are little removed from the animal, and that before many centuries have passed we shall be held in almost universal contempt.

If that does not stir us to do our best, we are indeed a nation of shopkeepers. But even the proprietor of the meanest store relies on his powers of prophecy for his profits.

The science of wireless is but a few years old.[4] We know about it little more than our schoolboy sons, and in many cases not so much; let us therefore be open-minded if we are still ignorant.

Commercial invention trusts too far to mass thinking: an original mistake is very closely related to an accomplishment.

The Importance of Sound in Wireless

A few lines of history are desirable here. I do not mean the history controlled by the fact that William the Conqueror made many important appointments in A.D. 1066

2 When horses and carriages—and horse manure—were prevalent in cities, crossing sweepers would clear a path for the affluent to cross busy streets in return for a tip. Jo in Dickens' *Bleak House* is the best-known literary example.

3 The Roman Goddess Diana is the equivalent of Artemis in Greek mythology (see 191, n. 16). Her temple at Ephesus (about 50 km south from the modern port city of Izmir in Turkey) was one of the Seven Wonders of the World.

4 Guglielmo Marconi started experimenting with radio waves in the 1890s. In 1896 he demonstrated wireless telegraph to the English telegraph office and was awarded the patent. The first successful transatlantic messages were received in 1901–1902.

or that Stephen was particularly busy in A.D. 1100.[5] I mean the history of wireless, for, although Radio Science is new, it has a history; all time is relative, and we ourselves are functions of that phenomenon.

Only a few years ago the efforts of wireless experimenters were entirely directed to the converting of the extremely delicate wireless oscillation, still but little understood, into a mechanical movement, in order that the motion of electrons in a problematical aether (which may be nothing but a thought projection and which may exist in many different forms) might be altered into something readable by a man with a check waistcoat and a stock and share list in his hand.

That particular use, and the information that one army is about to kill another could be transmitted to headquarters, naturally occurred to everyone as the first valuable applications of Radio.

The many devices, the electro-magnetic receivers, tape machines, coherers,[6] syphon recorders[7] and the thousand and one electrical machines produced at the time for these purposes, have practically all gone.

Even when to-day we want to send messages quickly, we record them upon a Dictaphone[8] and rely almost entirely upon the sense of hearing.

Sound, the regular oscillation, and noise, the irregular oscillation, of the air, are really the beginning and end of wireless as it is known to the public to-day.

I would go further when thinking of the public. They do not want to sit with a telephone upon their heads, even if their ears may be improved thereby. They require to walk into a drawing-room, and having stood for a moment upon the mat, they must be able to cross the room, touch a button in a fretwork cabinet, and by the movement of a lever be able to place themselves in touch with any part of the world. Paris, Hong-Kong, London, all must be one to them if we are to get their money for our art.

In other words, we are compelled to use what we now designate the "loud speaker." We have got to project a sound into the room before we can sell our instruments, and therein lies one great difficulty.

5 For the significance of 1066 as it struck contemporaries of the To-day and To-morrow writers, see W. C. Sellar and R. J. Yeatman, *1066 and All That: A Memorable History of England, Comprising All the Parts You Can Remember, Including 103 Good Things, 5 Bad Kings and 2 Genuine Dates* (London: Methuen, 1930). Count Stephen of Blois had fled from directing the Siege of Antioch in the First Crusade (1096–1099), and was in disgrace in 1100. His wife Adela (William the Conqueror's daughter) was so ashamed of him she refused to let him stay at home. The following year he joined the minor Crusade of 1101 (also known as the Crusade of the Faint-Hearted because so many involved had turned back from the First Crusade). He was captured and executed in 1102. His son Stephen became King of England in 1135.

6 A 'detector in the form of a glass cylinder containing metal filings, which cohere under the influence of an electromagnetic wave' (*OED*).

7 Electromechanical device invented by Lord Kelvin in 1867 as a receiver for messages transmitted by submarine telegraph cables.

8 The name 'Dictaphone' was trademarked in 1907 by the Columbia Graphophone Company, originally founded by Alexander Graham Bell and his associates, which soon became the leading manufacturer of sound-recording devices, which used a vibrating stylus to cut zigzag grooves in a rotating cylinder coated with wax.

In the first place we dare not exaggerate the movements of a delicate telephone very much or we shall spoil it—therefore we construct something which looks very much like a magnified telephone with a trumpet upon it. The mechanism is naturally rather heavy as regards the moving parts. In order to vibrate these heavy parts with the aid of our aetherial oscillations we have to amplify the available current, and during this process we naturally spoil the detail, or, in other words, we magnify it so much that electrical distortions occur through the whole range of various transformers and other items sold by every shop in the world—at double their value.

Most people are not content with a gentle sound: they find it necessary to express their joy at having reached their home by dancing; consequently they want plenty of sound, and they do not mind if it turns into noise.

They will tell you boldly that their wireless set with a couple of dozen foreign-made valves can be heard right across a large street, a street by the way in which we still permit as much nerve-shattering noise to occur as is thought necessary. This means that we must have quite a big movement on a diaphragm of large size, and a large diaphragm is made to move by the electrical oscillation, itself not very accurate; naturally, if it is heavy, like a poker or anything else, it has a will of its own, and therefore it continues to move when the wireless oscillation has told it to stop. It does not even commence to move when it is told to do so, as it would were it a thin delicate telephone diaphragm from which accurate music can be obtained.

This means further distortion, and so bad is it that a great many people say plainly that they will only listen to wireless concerts through a telephone and that they will only use crystals to obtain rectification because of the inaccuracies otherwise unavoidable to-day.

But this is not business, because do not forget we must have our cabinet with a fern upon it and beautiful music, if we are to be successful. Business always leads science, as we know.

Now think why it is that we need this big diaphragm moving so hard to get a big noise; let us neglect electrical details and consider what produces the noise; or sound, if we are lucky.

Sound is unfortunately purely a mechanical phenomenon as we chiefly understand it, and is produced by oscillations, alternate compression and rarefaction of the atmosphere. Unlike the aether, which sometimes oscillates only too readily, air is a heavy material and has great mass.

You will soon find this out if you put your head out of a railway carriage window, because the air is so heavy that we have got to really kick it and hit it hard before we can obtain a reasonable degree of noise.

When a speaker is standing at one end of a room, irrespective of what he says, the actual temperature-rise of the air can be measured, a fact which was used during the war for the inspection of sound.

Sound is a very complicated thing. It can be reflected in much the same way as light, and I suppose most school-boys know that if a concave mirror is at one end of a room and a similar mirror at the other with a watch hanging at its focus, the

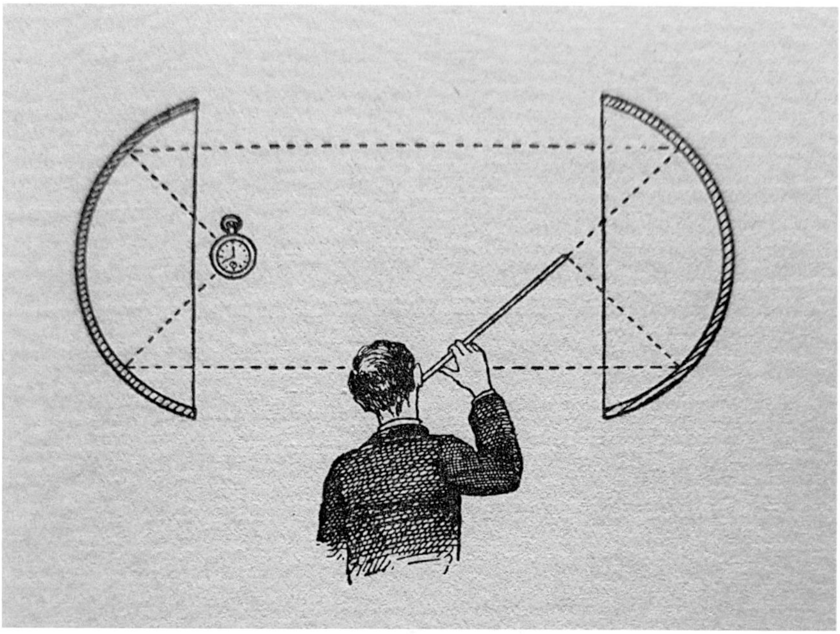

Figure 3.1 Sound can be focussed by a concave reflector and by this means the tick of a watch can be heard at a distance. The watch and the end of the wooden rod are placed at the focus of the reflector.

watch cannot be heard by an observer walking across the room, yet as soon as he places his ear at the focus of the other mirror he will hear the tick clearly, showing that sound is easily reflected. Everybody who has heard an echo should know this (Figure 3.1).

Sound travels also very slowly, and there is plenty of time for wind and different mechanical scraping effects to spoil the purity and partially absorb its delicacies.

Remember that if I am addressing a man by wireless who is one hundred miles away, someone who is listening on a telephone will hear my voice before I am heard at the end of the big hall where I am speaking, because the velocity of sound is only 1100 feet per second, and wireless, like light, travels much faster. Sound can also be actually refracted. Just as the old-fashioned jeweller used a globe of water to concentrate the light upon his work, so will a collodion balloon[9] filled with carbon dioxide, the ordinary gas product of average combustion, act as a lens for sound, which can be actually focussed by these means.

This exemplifies the complication of our subject, and indicates that the heavy diaphragm and other details of the loud speaker must produce serious distortion.

9 Collodion is a 'solution of gun-cotton in ether, forming a colorless gummy liquid, which dries rapidly in the air, owing to evaporation of the ether' (*OED*). It can be used to a produce very light, thin film, as here in balloon form to enclose a gas.

Let us be honest at once. We can only hear such distant places as America by the grace of heaven. Even Sunlight can tune sweet song into vague cracklings. Until true tuning can be obtained we are largely at the mercy of the reproducing instrument, which too often exaggerates every fault and gives the impression that wireless and music are in no way related. No loud speaker of to-day really produces voice and song which sound exactly like voice and song. It all too much resembles a bad gramophone, but without the advantage of the user having the choice of the music.

If user and manufacturer would concentrate upon obtaining purity, if they would try the effects of damping upon loud speakers, which are easily obtained; if they would realise that the horn of the loud speaker should be without resonance, that it should be also damped and pocketed, that its goose-necked shape is not adopted without an object, and if they would aim at the delivery of true music instead of noise—then, we should make a great advance. The average loud speaker can often be greatly improved by padding the horn with some kind of tape, and as an example of the great difficulties of proper transmission let it be made quite clear that with most cases of wireless communication the sending is nearly perfect. Reception is greatly at fault: it is the reception that mangles the sound and makes it too often almost unbearable to anybody of reasonably sensitive hearing.

At some large transmission stations it was at one time quite common to use three separate microphones for the modulation. One received notes of high pitch, one of low, and a third attempted to obtain the "S" sound with the result that, when this "S" microphone was adjusted for a man who did not say his "S" very loudly and someone appeared who did, it sounded exactly as if the speaker had dropped his false teeth.

All this is now avoided. The ordinary diaphragm is no longer in use, but a very small coil of aluminium wire is suspended between the poles of an electro magnet, allowed to rest against an ordinary pad of cotton wool, and that is all!

The infinitesimal movements of this aluminium coil will reproduce speech up to about 40,000 periods per second in oscillatory speed, yet speech is well recognisable if all frequencies over 4,000 per second are gridded out.

How difficult it is going to be to make a large, heavy, and rapidly moving diaphragm reproduce accurately when we have had to take all these precautions to obtain accuracy of transmission! It is not impossible; it will come one day. Now let us see what is the result of our sound troubles. We are told that before long it will be quite easy to hear birds singing in trees and the waves beating against the seashore. Quite right, quite easy to do it now, but if a bird singing in a tree sounds like a man moving his condenser or walking about with a pair of squeaky boots, is it progress?

Perhaps it is. The whole point of wireless is that it brings a man into your room, but it must sound like the man himself if it is to be really effective; it is this pitiful quality of reproduced sound that has wrecked the talking cinema.

It is very easy to photograph sounds and to reproduce them simultaneously with the projection of a picture (Figure 3.2), but to reproduce all the sounds of a cowboy scene accurately is, at present, almost impossible. If a hero says "Good-bye" to a

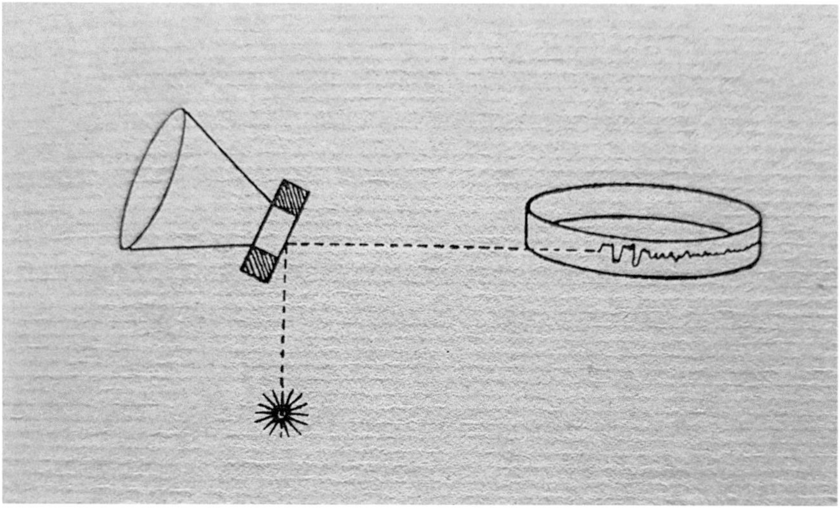

Figure 3.2 If light falls on a very thin diaphragm carrying a mirror, the sound waves being directed into it by a trumpet can be photographed upon a revolving sensitised film.

heroine with a kiss like a creeking board, in the middle of a twenty-reel drama, instead of improving upon the effect of your imagination, which tells you that it is real, and which acts the scene for you better than it can be shown in life or sound, it would be like putting up a blackboard across the screen with the words written upon it, "this is not real—it is only a fake." That is what is wrecking the talking cinema. As a scientific proposition it is easy, but the results are not good enough at present, and, if we can improve, let us first consider the loud speaker.

The talking cinema will come as a matter of course.[10] It is so easy to record sounds upon the film by photography with reproduction by the selenium cell or the neon tube; it is easy to photograph the wave; it is easy actually to impress the sound wave upon the film with the picture, or to use a gramophone; but reproduction is not like a human voice. Neither as yet is any reproduction ever like the voice itself. Let that painful fact be remembered.

Unfortunately, from the business point of view, the long distance reproduction effect is usually satisfactory even when re-broadcasted upon telephones, and for

10 When Low was writing, commercial cinema was still silent. There had been experiments to synchronize sound with motion pictures before the First World War; but amplifying the sound accurately to fill a cinema presented a problem. Speakers and amplifiers improved gradually. Warner Bros. were competing against the large corporations who controlled distribution and exhibition, and felt sound would increase profits. Their 1926 silent feature, *Don Juan*, included a sound track with orchestral music and clanging sword-fights; but few movie theaters had their Vitaphone equipment to play it. It was their fourth Vitaphone feature, *The Jazz Singer* (1927), that signaled the arrival of the 'talkies,' which became the norm by the end of the 1920s. Its 354 or so words of spoken dialogue caused a greater impact than Al Jolson's six musical numbers.

business purposes it is immaterial whether the voice that records the profit or loss is harsh or pleasant.

So the greatest effects we shall soon see from wireless and sound are these: we shall be able to speak to people all over the world by relaying and a combination of land-line and radio; we shall easily be able to connect our office with a wireless station on the coast, radio across the Continent, and then connect by land-line to another office on the other side of the Atlantic.

All this is so easy that no one can doubt that we shall soon listen in to native jamborees; no one can doubt that we shall hear the strange cries of partisans at a baseball match taking place a few miles from New York; no one can question these things, and when reproduction becomes so accurate that the very nature of the people is revealed to us through their speech, surely we might be a little more neighbourly even with those whom we now pretend to love? Relations are notoriously quarrelsome.

If you are in a concert hall and the number of people is varied, it will alter the effect of the sound. You have only to look at a sound-wave photograph produced from a violin to realise from its dainty intricacies that the least variation of any of its harmonics or the very exact shape of its wave beats will reveal all the difference between a beginner and the finest musician in the world. But these things are seldom noticed in wireless.

It is quite easy to photograph a sound, by means of a diaphragm beside which a soap bubble is thick, and to compare wireless sound with the original; even then we have the great difficulties of resonance, and a diaphragm cannot reproduce properly. How, therefore, dare we neglect the dreadful sounds we hear in the name of radio music?

If a piece of silvering, one thirty-secondth of an inch, be scraped from the back of a mirror and fastened to the outer part of a celluloid diaphragm (made by water-floating a drop of amyl acetate in which celluloid has been dissolved), it only requires a horn and a beam of light to render visible the waves of sound. A diaphragm movement of a millionth of a millionth of an inch is sometimes audible.

It is the science of wireless that is beautiful; it is the possibilities that are wonderful; but to talk of pure sound and to judge of it by the human ear which varies after every meal, is like measuring the amount of current passing through an electric-light bulb by feeling its heat with the hand. It is not generally known that, during the War, experiments were made with a sound-reflector for listening to different types of aeroplane and submarine, by means of a microphone placed at the centre of a concave mirror. The difficulty was that of distortion, which is the whole source of trouble with sound producers to-day. Distance is no difficulty and when we can obtain purity and realism as well as distance, the latter is no difficulty at all; then only will be the time when we shall have that spontaneous mental realism of vision that will help radio to alter the world.

In a few years time we shall be able to chat to our friends in an aeroplane and in the streets with the help of a pocket wireless set, and be able to do practically everything by the aid of radio that we now do with our voice. The only thing that will seem intensely strange will be that these comforts never existed before!

Wireless Inaccuracies

I have often wondered whether people realise that broadcasting, at present, is only possible or, shall I say is only popular, because of its extreme impracticability for most forms of secret communication.

Supposing two people had been able to converse privately and with absolute secrecy from other "listeners in," then we should not mind trusting all our messages to Radio. At present, what can be coded can be decoded, and we are not entranced by the idea of entrusting our pennies to the winds of Heaven and the vagaries of a thunderstorm.

If wireless had really been selective[11] in the first instance, Broadcasting would not have been its initial phase.

Wireless at present is excessively inefficient; a few yards from a large broadcasting station the power is measured in millionths of a horse-power, is disseminated in all directions, and is almost without definite selectivity.

When the day comes when we can tune with absolute accuracy; when we can combine waves with accuracy and obtain a directional beam with the shortest waves for re-broadcasting purposes, then we shall obtain real happiness from the results.

Parliament must have its special wave length, the divorce courts of the future will be broadcasted to prevent people from catching cold by waiting outside. It will be quite easy for the Judge, at a doubtful passage, to press a switch and to say, "I think we will cut that out."

One can imagine broadcasting of the future linking up every city from China to London; one can see special wave lengths for men, and equally special wave lengths for women. And we shall forget the time when ships at sea with ancient sets interfere with the murdering of music by the local amateur.

It has been said that, at present, those in authority find it necessary to choose special voices for the wireless broadcast-delivering. What an idea! The public want to hear everybody. They want to have local events broadcast, irrespective of the operator. They do not want a perfect voice, they want a perfect personality, and it is rather the wireless that must be altered to take any reception than the human voice whose very characteristics delight us.

We are too accustomed to relying upon our senses. We are apt to think that the ear is most delicate. It is nothing of the kind; it cannot even hear notes that delight the heart of a dog, and if one pictures life with the brain of a man, the ear of an antelope, and microscopic eyes, together with the nose of a dog, some little idea of the inefficiency of those few senses which we slightly understand can be obtained.

To live in any town would be impossible: the smells of Bond Street instead of pleasing the dog would tell us of rotting animal matter and alarm us to distraction. We could never sit down upon a beautiful piece of grass without listening to the worms and imagining ourselves with them. We could not bear to drink water for the peculiar bodies we would see in it. The wind in the trees, the people walking down our street or into our rabbit-warrens of flats, would sound like a battle from afar.

11 Selectivity is the 'ability of a receiver to tune separately to signals of adjacent frequencies' (*OED*).

It is only a question of relative senses, easily tested by anyone who has the patience to fit an effective microphone to the amplifier purchased in mahogany case at the local "store."

Wireless inaccuracies abound; anyone who hears its music will agree, but what of their effect upon our bodies!

The air, popularly speaking, must now be full of radio oscillations, and if you tell me that they are negligible in effect I may believe you, but if I hear there is no effect at all, I know that it cannot be true.

It is more than likely that, in the far future, the proper study of oscillatory theories, the proper investigation of the spectrum[12] only very partially explored by a few, will lead us to a better understanding of the nature of life, and will help us to appreciate the theories of electrical sonics. Theories of preventing local thunderstorms, of growing babies and wheat effectively, by electrical or other similar oscillatory means, of helping ourselves to see by wireless and of affecting our health at the end of many generations for the better, may all be developed in the time to come.

If some health effect is produced, why should we not try to render it beneficial. A small effect can be very cumulative in nature. One has only to inspect a human nail to agree with that statement.

Radio Television

It has been said of sound that a bell never ceases to echo and that the human voice never ceases when once words are spoken, truly, it is an alarming thought when the nature of most of our sayings is realised.

Not long ago it was claimed that by means of a delicate microphone the sayings of Henry VIII had been investigated—though nature of the subject was, with not less delicacy, omitted.

Much the same basic ideas apply to light except that we are dealing with a very much more interesting phenomenon, one indeed which is not apparently too material and a sense which gives us nearly all our non-physical sensations.

We actually transmit very few senses: we merely convert their nature by utilising different portions of the spectrum. Light has undoubtedly its tone values, as in the case of sound, and it has not yet been definitely established with what portion of the body vision is actually obtained. It is likely that light is projected along the electro lines of force by the movement of electrons but whether the ether consists of electrons themselves, whether it exists in many forms, or is merely a result of the mass effect of thought, we do not yet know. Light very possibly proceeds from the eye as well as from the luminous body concerned.

The science of Radio has taught us something of light, but only to a modest extent, for light yet remains one of the most inefficient factors of a civilisation which almost entirely depends upon it for existence.

12 The electromagnetic spectrum, ranging, with increasing frequency from radio waves, to microwaves, infrared, visible light, ultraviolet, X-rays, and gamma rays.

There is a strange factor which we may call the "Law of Supply and Demand." This strongly implies the faculty of invention, a facility of "wishfulness to improve"; something far better than the necessity for invention. Let us remember that our clothes are not necessities; they are merely comfortable, and it is comfort that distinguishes us to-day just as it is convenience that will in the future give us a life which will be better by far than that experienced by the kings and princes of to-day.

Civilisation has depended almost entirely upon the speeding up of communication. We can travel fast; we can convey our thoughts at great speed, but, unfortunately, although all these means of intercommunication are devised with the one idea of preventing physical work and of obviating the movements of our gross bodies, our senses are very closely combined. It is consequently not possible to ring up somebody on the wireless telephone, a fact itself easy of accomplishment, and to impress our personality upon the listener. This is simply because we require a combination of senses for hearing, seeing, smelling, and other reactions, in order to convey our whole personality.

Vision at a distance is, therefore, very necessary as our inclination for travel decreases and its comfort increases.

It is also important from the point of view of "speeding up," which we have no reason to suppose will cease. All operations have steadily increased in speed for many generations.

There was a time when we made appointments to meet our friends at the full of the moon, but now we say at "10 o'clock, and I can only give you two minutes." In the future we shall probably say, "Meet me at 10.2.1–5 secs., and do not keep me waiting." To do this we must have radio sight.

Many years ago, when experiments were made on the subject, the usual cry appeared from what I always mentally typify as the "Flat Earth Brigade"; they said, "Impossible." What would our forebears have said of talking to a man in an aeroplane? "Impossible!" It is a foolish word. Now all over the world experiments are being conducted, many of them with success and some with the guarantee of reasonable success in twenty years or less.

Now wireless, if I may apply the word here, is very like light in many ways; it is capable of refraction and shadow effects; it travels at the same speed, and if the wave-lengths of wireless could be sufficiently shortened to become visible we should probably find ourselves with a new, and possibly effective, method of transmitting wireless light and even power.

Radio is a phenomenon of the spectrum like ordinary photographic light, X-rays, and so on. It is effects which determine the difference to our eyes of things invisible, solid, and transparent.

It may well be that, when we succeed in inter-planetary communication, we shall discover that the inhabitants see by the X-ray, by wireless, or by heat.

It is not difficult to obtain a proportionate interchange of radio and light oscillations. Even sunlight affects wireless telegraphy, and experiments which have been conducted upon the carrying and directional power of certain other rays and oscillations have not been entirely without results. We may, one day, obtain far greater sensitivity of direction, greater carrying power, from small initial output with a degree of selectivity almost infinite, in comparison with modern working.

To use a light beam along which we can talk, to use a light beam initially and to turn it into light when required, is by no means difficult; it suggests the direct method of wireless vision, but from the mechanical aspect the problem is still less complicated. The difficulties of Radio Television to-day are constructional; in the far future it may be a question of pure physics.

There is, at least, one simple method of sending photographs by wireless with a reasonable degree of accuracy. Distance, re-broadcasting, relaying are, none of them, of any great technical importance. Interference is certainly a difficulty, for in the case of a picture the eye cannot distinguish between faults so easily as the ear can automatically separate unpleasant noises from music.

If an ordinary photograph is transferred to a copper plate, either flat or round, and a contact finger is allowed to pass over it, clearly the resistance between the plate and the finger will vary with the thickness of the photographic film. If this resistance is used to modulate the transmission in place of an ordinary microphone for speech, the current at the receiving end can be picked up, amplified, and used to mark darkly, lightly, or not at all, upon a prepared piece of paper which is affected by the passage of an electric current.

By these means good photographs can be reproduced, and doubtless in the future we shall be able to sign our cheques by the rapid transmission of motion; we shall be able to trace criminals, send out their finger-prints, and carry on very many classes of business which, at present, require our bodily attention.

What a help to the man who objects to a large city. Why could he not conduct his business from his house in comfort instead of having his spats[13] washed every week in order to maintain his financial reputation?

There is a still more rapid method of transmitting a photograph: it is to allow the light from an ordinary lamp to pass through a spot upon the negative and then to a selenium cell (Figure 3.3). Selenium is so constituted that its resistance to the passage of electricity varies with the amount of light to which it is exposed. This property has been used to light up and to extinguish ordinary street lamps, for demonstration purposes.

If a selenium cell is used in place of the ordinary broadcasting microphone, the transmission can be modulated in accordance with the passage of the light through a black spot on the negative, such as part of a top hat, or a white spot, such as a white face or part of it.

The received current is picked up and amplified in the ordinary manner, but instead of operating a diaphragm to produce speech, it is taken to a kind of electrically operated venetian blind, which allows light to pass through it or not to pass through it, in accordance with the transmitting current.

13 Spats, short for spatterdashes, or spatter guards, were cloth accessories which buttoned over the instep and heel to protect shoes and socks from mud.

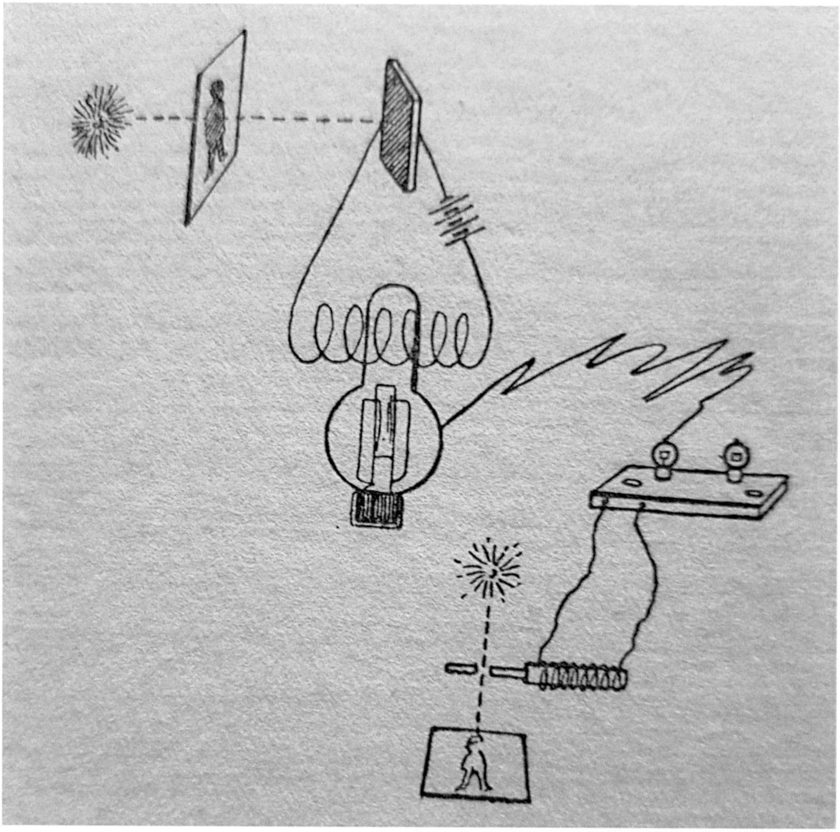

Figure 3.3 If light passes through the negative to be transmitted on to a selenium cell which modulates in place of a microphone, the wave can be picked up, amplified, and made to open or close a shutter. This permits another light to record, spot by spot, the reproduction of the original photograph.

It requires little imagination to see that, if a beam of light is allowed to pass through each point of the original negative in turn, the final picture can be built up from "spots" somewhat in the manner of a half-tone block.[14]

It takes a long time, is rather patched, and is liable to interference; but the whole process is perfectly simple. Consider the great importance of this experiment to Radio Television.

The human eye sees only one point at a time but in the fact that instantaneous vision of a complete picture is not necessary lie our hopes of television to-day.

The eye is a very defective piece of mechanism considered from an optical standpoint. The pointed rays which appear to come from stars show one example

14 Halftone in printing uses dots, varying either in size or in spacing, to reproduce different colors and shades.

of faulty optical construction, however wonderful may be the whole structure. Another property, and a feature of great importance from the aspect of television, is that of retentivity.

We all know that when a lighted cigarette is whirled round in the hand the result appears to be a ring of fire. Our brain assures us that the eye is telling lies and that it is really a moving point. This is because the image is impressed and actually lasts upon the eye or its retina.

This phenomenon is used in every cinematograph;[15] without it the ordinary film would not be practicable. Each picture of an arm about to light a cigarette shows the arm constantly closer and closer, and before one picture has had time to die out the other is thrown upon the screen. The result is an illusion of motion.

To return to the transmission of a photograph, let us imagine that it is sent in a series of spots beginning in the top left-hand corner at 12 o'clock: the bottom spot will probably be completed, at modern sending speeds, by about 12.15, in the case of a picture two inches square.

Clearly all we have to do is to reduce this time to 4/5 of a second altogether, and we shall be again sending the first spot before it has had time to die away from the apparent vision of the observer. In other words, we will see by wireless (Figure 3.4).

The obvious method of assisting in this speeding up of sending the thousands of spots, would be to graduate them by some means of rotary conversion or to decrease the number of spots. The latter is one method by which practical television can be accomplished to-day.

It would be quite easy to fix up an apparatus by means of which we could show whether an office in New York was lit up or not, the observer being situated in London. This is a form of energy or combination of phenomena which amounts to wireless sight, but it does not help us to see shapes or forms or to say if the light is from a candle, the sun, or an arc lamp.

By increasing the number of cells from one to, let us say, twenty, we could possibly indicate the difference between the moving shadow of a cross or a circle, but to radiate detail is a very difficult problem, which doubtless will be partially solved within the next few years. The electro-magnetic theory of light and the phenomena exhibited by the neon tube,[16] present many opportunities.

What an excellent invention this will be! It means that a telescopic camera could be attached to an aeroplane and the views seen by thousands in a cinematograph

15 The term was used for film projectors, as here, but also for cameras and complete systems which also printed the film.

16 Neon tube lighting designed by French engineer Georges Claude had made its first appearance at the 1910 Paris Motor Show. A similar technology using vacuum tubes produced cathode rays (electron beams). In 1906, Lee De Forest had used a modification to produce the first triode valve as an amplifier. Low was right that these developments heralded the future. Diodes (the wireless valves mentioned in the next section) and Triodes became fundamental components of electronics until the invention of the much more compact and reliable silicon transistor in the late 1950s.

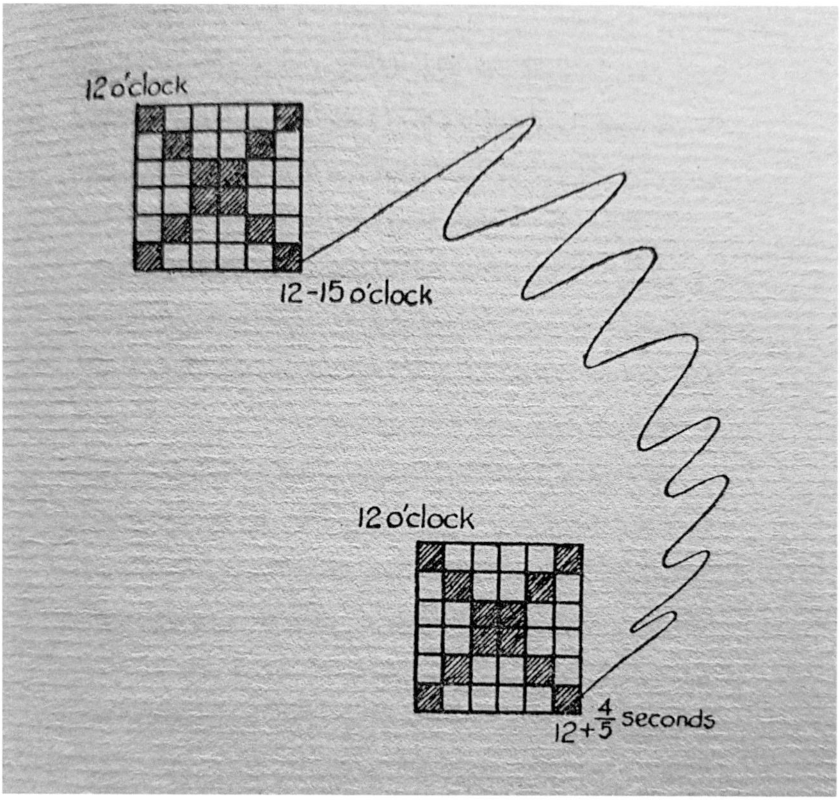

Figure 3.4 If a photograph is divided into spots and the last be sent only 4/5 sec. after the first, television can be obtained; but if the spots must be large to do this, only such items as the difference between a cross and a circle can be observed.

theatre who may have the pleasure of witnessing the finish of a horse-race and knowing without loss of time how much money they have lost.

It would mean that the crew of a ship, a submarine in difficulties, or the passengers in an aeroplane, might be visible to people many miles away. It could not yet occur without their wish, for the transmitting apparatus must first be put into operation.

The senses of seeing and hearing are possibly amongst the most important of all, and, if we can convey both of them to a distance, it means that we can call friends, nations, music, and personalities to our fireside, by the touching of a button.

Such possibilities need no enlargement. Wireless may prove a far more rapid link than the ordinary increase of travelling speed and may help nations to intermingle to the common good.

The question of seeing in colours has hardly yet been considered, but that also will come to us, however great the difficulties may appear to-day.

Certainly Leagues of Peace will have more arguments, and Generals will have more weapons.

The laziest millionaire to-day, in a physical sense, will be hard-worked in comparison with the fortunate individual of the scientific future. We will travel in the best possible manner and in such comfort that the mind will be free to receive impressions. Our main objective will be to train it for that purpose.

After all, what more can we do now?

Wireless and War

The subjects of War and Wireless cover a multitude of closely allied ills.

It is only natural that wireless should first have been applied to Love and War. I remember well one of the most remarkable applications of wireless mentioned in the press in the early days was that of a cable sent to an unfortunate man in mid-ocean, informing him that an all too successful arrival of twins had taken place.

War is, of course, a natural process a little less educated, and more unkind, in consequence, than birth control.

Most inventions are first applied to the science and art of warfare. Perhaps we should not regard this as all to the bad, for War has a remarkable capacity for acceleration.

Development of the wireless valve was greatly assisted by the War: the aeroplane, the art of plastic surgery, and many other human benefits have arrived more rapidly from the same cause.

Let us see, therefore, what wireless can do now, and what it may accomplish for the future of organised destruction.

Mentally, the fittest should survive, in both the realms of invention and physiology. It is only a few years since wireless was of no intrinsic value for ordinary land warfare, by virtue of the fact that interference was extremely easy, and that any coded message could be so easily decoded.

At present wireless messages are chiefly of service where secrecy is not of such importance as speed; but an enormous number of experiments are being conducted upon beam wireless, directional wireless, and in the combination of the Radio oscillation with some other oscillations such as those of visible or invisible light. By these means secrecy will be obtained when we discover how to use small powers for long distance, but at present Radio is chiefly of value as a time-saver.

The pilot in an aeroplane can talk to his base: he will soon be able to write and transmit vision from a plane which could be controlled by wireless. The time will come when low-flying wireless planes will explore, and render visible at many miles distant, places where no human pilot could remain for any length of time in safety.

It is not long ago that we rejoiced because a damaged ship was able to call for help by wireless, but we have only to look back to a recent war to remember an occasion when one ship was totally unable to call assistance because its wireless was jammed. In other words, enemy interference was possible.

This should show us how far we have yet to go in an utterly new and very little understood science.

We began with sparks, we progressed to coherers, and now we have valves; but let it not be thought for a moment that the valve represents finality to any thinking being.

Broadcasting at present has really become so universal only on account of the exceedingly public nature of wireless,[17] for, when we are able to obtain accuracy of tuning and direction, we shall not only use the latter to guide ships at sea, but we shall have correspondence which can be conducted with a reasonable degree of secrecy. We shall have special wave-lengths for the Government, special wave-lengths for Parliamentary debates, and the Divorce Courts. We shall not conduct our conversations in such a manner that any schoolboy with a piece of wire, a needle, and some sugar, can promptly listen in.

The very idea suggests a new "Peeping Tom."

As far as communication is concerned, we shall have whole armies in instantaneous touch with each other: it may indeed make real secrecy more difficult. It should always be recollected that when we refer to wireless speech, wireless control, and Radio Vision, we do not necessarily mean the same form of electrical wave by which we now broadcast a comic opera.

It is with oscillation that we are really concerned, and we may discover many forms of electronic vibration at present occupying portions of the so-called spectrum which are as yet very little understood.

It may be impossible for the Commander-in-Chief of the future to conceal a document from the eyes of wireless; and who knows but that the electrical operation of thought may be reduced to a science so that our very ideas are not secret without protection?

How many of us to-day could risk all our thoughts being known? It would probably improve moral standards if they were published: science tends to effect an average improvement.

We have never yet really seen the extraordinary value of wireless in war. If we had solved the problem of selection, the transference of speech by phonograph records dropped from aeroplanes would never have arisen.

Undoubtedly, we shall see wireless controlled tanks, submarines, and torpedoes on both land, air, and water. All will be accurately controlled, and they will possibly be able to find their way home and to operate from a distance while out of sight.

Even to-day it is possible for an aeroplane to operate a torpedo, to steer it properly, to slow it down; and for a pilot of an aeroplane many miles away to work his will upon it with a reasonable degree of accuracy and with the help of a gyro control.

The day will undoubtedly come when the problem of defending an island is not that of the mainland itself but of all its dependencies.

No large town could live for long if it were bombed from a distance by wireless, if gassed and poisoned from a distance, were it not for the balance of protection and defence which is usually maintained by nature.

17 The British Broadcasting Company had only been formed less than two years earlier, in October 1922 (it became a 'Corporation' in 1927).

We shall in the future, see forms of electric death and heat-rays which may materialise not as a direct projection of heat but as some form of oscillation which produces heat only when striking a metallic object.[18]

We have been so often told that power can be transmitted by Radio that we are apt to look upon this statement with contempt. This is quite wrong: power will one day be transmitted by wireless; power can at present be inductively sent over quite a large air gap, though the energy available quite close to any wireless station is practically negligible to-day.

When motor-cars and ships are controlled or stopped by wireless, it is not the wireless which does the work; the etherial oscillation merely sends signals to the ordinary operative mechanism.

Much excitement has been caused by the alleged injury of aeroplanes and motor-cars by wireless, but how is it that they can afterwards proceed? Do we forget that the petrol engine has to be restarted, and that, if allowed to fire when a car was in gear, it might be damaged and would probably not operate the moving parts?

If wireless power could be directed in such a form that it could be conveyed along a wave of "atomic" oscillation, many more valuable ends might be served than the enforced landing of aeroplanes.

Our clocks could be corrected by wireless, experiments could be conducted upon the nature of light and ether in various forms. We might decide the mode of propagation of light and thought, and investigate the apparent motion of the electron along the electro lines of force.

What an opportunity for study to the man of medicine! What a chance to find out how the oscillations of life are connected with those we partially understand. What a chance for the burglar to discover the presence of hidden spoons as a mass of metal by means of wireless; what a chance for the surveyor and the seeker after oil to use this all-prevailing sense of oscillation and even to discover the meaning of radiation.

Oscillation—that is all we mean by Radio; and oscillation is at the base of life itself. It will not be long before travellers by air, land, and water, will be no longer alone.

That they will be able to converse with their homes may seem no advantage, but that they can remain in touch with the rest of mankind is most obviously desirable.

If this were understood to-day, I should not need to make noises with my lips or require the simulacra of these noises to be produced upon paper to convey my thoughts. If thought is a process of energy-conversion—and who will deny it?—what form of screening prevents its use, and why should its reception be confined eventually to life upon this particular and very troublesome planet?[19]

18 This idea may be inspired by the death rays of H. G. Wells's Martians in *The War of the Worlds* (1898). But it also accurately predicts today's developments in laser weapons.

19 This passage anticipates J. D. Bernal's speculations in the 'Flesh' section of *The World, the Flesh and the Devil* on space travelers' minds being connected through wireless; p. 115, 119–22.

It is remarkable how little is known of wireless: the very simplicity of its painfully standardised features is a trap for the unwary. It is a universal science, but we do not yet know the correct diaphragm size for a loud speaker, nor how damping should be employed. The finest apparatus is available to all, and yet we do not understand the fullest range of wave-lengths. The study of radio-active materials and short wave radiation may in one day produce the cold-emitter valve, abolish the outside aerial, and bring to our closer understanding some of the many senses now so atrophied in mankind, that we can only speculate as to their existence. I doubt much if the schoolboy of the future will greatly esteem the radio expert of this century.

Archimedes[1]; *or, the Future of Physics*

L. L. Whyte (1927)

Headnote

Lancelot Law Whyte (1896–1972) had a varied career, split between being a theoretical physicist, financier, industrial engineer, and author. He later supported Frank Whittle in his development of the jet engine. He had deferred his university education to enlist in the Army in 1915, and had served as an artillery officer during the Battle of the Somme, winning the Military Cross. But he was a young Cambridge physicist of 30 or 31, who had been working under Ernest Rutherford, when C. K. Ogden recruited him to write *Archimedes.* In the later 1920s he spent time in Germany, where he met Einstein and discussed his work. Whyte was to continue his anti-dualistic futurological thinking in books such as *The Next Development in Man* (London: The Cresset Press, 1944) and *Everyman Looks Forward* (The Cresset Press, 1946); and his thought about psychology in *The Unconscious Before Freud* (New York: Basic Books, 1960). In his 1963 memoir, *Focus and Diversions* (London: The Cresset Press, 1963), 96–8, he looked back on *Archimedes*, noting wryly: 'It is clear that the impatient author had his time-scale wrong, for he predicted by 1940 a unification of physical theory which is not yet in sight.'

******

To Lotte

Chapter I: The Sciences Converge

One of the most fascinating features in the history of thought is that on several occasions an important new idea has come simultaneously to independent minds. Thus after Euclid's geometry had remained without a rival for two thousand years the conception of an alternative non-Euclidean system was reached separately by Gauss, Lobatschewsky, and Bolyai during the years 1820–30.[2] Bolyai's father,

1 The mathematician and physicist Archimedes (287 BCE–212 BCE) was a native of Syracuse, Sicily.

2 The German, Hungarian, and Russian mathematicians Johann Carl Friedrich Gauss (1777–1855); János Bolyai (1802–1860); and Nikolai Ivanovich Lobachevsky (1792–1856). Bolyai's father Farkas and Gauss had been close friends as students.

DOI: 10.4324/9781032625522-7

while ignorant of the fact that Gauss had already made the same discoveries, wrote to his son urging him to publish his results and used the following prophetic words:

> There is some truth in this, that many things have an epoch, in which they are found at the same time in several places, just as the violets appear on every side in the spring.

Another example of the simultaneous emergence of an idea in the minds of different thinkers is given by Darwin in his introduction to the *Origin of Species.* He there calls attention to the fact that in 1794–5 the broad idea of the evolution of species—though not its cause—was simultaneously formulated by Goethe in Germany, S[aint-]Hilaire[3] in France, and his own grandfather, Dr Darwin,[4] in England. Moreover Darwin himself had the remarkable experience of finding in an essay submitted to him in 1858 by A. R. Wallace a complete summary of his own unpublished theory of natural selection as the chief cause of the evolution of species.[5]

The last few years constitute another critical period of a similar kind, since an idea, which when made precise will transform scientific thought, has already come independently to many thinkers. Since 1922 many scientists have felt that in studying the emission and absorption of light physics has come near to the problem of life.[6] Others have proposed that in order to straighten out its atomic problems physics will have to take a hint from biology, but what this hint should be has not yet been indicated. The following pages suggest a definite line of advance for physics, and interpret these isolated flashes of intuition as evidence of a special feature in the present situation of the sciences.

We stand at the eve of a new epoch. Physics, biology, and psychology are converging towards a scientific synthesis of unprecedented importance, whose influence on thought and social custom will be so profound that it will mark a stage in human evolution. For centuries science has concentrated its highest genius on the study of inanimate matter; to-day the three great sciences are at last reaching the problem of life. For their researches on matter, life, and mind are now overlapping at one common issue: the nature of the fundamental electrical processes which underlie radiation and chemical combination.

Thus physics is at present occupied with the changes that occur when an atom emits either light or electricity. Biology is at the same problem in studying the electrical processes which are the basis of all organic behaviour, whether in primitive forms of protoplasm or in the highly developed central nervous system of man. Meantime

3 Étienne Geoffroy Saint-Hilaire (15 April 1772–19 June 1844).

4 Erasmus Darwin FRS (1731–1802) was also a slave-trade abolitionist and a poet.

5 Essays by Charles Darwin and Alfred Russel Wallace OM FRS (1823–1913) outlining their ideas about evolution were presented at a joint reading at the Linnean Society of London on 1 July 1858.

6 [Author's note] Whitehead, *Science and the Modern World.* Eddington comes near to the same idea in an essay in *Science, Religion, and Reality,* 1925. See also [Hermann] Weyl, *Was Ist Materie?* 1924, p. 84. It has also been expressed by others quite independently, though I do not know of other published references.

psychology is dealing with an identical process when it analyses the structure of mind, and considers the elementary changes of consciousness which are produced when light of a given colour falls on the retina and sends its influence to the brain.

As the result of these convergent researches, life and consciousness will soon be subject to the first stages of a theoretically-grounded control, compared with which the present tentative efforts of medicine and psychology will be looked back on much as we remember the haphazard work of the alchemists before the foundation of chemistry. But this development of human knowledge and powers will carry with it great responsibilities, and scientists have to prepare themselves for the new tasks that will very soon fall to them. By indicating the main ideas through which this broad scientific synthesis may come about, this essay aims at showing that this possibility has to be taken seriously. We shall first examine the situation in physics and then turn to consider the influence which future developments of physical theory may have on biology and psychology.

Two main types of process defy interpretation within the present scheme of physical conceptions: life itself, and the atomic processes of radiation and the building up of stable compounds. In organic processes on the one hand, and the energy-interchanges of atoms on the other hand, we find something happening which cannot adequately be explained as a change in the *structure* of the system considered. By structure is meant a spatial pattern of particles, which are supposed to be permanent and to move about like cricket balls or planets. Systems with a structure of this kind could not display the purposive quality of organic behaviour, and when we try to make a structural model of the atom we find that it fails to explain why the atom radiates energy in the abrupt packets which are called 'quanta', instead of in a continuous wave. We shall return presently to the question of organisms, after making an endeavour to discover why the atom cannot be described in terms of a particle structure.

In 1911 Rutherford[7] achieved remarkable success in accounting for the results of his own researches in radioactivity by adopting a model of the atom as a miniature solar system, with planetary electrons rotating rapidly around a nucleus. But in order to explain the fact that the spectrum of the light emitted by an atom shows a characteristic series of lines, Bohr suggested that an electron inside an atom could emit light only by making a discontinuous jump from one possible orbit to another quite distinct orbit. This apparent discontinuity in the motion of electrons has intrigued physicists for more than ten years, and the following interpretations have recently been offered for this puzzling behaviour:

1 Nature is made up of electrons, but neither space nor time is fundamentally discontinuous. The electron appears to have some freedom of choice, and to be able to reappear unexpectedly at forbidden places.

7 Ernest Rutherford (1871–1937); New Zealand physicist awarded the Nobel Prize in Chemistry in 1908 'for his investigations into the disintegration of the elements, and the chemistry of radioactive substances.' As Director of Cambridge's Cavendish Laboratory from 1919, he would go on to lead teams which discovered the neutron and split the nucleus in 1932.

2 Nature is not discontinuous or arbitrary, but nevertheless something prevents us determining all the things we should like to know about an electron. For instance, if we try to determine exactly where it is, it behaves so that we cannot simultaneously measure its exact velocity. (Heisenberg.[8]) This view may perhaps be interpreted to mean that we have made the atom model more complex than the atom itself is, and that consequently we have been using more quantities than are necessary for describing all we can observe of its behaviour.
3 Nature is not made up of electrons, but of waves. The atom must be considered as a system of electric waves spread over its whole volume. 'Electrons' are merely an inaccurate way of describing some of the properties of these waves. The wave picture of the atom is, however, to be considered only as a temporary expedient to be used until some better description of the atom can be invented, in which both the wave and the corpuscular properties of atoms will appear as aspects of some more profound physical property. (Schrödinger.[9])

The first alternative is a mere cry of despair, since it does not propose any line of advance. But the other two suggestions may be combined thus:

4 The view of the atom as a structure of Newtonian particles is wrong since it gives rise to discontinuities, and provides more quantities than we at present need. A new formulation of atomic processes must be found using fewer quantities which will explain why we find wave properties, and why sometimes the electron does behave like a small billiard ball though really it is some different sort of thing.

Now since the Newtonian mathematics of moving particles is inadequate for describing the changes that go on in the atom—just as it is for describing organic processes—there must be some assumption implicit in Newton's laws which is valid neither for atom nor for organism. Such an assumption can be found very easily, though physics has never given it much attention. It is that the elementary processes in nature are *reversible,* or would be if they could be isolated. By reversible is here meant that the laws governing the process remain unchanged when the direction of time is reversed, i.e. when -t is substituted for +t. If the law is changed by this substitution so that the reversed process never occurs or is recognizably different, then the process is called irreversible. An irreversible process can therefore be used to yield an objective criterion of past and future, when these terms have been once defined. To take an example. If I am standing behind a hedge and take a cinematograph film of a stone which suddenly rises in the air and disappears from sight, I could not tell from an examination of the film which way to wind it. Thus

8 Werner Karl Heisenberg (1901–1976) introduced the key quantum mechanics concept of the 'uncertainty principle' in a paper of 1927; the same year as *Archimedes*. He was awarded the Nobel Prize in physics in 1931.

9 The Austrian physicist Erwin Schrödinger (1887–1961) won the Nobel Prize with Paul Dirac in 1932 for their work on quantum physics. He coined the phrase 'quantum entanglement.'

if it is wound one way the stone appears to rise, and if wound the other way to fall from the sky. To tell which was the right way I should have to use my subjective sense of the direction of time, i.e. remember the fact that I saw the stone low in the air before I saw it high up. This case, like every gravitational process, is reversible, and motions of this kind have provided the basis for modern physical conceptions.

But suppose that instead I had taken a film of a cup of tea as it was cooling. One end of the film would show the steam above the cup and the spoon changing in length as it changed in temperature. Passing along the film these effects would grow less marked until the successive photos showed no variation when the temperature of the tea was nearly that of the surrounding air. It would be obvious which way to wind this film, without using any subjective criterion supplied from memory of the individual process which had been photographed. This process is irreversible, but physics has hitherto assumed that all such processes are merely the statistical result of a chaos of molecular motions each of them perfectly reversible.

The assumption of reversibility seems to some physicists so fundamental that they think there could be no science without it. But that is a mere prejudice arising from the fact that Newton conceived one particular way of giving mathematical formulation to the measurable features of physical processes. By suggesting that all the laws of nature might take a form similar to his law of gravitation, he made the implicit assumption that all elementary processes were reversible. Gravitational motions are so, at any rate within the accuracy of Newton's law, and as a consequence of the confirmation of his law and the fact that it has been taken as a model for the whole system of modern physical conceptions, the latter are only appropriate for reversible processes.

Apparent irreversibility, such as the cooling of a cup of tea, is attributed to statistical effects, and the second law of thermodynamics,[10] which asserts that temperatures tend to uniformity, is treated as merely a statement of what is highly probable. This is probably quite legitimate, but even where no statistical effect can enter and the process is clearly irreversible physics usually adopts any measure rather than assume that a fundamental elementary process is irreversible.[11] We cannot be surprised at this, since if physics once admitted that any elementary process was irreversible it would have to give up the whole system of Newtonian conceptions. Matter, force, energy, action, and wave properties are all unsuitable for the treatment of irreversible effects since they all ultimately depend on Newton's reversible law.

An entirely new set of ideas is necessary for describing processes which necessarily proceed in one direction, so that one particular state of the system must

10 The physical law stating that heat always flows from hotter to colder regions of matter. It follows that a system containing different temperature zones tends towards a uniform temperature; a process known as entropy.

11 [Author's note] E.g. the irreversible motion of an electron in the field of a bar magnet is rendered formally reversible by the assumption that the magnetic field is due to moving electrons. Yet this assumption is highly artificial since it postulates electronic movements that have never been observed. In other cases irreversibility is eliminated by the choice of special co-ordinate systems. Some physicists now hold the view that irreversibility may be inherent in atomic as it is in organic processes.

precede another state. It appears conceivable that an alternative set of conceptions to replace the Newtonian might be established by demanding the irreversibility of all natural laws, as well as the demands hitherto made by physics, i.e. the permanence of matter and the conservation of energy.

The question of the reversibility of natural processes provides the key to a great intellectual struggle which is now in progress behind the complexities of philosophic and scientific thought. The issue can be formulated thus:

Is there a real temporal process in nature? Is the passage of irreversible time a necessary element in any view of the structure of nature? Or, alternatively, is the subjective experience of time a mere illusion in the mind which cannot be given objective expression? These are not metaphysical questions that can still be neglected by science with impunity. For just as Einstein made his advance by analysing conceptions such as simultaneity, which had been thought to be adequately understood for the purposes of empirical science, so the next development of physical theory will probably be made by carrying on the analysis of time from the point at which Einstein left it. Moreover, the above questions may be put into precise scientific form by asking if the causal relations which are studied by science are symmetrical and reversible so that we cannot obtain from them any criterion by which to distinguish past and future. If, on the other hand, they are asymmetrical and irreversible, the laws of nature lead us on necessarily from what went before to what comes afterwards.

Chapter II: A Modern Duel: Einstein and Eddington v. Bergson and Whitehead[12]

In this battle over the importance of time and process great names stand out as representatives of the two opposed views: Einstein and Bergson, with their lieutenants, Eddington and Whitehead. The two leaders use very different methods. Einstein, as mathematical physicist, suggests that physical laws can best be expressed if we assume that space and time are so similar that physics can make no absolute distinction between them. Thus in relativity theory the symmetry of space involves the symmetry of time, and therefore the reversibility of physical laws, as has been shown by Birkhoff.[13] Bergson, as biologist and philosopher, denies that the view of

12 Sir Arthur Eddington, OM, FRS (1882–1944), theoretical physicist, astrophysicist, and writer of popular science books such as *The Nature of the Physical World* (1928) and *Science and the Unseen World* (1929). French philosopher Henri-Louis Bergson (1859–1941) was the leading theorist of time and process in the period, known for his major works *Time and Free Will* (1889); *Matter and Memory* (1896); and *Creative Evolution* (1907). Mathematician and philosopher Alfred North Whitehead (1861–1947) wrote *Principia Mathematica* with Bertrand Russell. His opposition to Cartesian dualism in *Science and the Modern World* (1925) was influential on many of To-day and To-morrow's contributors. His *Process and Reality* (1929) is a foundational text of the kind of process philosophy Whyte engages in.

13 American mathematician George David Birkhoff (1884–1944) contributed to the development of general relativity, quantum mechanics, and proved the possibility of a black hole being produced if a spherical star had sufficient mass. He published *Relativity and Modern Physics* (1923) with R. E. Langer, and later became best known for developing the ergodic theorem.

time which is implicit in relativity mathematics is adequate when a wider range of experience is taken into account.

Einstein starts by excluding all but a very narrow range of physical experience, and finds that he can make successful predictions about light and gravitation by treating the irreversibility of the passage of time as of no importance for scientific measurements. Bergson, by studying a wide range of biological and subjective experience, comes to assert the existence of a creative process, though the inherent limitations of the intellect and of science may leave the essence of this process outside their reach.

Both protagonists have left their flanks exposed, by omitting to present their view as a consistent logical system, Einstein because he is concerned only with the equations that can be empirically tested, and Bergson because his chief interest is non-intellectual. It is here that their lieutenants step forward to develop the two points of view, and hence to intensify the conflict.

Eddington provides a logical basis for the theory of relativity and reveals that the significance of physical laws is not quite what we used to think. They are, he argues, identities which the human mind discovers in its search for something permanent that it can call *matter* beneath all the changing appearances of the world. We have made matter the real thing by demanding permanence or indestructibility as the basis of physical reality. Now that we know that we have done this it need not trouble us too much to find that absolute unchanging matter doesn't exist, since this merely means that we started out with a demand that nature cannot fulfil. Unfortunately Eddington doesn't discuss what alternative demand we might now make in order to build up a more satisfactory system of scientific ideas. But in spite of his enthusiastic support of Einstein's theory, with its implicit assumption of reversibility, Eddington hesitates at least once in his advocacy of reversible laws, for facts are turning up which suggest that this undiscussed presupposition may not prove valid.[14]

Meantime Whitehead has been at work on the other side, and by sharpening his logic till few can understand him has made the idea of temporal process the basis of all intellectual and scientific thought, whereas up to now process has always presented many difficult problems for the intellect. He proposes that since the conception of matter has been found to be unsatisfactory we must start from the basic idea of process in building up a new physical theory. As a consequence of his line of thought, Whitehead found it necessary to reject some of Einstein's arguments and to show that Einstein's law could be reached from quite different

14 [Author's note] *Internal Constitution of the Stars* 1926 p. 56. Compare note on p. 44.

It may be convenient here to summarize the processes that give at any rate superficial evidence of their irreversibility: processes involving heat changes, or the radiation of light, or mass; the production of energy in a star, the motions of electrons in magnetic fields, certain types of atom-ion collision in mixed gases, processes dependent on retarded potentials, radioactivity, organic growth and evolution, and consciousness itself. Eddington deals only with the case of the emission and absorption of light, but suggests that the direction of time can only be deduced from statistical processes. This is the orthodox view, though it is very doubtful if it is valid now that the quantum processes are receiving formulation. In this connection, see note [15].

postulates. For instance, Whitehead assumed that the motion of light was irreversible, and that light did not travel with the same velocity in the two opposed directions.

So much for one aspect of the conflict, its logical and philosophical basis. But the issue must be decided by appeal to experimental confirmation over the widest range of phenomena. Orthodox physics still assumes reversibility, and has on its side the explicit statement made by Einstein in 1925,[15] but by doing so it excludes at the start any reference to organic processes. Conceptions based on this assumption could never be legitimately applied to life, and all attempts made hitherto to explain the central controlling processes of organisms in terms of classical physics have necessarily failed. We know now that this failure could have been foreseen.

The same objection cannot be made against the basic ideas of Bergson and Whitehead, nor against the new atomic physics as interpreted by Born,[16] as we shall see in a moment. To Bergson and Whitehead, as to many others amongst whom Lloyd Morgan[17] must be mentioned, the process of nature is creative, i.e. it involves the coming into being of the new, the appearance of new combinations essentially precluded before. This probably means that the laws of physics which are to describe what is actually happening in the world must be given irreversible form. For reversible equations make no distinction between to-day and to-morrow, and cannot express the fact that at later moments new forms may emerge, either in the evolution of organisms or of stars. On the other hand irreversible laws can be arranged so as to display time as an active factor in causation, i.e. to emphasize the fact that a certain period of time necessarily has to pass before some new combination can be attained.[18]

The upholders of a real process in nature can appeal to the facts of organic life, human memory, and to biological and stellar evolution. But their case is still weak because fundamental irreversibility has not yet received explicit mathematical formulation suitable for experimental test. When this has been done the intellectual battle will be brought to its decision, and if irreversibility wins the day biology and

15 [Author's note] Einstein. Berlin Akad., *Sitzungsberichte,* 1925, p. 418. But Einstein's view must be revised in view of recent experimental results (e.g. Harnwell, *Phys. Rev.,* Vol. 29, 1927. pp. 683 and 831), if these have been correctly interpreted. See Born, *Zeitschr. für Physik,* Vol. 40, pp. 177–8; and Jordan, *Naturw.* 1927, p. 792.

16 Max Born FRS, FRSE (1882–1970); German-British physicist and mathematician who worked on quantum mechanics with Heisenberg and was awarded the Nobel Prize in 1954.

17 Conwy Lloyd Morgan, (1852–1936), British comparative and animal psychologist and philosopher.

18 [Author's note] The idea that time may be an active factor in causation has the mathematical significance that 't' (for the system in question) must appear explicitly in the formulation of the law, and not merely as the square of a time-differential found convenient for the correlation of a standard clock with a reversible process which is being observed. A law whose mathematical formulation involves 't' measured from some moment in the history of the system, gives an entirely new meaning to 't', though one consistent with the properties of the reversible Newtonian differential 'dt'. Such a law may claim to express the fact of historic, irreversible, duration, a feature in nature which is neglected by laws involving only 'dt' squared.

psychology will find themselves in possession of a physical basis well suited to the facts with which they have to deal.

There is reason to believe that the decision will be made very soon. We saw that the implicit assumption of reversibility underlies all Newtonian conceptions. It may therefore be that the reason why we cannot interpret atomic behaviour in terms of particle motions is that electrical and radiational processes are essentially irreversible. Particle motion and wave propagation—the two ideas on which all modern theories of matter are based—are both represented by mathematical expressions which are essentially reversible since time enters only through the square of 'dt'. If the quantum processes should prove to be irreversible, we have already found a reason why the old conceptions of particles and waves must be inadequate.

This speculation may indeed be found correct, since Born, one of the leading experts in Quantum Dynamics, asserts that all quantum processes are irreversible and that the apparent reversibility of classical processes is only an approximation due to the fact that their irreversibility happens to be negligible.[4] We may therefore hope that the atomic physicists will soon formulate the quantum laws in a clearly irreversible form which admits of precise experimental test.

But this may take some years, and in the meantime we must look around and see how this issue is affecting current thought. We find the doubt about process presented by Mr Sullivan (in *Gallio*), who has not yet made up his mind to which side science will grant the victory. Thus on one page he writes: "it seems to be true that events do not really take place, we come across them" and suggests that process may be "a totally irrelevant idea when applied to reality". But later we learn to our surprise that "it seems likely that (in scientific theory) the world will have to be regarded as an evolutionary process, where patterns of value emerge". However, this inconsistency need not bother us, since we are told that "the teachings of science so far as the spiritual problems of man are concerned are merely irrelevant".[19]

These views reflect perfectly the uncertainty of the time, and will be looked back on as a precious record of the state of mind which preceded the scientific synthesis. Perhaps the most interesting feature of the essay is the indecision it displays with regard to the spiritual importance of science. This is a relic from the days when there were two worlds, the world of science and the world of religion and art. No one ever knew which of these worlds they were living in, and this is no wonder. For the division was made only because at one time it looked as though the scientific method could only deal with *quantities,* and therefore that science could have nothing to say about values or qualities. This view is no longer tenable. For instance, there is a quality in organic integration which most of us value, and without this and many other such conceptions biology and psychology could not get far.

Before proceeding any further it is necessary to correct a common misunderstanding with regard to the significance of Einstein's theory of relativity. This theory is mathematical, and is based on a series of postulates which rule out any claim

19 J. W. N. Sullivan's *Gallio, or the Tyranny of Science*, (London: Kegan Paul, 1927), was another To-day and To-morrow volume: pp. 38, 39, 92, 96.

to present an ultimate theory of space and time. One of these postulates[20] asserts that all our physical knowledge can be reduced to the space-time coincidences of pairs of point-events, or in other words the intersection of the world-lines of electrons. No respect for the supreme genius who predicted two experimental results and eliminated the chief discrepancies remaining in Newtonian theory should restrain scientists from pointing out that this postulate assumes something that has never been known to occur, and has no valuable reference to the world of physical experiment. The confirmation of Einstein's final equations cannot give any validity to this postulate. For it is difficult to think of any physical experience considered by theoretical physics which does not involve the perception of light or colour, and one cannot assume that the perception of light is a perception of coincidences. Light varies in colour and intensity; coincidence in space is too abstract to account for an effect which is subject to variation. Moreover all physical experience requires a certain amount of time, and this fact is neglected if perception is reduced to the recognition of instantaneous coincidences. Even if these two criticisms are left on one side we still have to notice that Einstein's postulate rules out from the range of physics the important fact that many processes are irreversible. For instance, if we accept Einstein's definition of physical experience, then the interesting fact that radioactivity is only observed in the form of disintegration, and not also as the reverse process of a spontaneous building up of heavier elements from lighter, has to be left over by physics to be dealt with by some other science.[21]

It almost always happens that the formulations of genius are exaggerated and form the basis of a pernicious orthodoxy, and it has certainly happened to relativity theory. Against a tide of exaggerated praise Whitehead, Larmor,[22] and Bridgman,[23] as well as some Continental astronomers, have debated the general assumption that the theory of relativity is adequate to its task, but those in whose hands the power of orthodoxy lies have not yet answered their criticisms in print. Neglect has always been the weapon by which orthodoxy has unknowingly hindered the advance of new ideas. But while this neglect is easy to understand, it is really remarkable that the postulates of relativity theory were not subjected to closer examination before it was made the basis of wide philosophical speculation. The experimental confirmation of Einstein's law of gravitation does not guarantee his postulates, since Whitehead has reached a similar law (identical within the accuracy of the observations) from different assumptions.

Einstein's profound creative intuition and use of a difficult technique compel our deepest respect, but his work should never have been regarded as a *general*

20 [Author's note] Einstein, *Annalen der Physik,* Vol. 49, pp. 776–777, 1916.

21 This is a surprising claim from a practicing Physicist, since Arthur Eddington had predicted nuclear fusion of hydrogen into heavier helium nuclei as the source of energy from stars: 'The Internal Constitution of the Stars', *Scientific Monthly*, Vol. 11, No. 4 (October, 1920), 297–230.

22 Sir Joseph Larmor (1857–1942), theoretical physicist specializing in electro-magnetism.

23 Percy Williams Bridgman (1882–1961); American physicist whose work on operationalism was more influential in the social sciences than the physical sciences for which he intended it. His book *The Logic of Modern Physics* appeared in the same year as *Archimedes* (1927). He won the 1946 Nobel Prize in Physics for his work on the physics of high pressures.

theory of time and space. Not only does he neglect the question of irreversibility but it is very doubtful if periodic processes can be made to fit into his scheme, as has been pointed out by Russell and Bridgman during the last year. Probably Einstein himself has never regarded his theory as more than a stage in the attempt to create a still wider physical synthesis, and we must not interpret in a broad sense his statement that one of the demands of his theory "takes away from space and time the last remnant of physical objectivity".[24] This could only be true if physical time shared the absolute symmetry of space, i.e. if physical processes were all reversible. But there are processes from which we can obtain an objective criterion of the direction of time, and hence time does retain an element of physical objectivity as distinct from the absolute symmetry of space. One of the most interesting features in the future of physics will be the explanation of the fact that Einstein reached a correct law from postulates of limited validity, and in this connection Whitehead's alternative derivation may prove to be of importance.

Chapter III: Time in Astronomy and Physics

The real discrepancy between the world of physics and that of life lies in the fact that physics has never recognized the irreversibility of time, while this is fundamental to life. We may even feel a doubt if the 't' of physics has the same significance as the time of biology, evolution, history, and human experience. The physical conception of time arose from the practical utility of clocks for describing natural processes, and finally took the form of defining astronomical time in terms of the rotation of the earth. The day was in fact taken as an absolute measure of time, and this remained quite satisfactory so long as the laws of physics were found to take a simple form with reference to the time so defined.

But then a complication arose. The study of the moon's motion suggested to astronomers that the earth's rotation was slowing down, i.e. to account for the apparent motion of the moon they had to assume that the day was increasing in length. The theory of the tides revealed a possible cause for this slowing down in the tidal friction on the bottom of shallow water basins, for instance the rush of the Atlantic tides into the Irish Sea provides an appreciable frictional force retarding the spin of the earth. In addition to this slowing down there appears to be a very slow periodic variation in the length of the day such as would be accounted for by a rhythmic expansion and contraction of the earth's crust.

The astronomers declare that our old measure of time is not only getting slower and slower, it is even varying rhythmically! It is clear that they have thrown over the earth as their definition of equal time intervals and have surreptitiously substituted something else. Yet one cannot discover any formal announcement of this, or find out if they realize that by doing this they have altered the theoretical significance of all physical measurements. In earlier days physics defined time in terms of a selected clock, and then set about finding the laws of nature. But the old ways

24 [Author's note] Einstein, *Annalen der Physik,* Vol. 49, pp. 776–777, 1916.

aren't good enough for the modern astronomer who gives us our time and sets the clocks of our physical laboratories. He has reasons for disapproving of the earth, and has almost reversed the procedure. In order to save the laws of inertia and gravitation in connection with the moon's motion—and to a lesser degree in the cases of the planets and the sun—he has made these laws his standard of equal time intervals in place of the earth's rotation.

It is a curious situation, especially in view of the fact that Einstein's law, which has superseded Newton's, is not very suitable for use as an astronomical clock, as has been pointed out by Larmor. Perhaps the physicist will soon be able to use the atom as the theoretical clock for physics, and we can go on using the corrected rotation of the earth as our practical standard. There is a faint chance, for instance, that if physics can invent some way of measuring the minute time intervals along the track of an electron, then electrons might be used as giving the fundamental measure of time. Thus if the velocity of an electron were first measured by some indirect method the electron itself might then be used as a clock. But in the meantime the astronomers should make a formal announcement to the Royal Society of what they have been up to. It then might be found necessary to appoint a commission to discover exactly what physics is now doing. For by using an astronomical clock of the new type it is assuming classical laws while researching on processes which are already known to undermine the absolute validity of these laws. Theoretical physics cannot hope to clear up its fundamental problems until it has considered exactly what is involved in this suspicious procedure.

Like most professions, physics includes a good deal of bluff, but unlike the others physics is now occupied on a campaign to get rid of all pretence. For instance, physical text-books have been filled for 20 years with phrases of this kind: "an electron with a velocity of so many cms per sec." Yet the professors omitted to tell their students the awful secret that this hypothesis of electron velocities is one that has never yet received direct experimental confirmation. To-day a reaction has set in and the demand is being made that physical theory shall not make use of conceptions that do not correspond to directly observed quantities. Thus the latest theories of the atom have eliminated the idea of electron orbits because it was realized that these were nothing more than a mathematical trick for calculating something quite different: the wave-length of the light an atom can emit. In place of the orbits it is hoped to substitute something which only makes use of the directly-observed features of the atom, but this new picture is not complete.

Yet physics still makes use of ideas that have not been adequately justified. For though the idea of moving electrons has been removed from the latest atomic model,[25] no substitute for it has yet been proposed for the case of electrons outside the atom. It therefore becomes very important for the experimental physicist to discover whether he can measure the distance travelled by an electron in a measured fraction of a second. As yet we have no proof that nature has not confused

25 The Bohr Model, or Rutherford–Bohr model, presented by Niels Bohr (1885–1962) and Ernest Rutherford in 1913, consists of a small, dense nucleus surrounded by electrons orbiting in discrete distances corresponding to individual quanta.

us by making electrons behave rather like moving particles, though really they are something different. In fact we have not yet made enough direct experiments to know even whether the dimensional system which is used for electrons is correct. Since no electron velocity has ever been directly measured we cannot be sure that the dimensions of the new constant 'h'—called Planck's constant—are really what we suppose, energy multiplied by time.[26] Until a way has been invented of making a direct measurement of some *time* involved in electronic motions, it is impossible for physical theory to know how it should deal with the quantum processes.

When we realize how uncertain are the conceptions on which the whole of electron theory is based, we may wonder what is really known about the atom itself. Yet it is possible that we know more about the atom than we think, and that what are talked about as facts concerning electrons and radiation may really be better viewed as information about individual atoms and the way in which they influence one another. The emission of light is an atomic process, and we only know about light when it has reached some atom and been at least partially absorbed. Some un-understood change of condition occurs in an atom when it radiates and passes this changed condition on to another atom. The absorbed energy may cause chemical change, as on a photographic plate. But if a human mind is to become aware of this change of condition, then sooner or later, directly or indirectly, its influence must be passed on to an atom in the retina. We know very little about this change of atomic condition, and though it is usually called a change of the internal electrical energy of the atom this supposes more than we really know until some electron velocity has been directly measured. The dimensions of electrical energy are taken as those of kinetic energy, i.e. mass times square of velocity, but we do not yet know if this describes atomic changes correctly. Since no one has ever measured a *time* involved in an electronic process, the scale of time in the atom might be quite different from that given by our calculations.

Our ignorance of what this change of atomic condition really signifies is so profound that some writers have begun to treat the atom as though it were an organism, alive when the atom is excited, and dead when in a state of minimum energy. Thus Whitehead proposes that we should call the atom an organism, though this of course may only muddle us since we know even less about life than we do about the atom.

Yet we do know one very interesting thing about this change which happens to atoms but cannot be reduced to a change of structure. When light reaches an atom in the retina, an electrical stimulus passes up a nerve and alters the condition of the protoplasm somewhere in the brain. This change in brain condition is known to us directly as the perception of colour. Therefore in one sense we know more about this change of atomic condition than we ever did about 'electric fields' or

26 Named after Max Planck (1858–1947), one of the founders of quantum theory, Planck's constant is a fundamental universal constant that defines the quantum nature of energy and relates the energy of a photon to its frequency. In the International System of Units (SI), the constant value is $6.62607015 \times 10^{-34}$ Joule-seconds. Planck received the Nobel Prize in 1918 for his discovery of energy quanta.

'gravitational potential' or any other of the mathematical conveniences used by physics in correlating observed quantities. The change in a sodium atom when we put salt in a flame is not a change in the consciousness of the sodium atom, because it is not part of a complex nervous system with the same high co-ordination as is found in the human being, and therefore the atom has no consciousness. But when an atom in the brain undergoes the same change we may become conscious of it, and the changes in matter which occur when light is absorbed are undoubtedly associated with the problem of consciousness.

Thus we are led to ask: how are single atoms built up into complex systems which have the characteristics of life, and finally into still more complex systems which have human consciousness?

Chapter IV: An Evolutionary Experiment

Questions are often made unnecessarily difficult by their being expressed in an abstract or theoretical form, and instead of asking What is life? it will be more valuable to put forward a practical issue for discussion: Could an infinitely wise physicist order the necessary chemicals to-day, and to-morrow put together a synthetic man? If not, why not? What are we really up against, that seems to put some aspects of life beyond our control?

Let us watch this ambitious physicist as he enters his laboratory. He has started quite easily and has in a moment prepared some simple molecules from their elements. Now he has completed the first colloid that he will require, and is starting on his first organic synthesis. But his infinite wisdom does not give him eternity within a minute, and we notice that he is getting on more slowly. While the actual combination of the first molecules took only about a thousandth of a second, once he had the apparatus ready, the simplest colloid took about a second. The organic colloid has taken him about a minute; it seems that nature won't work faster than that. She has her own rhythm and won't be rushed. If we wait patiently till the end of the day our friend may have his first speck of protoplasm,[27] and all the skill in the world would only have helped him to make more of it, not to have got any further in his game of evolution.

But look at him now! He is making a hasty calculation as though he had just realized some great secret of nature and knew that he could never create his homunculus.[28] We look over his shoulder and read:

Estimated minimum time required by the synthetic processes of nature to attain various evolutionary stages.

27 "The complex, translucent, colourless, colloidal material comprising the living part of a cell, including the membrane-bound cytoplasm, nucleus, and other organelles (but excluding food particles, secretions, large vacuoles, etc.)" (*OED* online).

28 The alchemist Paracelsus claimed that he created a false human being and referred to it as the homunculus (Latin for little human).

Starting from the elements, to	Minimum Time
Simple inorganic compound	1/1000 sec.
Simple colloid	1 sec.
Protein	1 hour
Primitive protoplasm	1 month
Simplest uni-cellular organism	10 years
Flagellate[29]	1,000 years
Mammal, including *Homo sapiens*	1,000,000 years

This highly speculative estimate is based on suggestive facts. A certain amount of time is necessary for two atoms to approach one another and form a molecule. The time required will be greater if many atoms have to settle down together into some special arrangement. For instance, the metal silver is normally crystalline, but if silver vapour is condensed too quickly the atoms will not have time to arrange themselves, and it is found that they pile up anyhow into an amorphous mass.

Colloidal processes require even longer periods, because great clumsy molecules have to arrange themselves on the surface of the colloidal particles. In elementary forms of protoplasm the molecular patterns are still more complex, and yet more time must be necessary to get the molecules properly adjusted.

It is probable that only our ignorance prevents us from building up protoplasm, but that we shall require rapidly increasing amounts of time for each successive stage of evolution. This will certainly be the case when we have reached organisms which can only be rendered more complex by controlling their environment while they reproduce themselves for many generations. A higher organism cannot be built up directly; the molecular arrangements in its body can only be reached through the synthesis of some simple form of life which must then be allowed to evolve through countless generations. Organic heredity resides in molecular patterns which can only be built up by this very slow process of repeated reproduction. Thus it is *shortage of time* that our ambitious scientist is up against in his haste to create a homunculus. Only the synthetic alchemy of time can build up organisms, each bearing within itself a long heredity.

The estimates given for the minimum time required in each case are about a thousandth of the actual time taken in a laboratory experiment or in the history of evolution as known from geological records. It may have taken a million years or more for the first mobile cells to have developed from inorganic materials and a thousand million years for the mammals. Yet perhaps these processes might have gone on more quickly. The times given are mere suggestions of a minimum time which may be necessary under ideal conditions. We waste a lot of time adjusting the apparatus in a laboratory experiment, and in evolution there may have been stationary periods with little or no new development. But it seems likely that when we know more about it we shall discover that a certain time is required for the formation

29 Cell with one or more whip-like appendages.

of organic systems of given complexity. In this sense we may say that then human spermatozoon and ovum carry within them the synthesis of at least a million years.

Only an International Institute of Evolutionary Research under the most stable of Leagues of Nations could hope to create an artificial man, and even then man could hardly take the credit, for Time would have done more than man. But with sufficient consistency of purpose man could do this, provided he learnt how to make use of every moment of the creative power of time, and never made a slip by which the accumulated treasure of the years (i.e. heredity) might be broken. How man would learn to value life, and how profoundly such an experiment might alter his view of human beings, each one a priceless miracle, fruit of a million years!

In 20 years' time scientific knowledge will be adequate for the beginning of this giant task, and we shall be subscribing our guineas for the foundation of the Institute. Time has created man; man may use time to create man once more. With a million years ahead of us before we reach the sensitive mammals, we need hardly fear criticism from the Society for the Prevention of Cruelty to Animals. We are simply going to allow life to evolve itself under ideal conditions with Switzerland as the State for Evolutionary Research.

It may happen that under such perfect conditions life will evolve more swiftly than it did on this rough-and-ready planet. But equally well we—or rather our descendants—may find that the Darwinian struggle for survival is essential for evolution, and then the nations would have to debate on the morals of reproducing the 'cruelty of nature' inside the World's Evolutionary Zoo. Perhaps a wrathful god will seek to punish mankind for attempting to build this ladder to the secret of life, this modern Tower of Babel, and amuse himself by watching the community of scientists stricken by a plague of inconsistency amongst their weights and measures.

The possibilities of such grand schemes have to be taken seriously. We are now highly self-conscious beings with a tremendous technique for research. Men with genuine creative imagination who reverence life must shoulder the responsibilities of the twentieth-century consciousness, and use scientific technique for creative not life-destroying purposes. One can imagine a growing fraction of the interest now given to war, other people's adultery, and greyhound racing, turned towards Switzerland, whence at critical moments wireless bulletins would announce that the first amoeba had just successfully taken nourishment. If we wish it, the future of science can be such as to recompense for its recent occupation with gunpowder. Governments would be powerless to make war if the physicists refused to make the guns and the Royal Society called upon scientists to go on strike until each war crisis had been settled by arbitration.

The problem of life may be seen in a new light if the speculations of the last section are accepted and we assume that a definite period of time is necessary for the building up of any living organism. For if this is so the laws which govern life must involve the age of the organism since some definite moment in its history. We might choose for this moment the instant when the parent spermatozoon entered the ovum in the case of a higher organism, or in the evolutionary experiment just described the age might be reckoned from the moment when the first elementary chemicals were combined into molecules. The point is that this whole evolutionary process must be described by laws which take into account the age of the system under consideration.

Let us take a very simple, indeed the simplest possible, example. If two hydrogen atoms having just the correct total energy for the formation of a hydrogen molecule have approached one another and combined, the law describing what has happened must indicate that at a definite moment the combination was complete and the process at an end. This is an example of an irreversible process, since the molecule does not *spontaneously* break up again. Moreover, the mathematical formulation of this process must include the definite age of the system at which the process was complete, this age being measured from some selected initial moment.

This process provides an interesting limitation to a principle put forward by Maxwell as the basis of physical science.[30] He suggested that the laws of physics must be considered to be eternal and unchanging and that therefore they must be expressed in a form which does not contain the time explicitly. This means that for physical laws there can be no difference between to-day and to-morrow. The laws are concerned with small changes which systems undergo in small time intervals, and need not express any fundamental distinction between one moment and another.

Such laws cannot express the fact that anything sudden ever occurs which makes an essential change in the system as when two systems become one, or when one system breaks up into two. The laws of organic growth or the evolution of individual systems must display the fact that at a certain age of the system special things happen, such as the combination of two hydrogen atoms, or the attainment of maturity by an organism. Maxwell's principle puts a limitation on the form of physical laws which precisely eliminates the laws that would be appropriate for organisms. But there is no reason why a broader physics should not try to frame this new type of law that would be applicable to the history and development of individual systems, and it is probable that if this could be done the reversible laws of Newton, Maxwell, and Einstein would appear as approximations which were valid when nothing of special interest was happening, i.e. when only spatial movements were involved without synthesis, disintegration or the emission of light.

Laws of the Newtonian type which Maxwell had in mind assume that one can adequately describe the present state of a system without specifying its past history. But we cannot say anything very precise about the inside of a living organism, and it is found far more efficient to describe what is known of its past history. We do not try to say where atoms are in an organism; instead we mention its species, age, etc. Organisms might be defined as systems whose future behaviour is more easily estimated from their past history than from what can be known about their immediate internal structure. The most convenient formulation of organic laws will therefore be expressed in terms of the age of the organism and take account of how its life has been spent. These laws are necessarily irreversible, since the assimilation of oxygen or food is always going on in a manner which can never be reversed.

30 James Clerk Maxwell (1831–1879), physicist. Whyte is probably referring to Maxwell's equations, which provide the basis for classical electromagnetism, optics, and electric and magnetic circuits. The equations feature two physical constants, known as the permittivity of space and the permeability of space.

Life is like a function which must always alter in one direction; when this development ceases life has disappeared.

The contrast of living and dead now appears less important than the following classification of natural processes:

1 Processes which are reversible and whose laws can be expressed independently of the age of the system, e.g. gravitational and mechanical motions which do not involve light or heat.
2 Processes which are irreversible, the laws being best expressed in terms of the total time which has passed since some initial state, e.g. chemical combination, growth, evolution, radioactivity, and all changes involving light or heat.

Physics has always asserted that processes of the first type were fundamental in nature, and astronomy provided the ideal example in planetary motion. It was this assertion that gave rise to the essential issue behind the conflict of mechanism and vitalism. But if Born is right, and the fundamental atomic processes are irreversible, then the situation is completely altered. There is no longer a question of life being an arbitrary irruption in a world of mechanical law, since the laws of gravitation and mechanics must then be looked on as the limiting case, when the irreversibility is vanishingly small, of a whole series of irreversible processes which constitute the most important examples of the fundamental order in nature. This series would include the atomic processes connected with heat, light, and electricity, chemical combination, colloidal effects, organic growth and evolution, and the highly co-ordinated electrical processes which form the physiological basis of consciousness. If this view is correct the atomic processes of radiation and chemical combination should be just what the biologist needs to build up organisms. Instead of a chaos of little particles obeying inverse square laws, the modern physicist offers to the biologist a new kind of atom with electrical and magnetic properties which cause it to build up stable compounds.

The biologist may reply: "Yes, but organisms have four chief characteristics, their behaviour is irreversible, and displays growth, memory, and purposiveness. If you tell me that your atoms obey irreversible laws, so much the better, because my organisms certainly do. But your crystals grow very differently from my cells and organisms, and you can't explain away the apparent purposiveness of all life."

To which the physicist may answer: "Suppose that two hydrogen atoms are some distance apart with the total energy necessary to make a molecule. If they begin to move towards one another under some attractive influence which they exert we display no surprise. But they are moving towards a final end, which is an end, even though they are of course unconscious of it; and provided that nothing interferes they will reach one another, form a molecule, and the process will be consummated. The atoms move under an irresistible law of attraction towards a final condition which is unavoidable unless outside influences prevent it. The system of the two atoms develops necessarily towards a consummation, and the process has in this sense a teleological quality though this need not mean that any god or man had consciously planned the end for these particular hydrogen atoms.

"This quality was not present in Newton's law of gravitation precisely because it failed to say what happens at the end of any process, for instance when a meteorite hits the earth. Newtonian laws avoid the responsibility of dealing with all the exciting events, like the wedding of the atoms or the death of the meteorite. On the other hand it appears probable that all irreversible laws can be interpreted as leading either from or to some critical end condition. Thus all heat processes tend towards an approximate uniformity of temperature, and chemical reactions also move towards a final condition.

"Such systems as these display the rudiments of unconscious purpose. One must imagine these systems made much more complex so that it takes a long time and considerable nourishment before their unconscious purpose is fulfilled, whether this be the instinctive reproduction of their kind or any other biological function. "

"Maybe. I like the unconscious purpose which you have revealed in irreversible physics, because I am troubled by colleagues who see conscious mind everywhere.

"But if I grant that your view of the atom, and hence of molecules and colloids, allows me two of the four features I find in life, i.e. irreversibility and unconscious purpose, you have still to deal with growth and organic memory."

"Yes. Growth and memory are things that physics has as yet little to say about. But we have at any rate reduced the problem of life to smaller proportions. It is no longer the question what is life? but, how do colloidal processes build themselves up into continuously-active, developing systems which can react to their surroundings so that some distant condition can ultimately be attained? This is a much less difficult question. Moreover, since the problem of radiation underlies all the chemical processes which are associated with the maintenance of life, we may expect considerable assistance when physics has cleared up this crucial problem."

Chapter V: Physics and Mind

If a psychologist who was not a behaviourist had been listening to this conversation he might break in:

"Does the physicist seriously propose that we should try to leave mind out of our picture of the human organism? Even if we can eventually explain the unconscious purposes of the lower organisms as ends towards which they are driven by physical laws, yet man has the supreme distinction of a conscious mind, he can select his aim, and if he likes renounce it again for something else. You must therefore allow in your picture for the emergence of mind at some point during the course of evolution."

"Wait a moment," replies the physicist. "Your whole outlook towards consciousness betrays not only an anthropomorphic standpoint, but one limited to a single stage in man's development. There is no single condition adequately described by the word 'conscious'. There are in fact a great many different states of awareness which may grade into one another, or may form a series of distinct conditions. We do not know much about them yet, but their variety is most striking. There is the dim sentience as we awake from chloroform, the awareness of the dreaming state, the passive experiencing that accompanies any intensely rhythmic activity such as running. Again, quite different states are known in day-dreaming, intellectual concentration and the delicately-balanced semi-consciousness of creative thought.

"Consider especially the states of awareness associated with love, or with the supreme creative activities of the mind. Free-will, or the deliberate choice of a purpose, is completely lost in a whole-natured falling in love, as it is also in the artist's need to follow some dimly-conscious intuition of a task he must attempt. At these important occasions free-will disappears before a sense of inner organic necessity.

"These examples seem to me to make it clear that 'conscious purpose' is not in any sense the ultimate or highest criterion of human behaviour, and that free-will need not be taken necessarily to mean the power to over-ride any laws of nature. In my view 'free-will' is simply the apparent characteristic of organic behaviour when no complete integration of the personality has been achieved and the mind seems to be able to oscillate from one purpose to another. We really have to deal in human beings with a whole series of forms of behaviour of increasing complexity and integration: reflex and instinctive actions, deliberate activity, and finally the intuitive whole-natured creative functioning which leads to ends which could not have been intellectually foreseen. To each of these must correspond a certain type of awareness, and in my view, a brain process of a definite degree of complexity. By analogy with our own experience of different modes of consciousness, we may be able to infer from the structure of the central nervous system of an organism what sort of awareness it can experience.

"Eventually we must expect to be able to give a complete scheme of all organic behaviour in terms of the organic processes and their laws, but none the less it will remain a great deal more convenient in some cases to refer to what happens to human beings by using words that suggest their conscious experience. The behaviourist denies the scientific significance of all but the very barest elements of conscious experience, but of course he has to start from the human perception of light and colour. Science cannot get on without ideas which obtain their whole meaning from the qualities of conscious experience, and hence the extreme behaviourist position merely arises from a prejudice which prevents clear thinking. But as a campaign to put more stress on the direct observation of what really happens to living beings in terms of physical movements, behaviourism can only do good by bringing more unbiassed knowledge about life.

"My own interpretation of the question may be put in this way. The thing that is given in nature is a process in time. According to its complexity and degree of co-ordination an organic process has different degrees of awareness. There is no one condition called human consciousness, because the human organism can function with different degrees of coordination, and if we ask if an atom in absorbing light is conscious, the question has no definite meaning. But in a few years those who are studying the physiology of the central nervous system will be able to indicate how many steps of synthesis and integration occur between the simplest cell and the creative thinker, and to each of these stages will be ascribed a mode of awareness. But below a certain degree of organic complexity this 'awareness' will cease to be anything that can be consciously imagined by man, e.g. below the dimmest sentience one might allow an undifferentiated knowledge of mere continuance, based in turn on the rhythmic pulsation of the elementary cells."

"Your scheme is of course still rather vague, but in its main outlines it appears satisfactory", replies the psychologist. "But tell me outright, can mind influence matter? If I understand you rightly, you suggest that matter certainly influences mind."

"On the contrary, I do not! You are back at the meaningless questions on which philosophers have wasted much time. To ask if mind can influence matter does not mean anything until you know what you mean by mind and matter, and to a scientist that means knowing the laws they obey. Now, on the one hand, relativity and modern quantum theory indicate that there is no matter in the old sense of particles made of some unchanging stuff, and physical science recognizes atomic and other *processes* as fundamental in the place of 'matter'. On the other hand, you really mean by 'mind' one particular form of conscious activity: the deliberate selection of a purpose. Therefore to give your question real meaning I have to ask instead 'Does the conscious selection of a purpose alter the physical processes going on in the human organism?'

"But that is an absurd question. It is like asking: Does a dint in the outside of a hat *cause* an alteration in the shape of the inside of the hat? To which the only reply is that the dint on the outside is merely another way of describing the dint on the inside. There is no *causing* of the one by the other any more than if you fold a bit of paper you can say that the crease on one side causes the crease on the other side. They are identical and the double method of description used in the question creates a meaningless problem.

"'Conscious selection of a purpose' is one way of describing a particular process, and after this process has occurred the brain will be different from before. The old theories of the correlation or interaction of mind and matter presupposed that they were separate things in themselves. The important questions become quite different when one realizes that mind and matter do not exist independently, but that they are both somewhat inadequate ways of describing certain *aspects* of one organic process. The spatial aspect of organic process is called the physical organism. The temporal aspect of organic process corresponds to the content of its consciousness. The physical body is a group of spatial characteristics. Consciousness is a system of temporal elements; memory, anticipation, deliberate repetition, creative longing, hope and fear are all things set in time.

"Professor Alexander has said 'Time is the mind of Space.'[31] He attempts to explain space and time by an anthropomorphic analogy. It is a very suggestive

31 The philosopher Samuel Alexander (1859–1938) wrote on the metaphysics of experience. *Space, Time, and Deity*, 2 vols (London: Macmillan, 1920), vol. 2, p. 38.

idea, though for the searcher whose goal is the nature of consciousness itself it is more valuable to put it the other way round: mind is the temporal aspect of process, body the spatial aspect. But it is very important indeed to notice that we have not yet found the adequate terms for describing these two aspects of process. Matter is unsatisfactory for the spatial aspect, because there are no unchanging particles. But nor is mind sufficient for the temporal aspect, because there is a temporal aspect to the combination of hydrogen atoms and to chemical and colloidal processes, and yet we must not speak of these as having mind. When the new words for these two aspects are invented they will form the foundation of the scientific synthesis which I am expecting."

To which the psychologist may answer: "Well, at heart I have always been a thorough-going determinist like you, at least in dealing with my patients. Moreover I find it works, because I have always included in any picture of the patient a life-impulse of some sort, which can be influenced by my personality. Thus if the behaviour of my patient is absolutely determined, the conditions which determine what happens to him include some inner life tendency, and also the effects produced on him by all the people he meets.

"But if one attempts to formulate such an absolute determinism, or to apply it to oneself, one gets into deep waters, and I haven't the courage to try it. It seems you must be right at bottom, but that only a god could believe it without its upsetting his mental balance or his sense of moral responsibility."

"There I agree," replies the physicist, "as long as one does not simultaneously revise one's whole view of life in terms of this new organic knowledge. That is a very big task, but I should like one day to attempt it. Two things especially would attract me to such a revision of human values. One is that people who ought to know better still go about making moral judgments about their acquaintances. Now that we know how profound is the influence on a child of the treatment it receives during its first five years of life, moral judgments become rather old-fashioned and only show that the person making them has himself not yet learnt to find emotional fulfilment in healthier ways. An analysis of human behaviour along the lines of organic determinism might do something to show that moral condemnations, whether of bolshevism or of the sins of one's children, are never effective unless immediately accompanied by positive example or creative suggestion.

"But there is another more attractive reason why I should like to attempt this transvaluation of values.[32] If organic determinism is valid, then the artist's aspiration to create is a natural consequence of some organic law. Creative aspiration may then be looked on as the natural destiny of certain human beings, though they no more know where they are going than did the two hydrogen atoms. But organic determinism allows us to understand why it is of no importance that the artist doesn't know what he is going to create before he does it. It seems that in

32 Nietzsche's phrase for the replacement of what he saw as decadent Christian idealism with a new life-affirming philosophy.

some matters our organic body is wiser than ourselves, or rather wiser than our very immature consciousness. When we have developed our consciousness by the discovery of the organic laws of our own natures we may be able to make human life more beautiful."

Chapter VI: The Future of the Sciences

The preceding pages have very broadly indicated the way in which current physical researches may influence the scientific outlook on the problems of matter, life, and mind. The view has been put forward that we are on the eve of a profound scientific synthesis of which the main outlines are already determined. These general suggestions will now be made more precise in order to offer to anyone who is interested the opportunity of testing for himself some definite prophecies regarding the future of scientific thought. The forecast made here does not involve any supernatural reading of the future, but is based on tendencies already inherent in the different departments of science. For convenience it is expressed in the form of separate assertions concerning the future of physics, biology, and psychology.

1. Before 1940 a very remarkable simplification will be made in atomic theory, which will indicate that in quantum processes physics has 'touched bottom' and that—for the time being—we may consider that nature is not infinitely complex within the heart of the atom. The proof of this apparent if not absolute limit to the micro-structure of nature will take the form of the discovery of simple relationships between the fundamental constants of atomic structure, e, m, M, c, and h. (The electronic charge and mass, the mass of the hydrogen nucleus, the velocity of light, and Planck's constant.) Such relations are already known but are considered to be of no significance since they are ruled out by the accepted theory of electrical dimensions.

   Yet this dimensional system is not based on direct observation, and the importance of these relationships will soon be recognized in consequence of experiments aimed at a direct determination of an 'electron velocity', in a curved track. 'Electron velocity' as calculated from deflection experiments will be found not to be the same as the directly measurable cms. per sec., and in the case of straight electron tracks, the measured velocity may be found to be always that of light, though this does not mean much since the velocity of light in one direction has never been measured.

   As the result of the study of individual radiation tracks, for instance in the reflection of electrons by crystals, and particularly of any *time* measurements that can be made, a new system of physical conceptions will be built up appropriate to irreversible processes, which will be substituted for the Newtonian reversible system. The new scheme will probably be based on the conception of the atom, with its radiating electron tracks, as a natural clock which not only can be used to measure out equal time intervals, but also to yield an objective criterion of past and future. In order to make this idea, or at least one part of it, capable of empirical test the following hypothesis is put forward: The time-interval between any two point-events on any electron track is a simple function of the length and curvature of the part of the track between the two points. This hypothesis contradicts the current interpretation of electron theory on a point which has never yet been subjected to experimental test.

The conceptions which will be built up on electron velocity experiments will very quickly bring within one simple theory the facts of chemical combination and colloidal processes. For these depend upon irreversible effects connected with radiation and electrons, and will therefore be amenable to treatment by the new conceptions for the very reason which necessarily puts them beyond the scope of Newtonian laws.

2. As the result of the alteration in physical conceptions biology will soon cease to draw a definite line between inanimate and living systems. The normal characters of life will be recognized as appearing in steps as one passes up the series atom, molecule, colloid, protoplasm, cell, and through further stages to mammal and man. In each class of organism a central controlling process will be discovered and its laws formulated with some precision, in terms of irreversible electrochemical processes. The process which in each organism represents the co-ordinating factor and is the life of the organism considered as a unit may for instance be described in terms of a quantity which we shall call 'f'. 'f' would be such that so long as 'f' keeps on increasing the organism is alive, while if 'f' stands still the organism dies. The rate of increase of 'f' indicates the tempo or intensity of the organism's life. In a simple case 'f' might be directly related to the intake of oxygen or food, and just as respiration and assimilation are irreversible, so is the change in 'f'. 'f ' must go on increasing, or else cease to represent any quantity in nature; as soon as it ceases to increase the process to which it corresponds cannot be identified any longer.

The most important factors which influence the life-function 'f' (i.e. which affect the central controlling process in any organism) will be known before about 1950, with the result that local rebellions such as cancer will not only be controllable, but easily prevented. Harmless methods for increasing the rate of change of 'f', i.e. for increasing the *élan vital*[33] of the organism, will be discovered, so that, for instance, the duration of child-birth will be reduced to a natural minimum. If child-birth sometimes takes very long nowadays, this is presumably because the woman's body is tired, exhausted, or partially poisoned by her mode of living, and by raising her vitality at the critical moment we may expect to be able to let the process go on at its natural speed. There must be some minimum time necessary for the act, since a vast number of complex organic processes have to complete themselves in a certain order, but probably this time is considerably shorter than that during which many women in this country have to suffer.

33 The term was introduced by Bergson in his book *Creative Evolution* (1907) to designate the life force driving evolution.

It is already known that the Mendelian *genes* which determine heredity are related to the rates of development of special processes in the organism, and a control over the life-tempo, or rate of increase of 'f', in any organism or group of cells within an organism, will provide a new method of tackling the practical problem of heredity. It is possible that hereditary tendencies to specific weakness or disease will be overcome by accelerating or retarding the rate of development of the human system at some special moment between conception and maturity.

Rejuvenation[34] will soon be safe and efficient, but not as a means for attempting immortality. It will be socially recognized as healthy and legitimate only when undertaken to compensate for premature ageing due to specific repressions, illness, or anxiety.

The elimination of known diseases by a genuine science of life does not mean that other diseases will spring up perhaps worse than before. A theoretical science of life will know the meaning of all disease, and will not prevent one in such a way as to give rise to another. Instead of making campaigns against influenza or any other one disease, it will determine the conditions in which no disease can survive, and thus gradually eliminate all the organic diseases which attack the body.

But this does not mean the attainment of a hygienic Utopia in which human life necessarily fulfils itself. A balance will be made to the disappearance of cancer and syphilis, not by the arising of other diseases but as a result of the consequent increase in the sensitiveness of the human brain.

The supremely difficult task of the next hundred years will be to keep the mind of the race healthy and stable through a period of critical sensitiveness. We are in a transition stage of violent instability, of intense cruelty coupled with compassion (America), of blended love of liberty and need of discipline, of emotional religions and of wars—but we must hope that it will lead to some mode of life with greater inherent stability.

3. Psychology is now occupied with the discovery that the human response to perceptions is not additive, i.e. that the effect made by a group of sounds or colours depends on the pattern in space and time in which they are arranged. (*Gestalt-theorie.*) For instance, the effect made on a man by the individual notes of 'God save the King' when played in the wrong order is negligible, and bears no relation to his response when he hears the tune played in a cinema, and it reminds him of 'patriotism' and the War. So far no scientific method has been found of describing when a group of elements is to be treated as a 'whole' for the purposes of psychology, and this is where the greatest advances may be expected.

Most scientific conceptions have been based on the method of spatial analysis, i.e. the reduction, where possible, of a thing to its smallest spatial elements. Physics,

34 See *Halcyon*, pp. 266–267.

biology, and psychology have all lacked the equipment to describe what makes the atom, organism, or the pattern function as a unit, and how we are to know if some group is a unit or not. The analytical method is fully developed, but we lack even the basis for a synthetic treatment. This leads some hard-headed scientists of the materialistic school who will 'stand no nonsense' to assert that there is no such thing as 'synthesis', that this is a mystical idea left over from primitive anthropomorphism. Yet to any mind that is guided not by prejudice but by a simple search for truth, the fact of synthesis is obvious, though not yet properly formulated.

Here modern physics can supply a clue. Analysis is the method required in a search for instantaneous spatial structure; the synthetic method which we need must deal with the temporal history and behaviour of systems. The fact that the human being reacts in the ways he does to a tune as a whole is evidence of something in his history, that he has heard the tune often under certain emotional surroundings. The unity of any synthesis, whole, or organism is not an instantaneous fact explicable in terms of structure, for we can recognize this unity only from a continued observation over a period of time.

Physics can invent one law to describe the approach of the two hydrogen atoms to form a molecule, and in doing so treats the two together as a unit. This suggests that the fact of organic unity is to be defined and formulated in terms of an irreversible law which governs the system as a whole. Thus a group of atoms, cells, or any other elements is to be called a unit when, and only when, one irreversible law can be found which expresses the behaviour of the different elements as contributing towards some common end, like the formation of the molecule in the case of the hydrogen atoms.

We can now draw a practical conclusion for the future of psychology, which is in great need of a moral principle to guide its treatment of disintegrated human personality. On the analogy of the two atoms, a human being is to be considered as a unity when his whole behaviour displays continuous co-ordination towards some end. But there is an important difference in the two cases: the atoms move towards an end which we know because it has already happened in history, whereas man's development is creative, that is it proceeds towards an end we cannot know exactly before it comes into being. Thus the parent or psychologist need not trouble if he cannot understand what his child or subject is aiming towards: so long as some consistency and harmony of functioning is apparent, the 'end' can be left to nature to look after, because such harmony *means* that the organism is tending towards some ultimate condition.

The psychologists of the future will therefore have to follow some principle such as this: their only legitimate aim is the maintenance and restoration of harmonious co-ordination of all the human functions, and no concern need be paid to ultimate intellectual or spiritual ideals. Of course if the person considered is apparently tending towards some degenerate condition, that is known to the onlooker because it is *not* new but a repetition of what many human beings have done before, then this tendency can be altered. At least, it can be altered if the onlooker can use his intuition to discover signs of repressed conflict which show that the immediate tendency is not whole-natured, but based on the repression of some more profound aspiration or desire. Then by bringing this repressed aspiration back into consciousness the degenerate tendency may be arrested. But this control over the lives of others can only be effectively exercized by the intuitive discovery that their present tendencies are not whole-natured.

Prophecy can never be scientific, and forecasting in the realm of science is perhaps the most dangerous form of intellectual acrobatics. Science must be thorough, and all vague speculation is its enemy. But there are moments when a profound revision is necessary, and amidst the responsibilities and rich appeal of daily life no one will undertake this task who does not believe that it offers an adequate reward to science and to man. To-day prophecy can call attention to unjustified limitations inherent in current scientific thought, and encourage the students of matter and of life to get together and try to discover the single system of natural law which we must believe covers both realms. It may even help them to find crucial experiments by which to guide their search.

The reward is certainly great. The indifference to the destruction of life which has marked recent years is no cause either for surprise or for despair after an epoch of orthodox and insincere religion coupled with an abstract science of matter. One thing only can guide humanity to a saner and richer life: the recognition and valuation of life. This can be assisted by science and art both revealing life in all its significant forms. But the roots of art have been destroyed by the domination of a science which had not recognized the significance of life within the realm of natural law. For great art can only arise from a profound reverence for life, whereas to the scientific mood of this period life appeared as an arbitrary impulse in continual conflict with the laws of matter.

Physics is now studying light. The radiant influence of light nourishes life and within human body forms the fabric of consciousness. We are alive and conscious, but our consciousness is immature for we do not yet know the laws that govern our own lives and thoughts. Yet it is certain that light, life, and consciousness are bound together by some, undiscovered law. This secret of nature's alchemy is still hidden from us within our own bodies. By revealing it physics will create a new hope for man.

The World, the Flesh and the Devil[1]

An Enquiry into the Future of the Three Enemies of the Rational Soul

J. D. Bernal (1929)

Headnote

John Desmond Bernal (1901–1971), FRS, was an Irish scientist who pioneered the use of x-ray crystallography to develop molecular biology. He had been appointed as the first lecturer in Structural Crystallography at Cambridge in 1927, and assistant director of the Cavendish Laboratory there in 1934. From 1937 he was Professor of Physics at Birkbeck College, University of London.

Bernal became a member of the Communist Party from 1923, and was a leading Marxist scientist and philosopher of science. *The Social Function of Science* (1939) remains an important account of how research is organized. His monumental four-volume *Science in History* appeared in 1954.

He was a friend of Pablo Picasso and Pablo Neruda, and his grasp of three-dimensional form was said to have been an inspiration for Barbara Hepworth.

During the Second World War Bernal worked on the effectiveness of bombing, and the planning for the D-Day landings. Although he never worked on radar, the atom bomb, or code-breaking, he was nevertheless described in one official history as the British scientist who did most to win the Second World War.

The World, the Flesh and the Devil was his first book. Like *Daedalus* and other volumes, it had begun as a university talk. An earlier version had appeared—also as 'The World, the Flesh and the Devil'—in Ogden's *Psyche*, 9:4 (April 1929), 3–26, with a note (on p. 3) saying: 'This rough draft of Mr. Bernal's forthcoming volume' was 'adapted from a paper read in Cambridge in March 1929.' Arthur C. Clarke described the book as 'the most brilliant attempt at scientific prediction ever made.' Robert Scholes called it a 'book of breathtaking scientific speculation' which 'is probably the single most influential source of science fiction ideas.'[2]

1 The title phrase from medieval theology is in Abelard and Aquinas, and gets into the Litany of the *Book of Common Prayer*: '[... F]rom al the deceytes of the worlde, the fleshe, and the deuill: Good lorde deliuer us.' It is traditionally taken to refer to the three main temptations to a Christian soul—the antithesis of the trinity. Dora Russell uses the expression in *Hypatia*, p. 197.

2 Clarke, *Greetings, Carbon-Based Bipeds* (New York: St Martin's Press, 1999), 410. Scholes, in Eric S. Rabkin, 'Bibliography III: Science Backgrounds', *Science Fiction: History, Science, Vision* (London: Oxford University Press, 1977).

DOI: 10.4324/9781032625522-8

In the Foreword, dated November 1968, to the second edition (London: Cape, 1970), Bernal wrote that the book had been written 'on the eve of the greatest discoveries and inventions of science': the atomic age, lasers, computers, and the discovery of the double-helix structure of DNA.

******

Chapter I: The Future

There are two futures, the future of desire and the future of fate, and man's reason has never learnt to separate them. Desire, the strongest thing in the world, is itself all future, and it is not for nothing that in all the religions the motive is always forwards to an endless futurity of bliss or annihilation. Now that religion gives place to science the paradisical future of the soul fades before the Utopian future of the species, and still the future rules. But always there is, on the other side, destiny, that which inevitably will happen, a future here concerned not as the other was with man and his desires, but blindly and inexorably with the whole universe of space and time. The Buddhist seeks to escape from the Wheel of Life and Death, the Christian passes through them in the faith of another world to come, the modern reformer, as unrealistic but less imaginative, demands his chosen future in the world of men.

Can we in any better way reconcile desire and fate? In the belief of the scientist the future can yield to an objective analysis only if he can put aside all desire of one future or another; and yet, in reaching for this unattainable understanding, by some mutual influence his desires and the events may grow more and more into harmony. Holding this hope, or better still, moved by a pure curiosity for things to come, how is it possible to examine scientifically the future? For in the science of the future observation is as impossible as experiment; and of the three methods there is left to us only prediction. In the other sciences prediction plays but a small part, and rightly so, for verification follows closely on its heels; but there are general methods in scientific prediction and we may try to apply them in dealing with the whole future state.

First and always, it is necessary to exclude as far as possible, illusion; for to most of us the future is the compensation and fulfilment of all that the present and the past have lacked; and the future being unknown and incontrovertible has been a fair ground on which to place all these hopes and desires. But in scientific prediction these desires are the most delusive guides. The opposite danger is as great and more insidious: in our lives we take the present for granted to an extent far greater than we can realize, so that even when we are thinking of the future we cannot separate the historic accidents of the society in which we were born from the axiomatic bases of the universe. Until the last few centuries this inability to see the future except as a continuation of the present prevented any but mystical anticipations of it. Luckily these complementary errors affect different parts of the future. It is in the near future where we are still sympathetically related to men and events that our desires have the most power to twist our appreciation of facts.

We care less about the more distant future, but to approach it at all we must divest ourselves of so many customary forms, that even the more enlightened prophets let their imagination stop in some static Utopia in despite all evidence pointing to ever increasing acceleration of change.

What positive ideas can be found to take the place of the naïve anticipation that the future will be like the present but more pleasant (or more unpleasant according to one's disposition)? The leading principle is that by which Lyell founded scientific geology:[3] the state of the present and the forces operating in it contain implicitly the future state and point the way to its interpretation. We have three disciplines of thought to help us to this interpretation. History (of which human history is only a minimal part) tells us how things have changed and how by inference they will change in the future. Strictly, prophecy should be treated as part of history, but, until history has found its laws, it must chiefly be used as a storehouse of illustrative facts; though one might say loosely that everything that will happen must conform with the spirit of History. The physical sciences, as far as we know them, give us the material of which future as much as past is built, and the manner of that building. The manner appears to us as physical law but it may well be found to be a tautology which we are congenitally too limited to grasp. Lastly there is the knowledge of our desires, but though the future according to our desires, is an illusion, our desires are, paradoxically, already tending to be the chief agent of change in the universe; it is only that the actual change is so rarely the desired change.

The initial difficulty in the general prediction of the future is its enormous complexity and the interdependence of all its parts; but this complexity is not completely chaotic and we can always attack it by considering it as a product of chance and determinism, chance where we cannot see relationships, determinism where we can. The events out of which so complicated a thing as the general state of the universe is built, form neither one indivisible whole nor a set of equally independent units, but consist of complexes (nebula, planet, sea, animal, society) of which the components are themselves complex parts. This hierarchy of complexes is not imagined to have any objective validity, it is only an expression of the modes of human thought, a convenient simplification which makes science possible. Inside each complex, development proceeds according to its own rules, determined by the nature of the complex; but these rules always include, if they do not entirely reduce to, what is, in effect, the statistical chance interaction of complexes of a lower order. The death-rate of a town, for instance, can be shown to be a function of the amount of money it spends on sanitary measures, but the individual deaths appear, from the point of view of the town, to be due to chance circumstances, though again for each individual concerned they are determined. We can always leave out the higher complexes when we are considering the lower. An atom of oxygen will respond to its environment in the same way in a nebula, in a rock or in the human brain.

3 Sir Charles Lyell (1797–1875). His *Principles of Geology* (1830–1833) explained the past development of the Earth as driven by the same processes discernible today.

Now the complex we are concerned with here is the human mind, and so we can fairly start with the assumption that the rest of the universe goes on its way determined by its physical, chemical and biological laws except in so far as man himself intervenes. Absolutely, we know hardly anything of these laws, but relatively to our knowledge of human behaviour we know them so well that the future they present—the astronomical, geological biological future—seems a fixed and stable thing.

In human affairs the immediate future reveals itself in the following of tendencies visible in the present; beyond that must come the applications and the development of present knowledge. This is the minimal basis for prediction; but our present knowledge carries with it the implication of still further advances in knowledge along the same lines. It is the applications of this new knowledge and the secondary results that flow from them that will chiefly concern us, because it is clearly impossible to go further and include unimagined discovery. Of course, there is a considerable chance that one of the unpredictable discoveries will be so important that it will turn aside the whole course of development. But to be deterred by this chance would be to abandon any attempt at prediction. Already the chance element comes in when we consider applications or developments of knowledge in more than one restricted field; because although we can predict the development in that field fairly well, we cannot predict the rate of development; and so the rates of development in differing fields, which are constantly reacting on each other, being unpredictable, the resultant future becomes more and more uncertain the farther we look forward. The only way to deal with this complexity is by separating the variables as best we can, by arbitrarily considering developments as proceeding in one field without any developments in any of the others, and then combining the results attained by applying this method in different fields. At the same time we must keep in mind that the state of development at any other period must be a self-consistent whole. Each line of development must have reached the level which is implied by the necessities of any of the other lines: for instance, the chemical control of life requires the development of chemical technique and apparatus of a very high order. On the other hand, whole sections of certain developments may become superfluous owing to developments in other fields; for instance, the manufacture of synthetic food and the industry connected with it would be unnecessary if blood were used directly as the motive power for animals.

Obviously we cannot proceed with this method in detail: if we could, we should not only be able to predict the future exactly, but to make it the present. For brevity, it is worth considering three fields only.

Man is occupied and has been persistently occupied since his separate evolution, with three kinds of struggle: first with the massive, unintelligent forces of nature, heat and cold, winds, rivers, matter and energy; secondly, with the things closer to him, animals and plants, his own body, its health and disease; and lastly, with his desires and fears, his imaginations and stupidities. In each of these divisions in turn we will make the arbitrary assumption that his progress in it will continue while in other respects he remains the same.

Chapter II: The World

FIRST, then, in the material world. Here prediction is on its surest ground, and is in the first stages, almost a business of mathematics. The physical discoveries of the last twenty-five years must find their application in the world of action—a process which has hardly begun, but the nature of which can easily be seen. So far we have been living on the discoveries of the early and mid-nineteenth century, a macro-mechanical age of power and metal. Essentially it succeeded in substituting mechanism for some of the simpler mechanical movements of the human body, with steam and later electrical power in the place of muscle energy. This was sufficient to revolutionize the whole of human life and to turn the balance definitely for man against the gross natural forces; but the discoveries of the twentieth century, particularly the micro-mechanics of the Quantum Theory which touch on the nature of matter itself, are far more fundamental and must in time produce far more important results. The first step will be the development of new materials and new processes in which physics, chemistry and mechanics will be inextricably fused. The stage should soon be reached when materials can be produced which are not merely modifications of what nature has given us in the way of stones, metals, woods and fibres, but are made to specifications of a molecular architecture. Already we know all the varieties of atoms; we are beginning to know the forces that bind them together; soon we shall be doing this in a way to suit our own purposes. In fact, Professor Goldschmidt[4] of Oslo has already made many model structures in which existing substances are closely copied in different atoms, so as to make a new substance, softer or harder, or more or less fusible. Sulpho-nitrides with silicate structures will be harder and more infusible than anything on earth. A similar substance—carboloy—which is already on the market—combines the strength of steel with the hardness of diamond, and is capable of working glass like metal. There are similar possible model structures for organic substances; the complexities are greater but the results will be more far-reaching. The linked molecules that make fibres and elastic substances such as rubber or muscle, are already yielding to X-ray investigation; the proteid bodies of living matter must have an analogous but more complex structure. After the analysis will come the synthesis; and for one place in which we can imitate nature we will be able to improve on her in ten, and furnish models of organic materials with more varied properties and capable of withstanding more rigorous conditions. The result—not so very distant—will probably be the passing of an age of metals and all that it implies—mines, furnaces and engines of massive construction. Instead we should have a world of fabric materials, light and elastic, strong only for the purposes for which they are being used, a world which will imitate the balanced perfection of a living body.

At the same time, much that we require for the purposes of modern life would become no longer necessary. With improved systems of chemical manufacture our food and our clothing will be made with much less expenditure of energy in

4 Victor Moritz Goldschmidt (1888–1947); Norwegian mineralogist who (with Vladimir Vernadsky) founded modern geochemistry and crystal chemistry.

manufacture and transport. And the development of mechanism will not cease: it should turn into more refined forms—heat-engines capable of working at lower and lower temperature differences, engines of higher and higher speed, electrical machines of high potential and high frequency—and should lead to the solution of two most fundamental problems, the efficient transmission of energy by low frequency (wireless) waves, and the direct utilization of the high frequency (light) waves of the sun. On the chemical side the problem of the production of food under controlled conditions, biochemical and ultimately chemical, should become an accomplished fact. In the new synthetic foods will be combined physiological efficacy and a range of flavour equal to that which nature provides, and exceeding it as taste demands; with a range of texture also, the lack of which so far has been the chief disadvantage of substitute food stuffs. With such a variety of combinations to work on, gastronomy will, for the first time, be able to rank with the other arts.

All these developments would lead to a world incomparably more efficient and richer than the present, capable of supporting a much larger population, secure from want and having ample leisure, but still a world limited in space to the surface of the globe and in time to the caprices of geological epochs. Already ambition is stirring in men to conquer space as they conquered the air, and this ambition—at first fantastic—as time goes on becomes more and more reinforced by necessity. Ultimately it would seem impossible that it should not be solved. Our opponent is here the simple curvature of space-time—a mere matter of acquiring sufficient acceleration on our own part—which, sooner or later, must be practicable. Even now it is possible to imagine methods of accomplishing it, based on no more knowledge than we already possess. The problem of the conquest of space is one in which all the difficulties are at the beginning. Once the earth's gravitational field is overcome, development must follow with immense rapidity. Without going too closely into the mechanical details, it appears that the most effective methods is based on the principle of the rocket, and the difficulty, as it exists, is simply that of projecting the particles, whose recoil is being utilized, with the greatest possible velocity, so as to economize both energy and the amount of matter required for propulsion. Up to the present all forms of rocket depend on the movement of masses of gas in which the individual molecules are moving at high velocities in perfectly random directions, and use is only made of the average velocity in the desired direction. What is wanted in the first place is a form of Maxwell's Demon[5] which will allow only those molecules, whose velocities are high and in the direction opposite to the trajectory of the rocket, to escape. The next difficulty is that to set in motion any large rocket the mass of gas required is the same of the same order as the weight of the rocket itself, so that it is difficult to imagine how the rocket could contain enough material to maintain its propulsion for any length of time. When the radio-transmission of energy is effected half the difficulty will be removed and the

5 A thought experiment by James Clerk Maxwell imagining an entity 'allowing only fast-moving molecules to pass through a hole in one direction and only slow-moving ones in the other direction, so that if the hole is in a partition dividing a gas-filled vessel, one side becomes warmer and the other cooler, in contradiction of the second law of thermodynamics' (*OED*).

projection may very well ultimately be effected by means of positive rays at high potential. It may be that both the problem of space travel and the ethereal transference of energy have already been solved by Professor Japolsky's magneto-fugal waves.[6] These are a type of magnetic vortex ring, propagated through space, which, instead of spreading as ordinary electromagnetic waves, remain concentrated along the axis of propagation. Apart from its mode of projection, the construction of the space vessel offers little difficulty since it is essentially the same problem as that of the submarine. Naturally the first space vessels will be extremely cramped and uncomfortable, but they will be manned only by enthusiasts. The problem of landing on any other planet or of returning to earth is much more difficult, mainly because it requires such a nice control of acceleration. Probably the first journeys will be purely for exploration, without landing, and the travellers, if they return to earth at all, will have to abandon their machine and descend in parachutes.

However it is effected, the first leaving of the earth will have provided us with the means of travelling through space with considerable acceleration and, therefore, the possibility of obtaining great velocities—even if the acceleration can only be maintained for a short time. If the problem of the utilization of solar energy has by that time been solved, the movement of these space vessels can be maintained indefinitely. Failing this, a form of space sailing might be developed which used the repulsive effect of the sun's rays instead of wind. A space vessel spreading its large, metallic wings, acres in extent, to the full, might be blown to the limit of Neptune's orbit. Then, to increase its speed it would tack, close-hauled, down the gravitational field, spreading full sail again as it rushed passed the sun.

So far, those who have considered spatial navigation have regarded it from the point of view of exploration and planetary visitation, but the vast importance of escaping from the earth's gravitational field has been almost entirely overlooked. On earth, even if we should use all the solar energy which we receive, we should still be wasting all but one two-billionths of the energy that the sun gives out. Consequently, when we have learnt to live on this solar energy and also to emancipate ourselves from the earth's surface, the possibilities of the spread of humanity will be multiplied accordingly. We can imagine this occurring in definite stages. When the technicalities of space navigation are fully understood there will, from desire or necessity, come the idea of building a permanent home for men in space. The ease of actual navigation in space together with the difficulties of taking-off from or landing on planets like the earth with considerable gravitational fields will in the first place lead to the necessity for bases for repairs and supplies not involving these difficulties. A damaged space vessel would, for instance, almost be bound to be destroyed in attempting earth landing. At first space navigators, and then scientist whose observations would be best conducted outside the earth, and then finally those who for any reason were dissatisfied with earthly conditions would come to

6 'Magnetofuge' was N. S. Japolsky's name for an electric linear induction motor (as used in wheel-less maglev trains, for example, where electromagnetism both causes the vehicle to hover and propels it).

inhabit these bases and found permanent spatial colonies. Even with our present primitive knowledge we can plan out such a celestial station in considerable detail.

Imagine a spherical shell ten miles or so in diameter, made of the lightest materials and mostly hollow; for this purpose the new molecular materials would be admirably suited. Owing to the absence of gravitation its construction would not be an engineering feat of any magnitude. The source of the material out of which this would be made would only be in small part drawn from the earth; for the great bulk of the structure would be made out of the substances of one or more smaller asteroids, moons, rings of Saturn or other planetary detritus. The initial stages of construction are the most difficult to imagine. They will probably consist of attaching an asteroid of some hundred yards or so diameter to a space vessel, hollowing it out and using the removed material to build the first protective shell. Afterwards the shell could be re-worked, bit by bit, using elaborated and more suitable substances and at the same time increasing its size of by diminishing its thickness. The globe would fulfil all the functions by which our earth manages to support life. In default of a gravitational field it has, perforce, to keep its atmosphere and the greater portion of its life inside; but as all its nourishment comes in the form of energy through its outer surface it would be forced to resemble on the whole an enormously complicated single-celled plant.

The outermost layer would have a protective and assimilative character. The presence of meteoric matter in the solar system moving at high speeds in eccentric orbits would be the most formidable danger in space travelling and space inhabitation. Certain meteorite swarms could be avoided altogether by keeping out of their tracks; larger meteorites could be detected at a distance by visual observation or by the effect of their gravitational fields. These might be avoided by charging changing the course of the globe or deflecting the meteorites by firing high-speed projectiles into them. Smaller meteorites would be impossible to avoid. The shell of the glove would have to be made strong enough not to be penetrated or cracked by them, and would have to possess regenerative mechanisms for repairing superficial damage. Possibly the function which our atmosphere performs for the earth could imitated by jets of high-speed gas or electrons which, projected at meteorites would vapourize them and thus prevent them doing any damage. At the same time meteoric matter might be the chief source of the material required for growth or propulsion of the globe if a method of assimilating it could be found.

The outer shell would be hard, transparent and thin. Its chief functions would be to prevent the escape of gases from the interior, to preserve the rigidity of the structure, and to allow the free access of radiant energy. Immediately underneath this epidermis would be the apparatus for utilizing this energy either in the form of a network of vessels carrying a chlorophyll-like[7] fluid capable of re-synthesizing carbohydrate bodies from carbon dioxide, or some purely electrical contrivance for the absorption of radiant energy. In the latter case the globe

7 Chlorophyll is the green pigment in plants, algae, and cyanobacteria. It is vital for photosynthesis, enabling them to absorb energy from light.

would almost certainly be supplied with vast, tenuous, membranous wings which would increase its area of utilization of sunlight. The subcutaneous circulation would also have the necessary function of dissipating superfluous heat, in as low temperature radiation as possible. Underneath this layer would probably lie the main stores of the globe in the form of layers of solid oxygen, ice and carbon or hydro-carbons. Inside these layers, which might be a quarter of a mile in thickness, would lie the controlling mechanisms of the globe. These mechanisms would primarily maintain the general metabolism, that is they would regulate the internal atmosphere and climate both as to composition and movements. They would elaborate the necessary food products and distribute mechanical energy where it was required. They would also deal with all waste matters, reconverting them with the use of energy into a consumable form; for it must be remembered that the globe takes the place of the whole earth and not of any part of it, and in the earth nothing can afford to be permanently wasted. In this layer, too, would be the workshops and laboratories concerned with the improvement of the globe and arrangements for its growth.

Inside the mechanical layer would be the living region and here imagination has a more difficult task. It would, of course, not be necessary to have either houses or rooms in the same sense in which we have them on earth. The absence of bad weather and of gravitation makes most of the uses that we have for houses superfluous. Perhaps we can safely assume that a certain number of cells closed by thin, but sound-proof, partitions would be necessary for work requiring special isolation, but the major part of the lives of the inhabitants of the globe would be spent in the free space which would occupy the greater portion of the centre of the globe.

The three-dimensional, gravitationless way of living is very difficult for us to imagine, but there is no reason to suppose that we would not ultimately adjust ourselves to it. We should be released from the way we are dragged down on the surface of the earth all our lives: the slightest push against a relatively rigid object would send us yards away; a good jump—and we should be spinning across from one side of the globe to the other. Resistance to the air would, of course, come in, as it does on the earth; but this could be turned to advantage by the use of short wings. Objects would become endowed with a peculiar levity. We should have to devise ways of holding them in place other than by putting them down; liquids and powders would at first cause great complications. An attempt to put down a cup of tea would result in the cup descending and the tea remaining as a vibrating globule in the air. Dust would be an unbearable nuisance and would have to be suppressed, because even wetting it would never make it settle. We should find in the end that all these things were great conveniences, but at first they would be extremely awkward. The possibilities of three-dimensional life would make the globes much roomier than their size would suggest. A globe interior eight miles across would contain as much effective space as a countryside one hundred and fifty miles square even if one gave a liberal allowance of air, say fifty feet above the ground.

The activity of the globe is, of course, by no means confined to its interior. In the first place it would necessarily have a number of effective sense and motor organs.

Essentially the former would consist of an observatory which continually recorded the position of the globe and at the same time kept a look-out for any meteoric bodies of perceptible size which might damage it. On the whole the globe would not be designed for travel. It would move in an orbit around the sun without any expenditure of energy; but occasionally it might be necessary to shift its orbital position to a more advantageous one, and for this it would require a small motor of a rocket variety.

Yet the globe would be by no means isolated. It would be in continuous communication by wireless with other globes and with the earth, and this communication would include the transmission of every sort of sense message which we have at present acquired as well as those which we may acquire in the future. Interplanetary vessels would insure the transport of men and materials, and see to it that the colonies were not isolated units.

However, the essential positive activity of the globe or colony would be in the development, growth and reproduction of the globe. A globe which was merely a satisfactory way of continuing life indefinitely would barely be more than a reproduction of terrestrial conditions in a more restricted sphere. But the necessity of preserving the outer shell would prevent a continuous alteration of structure, and development would have to proceed either by the crustacean-like development in which a new and better globe could be put together inside the larger one, which could be subsequently broken open and re-absorbed; or, as in the molluscs, by the building out of new sections in a spiral form; or, more probably, by keeping the even simpler form of behaviour of the protozoa[8] by the building of a new globe outside the original globe, but in a position to set up an independent existence.

So far we have considered the construction and mechanism of the globe rather than its inhabitants. The inhabitants can be divided into the personnel or crew, and the citizens or passengers. With the first—except that their tasks would be more complicated and more scientific than those that fall to the officers and crew of a modern ship—we need not be concerned. To the others the globes would appear both as hotels and laboratories. The population of each globe would be by no means fixed; constant interchange would be taking place between them and the earth even when the greater portion of human beings were actually inhabiting globes. There would probably be no more need for government than in a modern hotel: there would be a few restrictions concerned with the safety of the vessel and that would be all.

Criticism might be made on the ground that life in a globe, say of twenty or thirty thousand inhabitants would be extremely dull, and that the diversity of scene, of animals and plants and historical associations which exist even in the smallest and most isolated country on the earth would be lacking. This criticism is valid on the initial assumption that men have not in any way changed. Here, to make globe life plausible, we must anticipate the later chapters and assume men's interests and occupations to have altered. Already the scientist is more immersed in his work and

8 Single-celled organisms.

concentrates more on relations with his colleagues than in the immediate life of his neighbourhood. On the other hand, present æsthetic tendencies verge towards the abstract and do not demand so much inspiration from untouched nature. What has made a small town or a small country seem in the past a narrow sphere of interest has been on the one hand its isolation, and on the other hand the fact that the majority of its inhabitants are at so low a level of culture as to prevent any considerable intellectual interchange within its boundaries. Neither limitation holds for the globes, and the case of ancient Athens is enough to show that small size alone does not prevent cultural activity. Free communications and voluntary associations of interested persons will be the rule, and for those whose primary interest is in primitive nature there will always remain the earth which, free from the economic necessity of producing vast quantities of agricultural products, could be allowed to revert to a very much more natural state.

As the globes multiplied they would undoubtedly develop very differently according to their construction and to the tendencies of their colonies, and at the same time they would compete increasingly both for the sunlight which kept them alive and for the asteroidal and meteoric matter which enabled them to grow. Sooner or later this pressure, or perhaps the knowledge of the imminent failure of the sun would force some more adventurous colony to set out beyond the bounds of the solar system. The difficulty involved in making this jump is probably as great as that of leaving the earth itself. Interstellar distances are so large that high velocities approaching those of light, would be necessary; and though high velocities would be easy to attain—it being merely a matter of allowing acceleration to accumulate—they would expose the space vessels to very serious dangers, particularly from dispersed meteoric bodies. A space vessel would, in fact, have to be a comet, ejecting from its anterior end a stream of gas which, meeting and vapourizing any matter in its path, would sweep it to the sides and behind in a luminous trail. Such a method would be very wasteful of matter, and one might perhaps count on some better one having been devised by that time. Even with such velocities journeys would have to last for hundreds and thousands of years, and it would be necessary—if man remains as he is—for colonies of ancestors to start out who might expect the arrival of some remote descendants. This would require a self-sacrifice and a perfection of educational method that we could hardly demand at the present. However, once acclimatized to space living, it is unlikely that man will stop until he has roamed over and colonized most of the sidereal universe, or that even this will be the end. Man will not ultimately be content to be parasitic on the stars but will invade them and organize them for his own purposes.

A star is essentially an immense reservoir of energy which is being dissipated as rapidly as its bulk will allow. It may be that, in the future, man will have no use for energy and be indifferent to stars, except as spectacles, but if (and this seems more probable) energy is still needed, the stars cannot be allowed to continue in their old way, but will be turned into efficient heat engines. The second law of thermodynamics which, as Jeans delights in pointing out to us, will ultimately bring this universe to an inglorious close, may perhaps always remain the final factor.[9] But by intelligent organization the life of the universe could probably be prolonged to many millions of millions of times what it would be without organization. Besides, we are still too close to the birth of the universe to be certain about its death. In any case, long before these questions become urgent it would seem impossible not to assume that man himself would have changed radically in this environment and the nature of this change we must consider in the next chapter.

Chapter III: The Flesh

In the alteration of himself man has a great deal further to go than in the alteration of his inorganic environment. He has been doing the latter more or less unconsciously and empirically for several thousand years, ever since he ceased being parasitic on his environment like any other animal, and consciously and intelligently for at least hundreds of years; whereas he has not been able to change himself at all and has had only fifty years or so to begin to understand how he works. Of course, this is not strictly true: man has altered himself in the evolutionary process, he has lost a good deal of hair, his wisdom teeth are failing to pierce, and his nasal passages are becoming more and more degenerate. But the processes of natural evolution are so much slower than the development of man's control over environment that we might, in such a developing world, still consider man's body as constant and unchanging. If it is not to be so then man himself must actively interfere in his own making and interfere in a highly unnatural manner. The eugenists and apostles of healthy life, may, in a very considerable course of time, realize the full potentialities of the species: we may count on beautiful, healthy and long-lived men and women, but they do not touch the alterations of the species. To do this we must alter either the germ plasm or the living structure of the body, or both together. The first method—the favourite of Mr. J. B. S. Haldane—has so far received most attention. With it we might achieve such a variation as we have empirically produced in dogs and goldfish, or perhaps even manage to produce new species with special potentialities. But the method is bound to be slow and finally limited by the possibilities of flesh and blood. The germ plasm is a very inaccessible unit, before we can deal with it adequately we must isolate it, and to do this already involves us in surgery. It is quite conceivable that the mechanism of evolution, as we know it up to the present, may well be superseded at this point. Biologists are apt, even if they are

9 On the second law of thermo-dynamics—that concentrations of energy tend to disperse, so that temperatures equalize—see *Archimedes*, 83, n. 10. In *Eos; or, the Wider Aspects of Cosmogony* (1928), his book for To-day and To-morrow published the previous year (and excerpted here), James Jeans had written: 'so far as present-day science can see, the final end towards which all creation moves, and at which it must, at long last, arrive' was when 'there would be neither sunlight nor starlight but only a cool glow of radiation uniformly diffused through space' (56).

not vitalists,[10] to consider it as almost divine; but after all it is only nature's way of achieving a shifting equilibrium with an environment; and if we can find a more direct way by the use of intelligence, that way is bound to supersede the unconscious mechanism of growth and reproduction.

In a sense we have already started using the direct method; when the ape-ancestor first used a stone he was modifying his bodily structure by the inclusion of a foreign substance. This inclusion was temporary, but with the adoption of clothes there began a series of permanent additions to the body, affecting nearly all its functions and even, as with spectacles, its sense organs. In the modern world, the variety of objects which really form part of an effective human body is very great. Yet they all (if we except such rarities as artificial larynges) still have the quality of being outside the cell layers of the human body. The decisive step will come when we extend the foreign body into the actual structure of living matter. Parallel with this development is the alteration of the body by tampering with its chemical reactions—again a very old-established but rather sporadic process resorted to to cure illness or procure intoxication. But with the development of surgery on the one hand and physiological chemistry on the other, the possibility of radical alteration of the body appears for the first time. Here we may proceed, not by allowing evolution to work the changes, but by copying and short-circuiting its methods.

The changes that evolution produces apart from mere growth in size, or diversity of form without change of function, are in the nature of perversions: a part of the fish's gut becomes a swimming bladder, the swimming bladder becomes the lung; the salivary gland and an extra eye are charged with the function of producing hormones. Under the pressure of environment or whatever else is the cause of evolution, nature takes hold of what already had existed for some now superseded activity, and with a minimum of alteration gives it a new function. There is nothing essentially mysterious in the process: it is both the easiest and the only possible way of achieving the change. Starting *de novo* to deal with a new situation is not within the power of natural, unintelligent processes; they can only modify in a limited way already existing structures by altering their chemical environment. Men may well copy the process, in so far as original structures are used as the basis for new ones, simply because this is the most economical method, but they are not bound to the very limited range of methods of change which nature adopts.

Now modern mechanical and modern chemical discoveries have rendered both the skeletal and metabolic functions of the body to a large extent useless. In teleological biochemistry one might say that an animal moves his limbs in order to get his food, and uses his body organs in order to turn that food into blood to keep his body alive and active. Now if man is only an animal that is all very satisfactory, but viewed from the standpoint of the mental activity by which he increasingly lives, it is a highly inefficient way of keeping his mind working. In a civilized worker the limbs are mere parasites, demanding nine-tenths of the energy of the food and

10 Vitalism: "The doctrine or theory that the origin and phenomena of life are due to or produced by a vital principle, as distinct from a purely chemical or physical force." (*OED*)

even a kind of blackmail in the exercise they need to prevent disease, while the body organs wear themselves out in supplying their requirements. On the other hand, the increasing complexity of man's existence, particularly the mental capacity required to deal with its mechanical and physical complications, gives rise to the need for a much more complex sensory and motor organization, and even more fundamentally for a better organized cerebral mechanism. Sooner or later the useless part of the body must be given more modern functions or dispensed with altogether, and in their place we must incorporate in the effective body the mechanism of the new functions. Surgery and biochemistry are sciences still too young to predict exactly how this will happen. The account I am about to give must be taken rather as a fable.

Take, as a starting point, the perfect man such as the doctors, the eugenists and the public officers between them hope to make of humanity: a man living perhaps an average of a hundred and twenty years but still mortal, and increasingly feeling the burden of this mortality. Already Shaw in his mystical fashion cries out for life to give us hundreds of years to experience, learn and understand;[11] but without the vitalist's faith in the efficacy of human will we shall have to resort to some artifice to achieve this purpose. Sooner or later some eminent physiologist will have his neck broken in a super-civilized accident or find his body cells worn beyond capacity for repair. He will then be forced to decide whether to abandon his body or his life. After all it is brain that counts, and to have a brain suffused by fresh and correctly prescribed blood is to be alive—to think. The experiment is not impossible; it has already been performed on a dog and that is three-quarters of the way towards achieving it with a human subject. But only a Brahmin philosopher would care to exist as an isolated brain, perpetually centred on its own mediations. Permanently to break off all communications with the world is as good as to be dead. However, the channels of communication are ready to hand. Already we know the essential electrical nature of nerve impulses; it is a matter of delicate surgery to attach nerves permanently to apparatus which will either send messages to the nerves or receive them. And the brain thus connected up continues an existence, purely mental and with very different delights from those of the body, but even now perhaps preferable to complete extinction. The example may have been too far-fetched; perhaps the same result may be achieved much more gradually by using the many superfluous nerves with which our body is endowed for various auxiliary and motor services. We badly need a small sense organ for detecting wireless frequencies, eyes for infra-red, ultra-violet and X-rays, ears for supersonics, detectors of high and low temperatures, of electrical potential and current, and chemical organs of many kinds.[12] We may perhaps be able to train a great number of hot and

11 Like other To-day and To-morrow writers, Bernal is alluding to George Bernard Shaw's *Back to Methuselah (A Metabiological Pentateuch)*, published in 1921, which imagined a future of extended human longevity.

12 Compare Low's speculation in *Wireless Possibilities*, 70, that 'It may well be that, when we succeed in inter-planetary communication, we shall discover that the inhabitants see by the X-ray, by wireless, or by heat.'

cold and pain receiving nerves to take over these functions; on the motor side we shall soon be, if we are not already, obliged to control mechanisms for which two hands and feet are an entirely inadequate number; and apart from that, the direction of mechanism by pure volition would enormously simplify its operation. Where the motor mechanism is not primarily electrical, it might be simpler and more effective to use nerve-muscle preparations instead of direct nerve connections. Even the pain nerves may be pressed into service to report any failure in the associated mechanism. A mechanical stage, utilizing some or all these alterations of the bodily frame might, if the initial experiments were successful in the sense of leading to a tolerable existence, become the regular culmination to ordinary life. Whether this should ever be so for the whole of the population we will discuss later, but for the moment we may attempt to picture what would at this period be the course of existence for a transformable human being.

Starting, as Mr. J. B. S. Haldane so convincingly predicts, in an ectogenetic factory,[13] man will have anything from sixty to a hundred and twenty years of larval, unspecialized existence—surely enough to satisfy the advocates of the natural life. In this stage he need not be cursed by the age of science and mechanism, but can occupy his time (without the conscience of wasting it) in dancing, poetry and love-making, and perhaps incidentally take part in the reproductive activity. Then he will leave the body whose potentialities he should have sufficiently explored.

The next stage might be compared to that of a chrysalis, a complicated and rather unpleasant process of transforming the already existing organs and grafting on all the new sensory and motor mechanisms. There would follow a period of re-education in which he would grow to understand the functioning of his new sensory organs and practise the manipulation of his new motor mechanism. Finally, he would emerge as a completely effective, mentally-directed mechanism, and set about the tasks appropriate to his new capacities. But this is by no means the end of his development, although it marks his last great metamorphosis. Apart from such mental development as his increased faculties will demand from him, he will be physically plastic in a way quite transcending the capacities of untransformed humanity. Should he need a new sense organ or have a new mechanism to operate, he will have undifferentiated nerve connections to attach to them, and will be able to extend indefinitely his possible sensations and actions by using successively different end-organs.

The carrying out of these complicated surgical and physiological operations would be in the hands of a medical profession which would be bound to come rapidly under the control of transformed men. The operations themselves would probably be performed by mechanisms controlled by the transformed heads of the profession, though in the earlier and experimental stages, of course, it would still be done by human surgeons and physiologists.

It is much more difficult to form a picture of the final state, partly because this final state would be so fluid and so liable to improve, and partly because there

13 See Haldane's *Daedalus* in this volume, 37–38.

would be no reason whatever why all people should transform in the same way. Probably a great number of typical forms would be developed, each specialized in certain directions. If we confine ourselves to what might be called the first stage of mechanized humanity and to a person mechanized for scientific rather than æsthetic purposes—for to predict even the shapes that men would adopt if they would make of *themselves* a harmony of form and sensation must be beyond imagination—then the description might run roughly as follows.

Instead of the present body structure we should have the whole framework of some very rigid material, probably not metal but one of the new fibrous substances. In shape it might well be rather a short cylinder. Inside the cylinder, and supported very carefully to prevent shock, is the brain with its nerve connections, immersed in a liquid of the nature of cerebro-spinal fluid, kept circulating over it at a uniform temperature. The brain and nerve cells are kept supplied with fresh oxygenated blood and drained of de-oxygenated blood through their arteries and veins which connect outside the cylinder to the artificial heart-lung digestive system—an elaborate, automatic contrivance. This might in large part be made from living organs, although these would have to be carefully arranged so that no failure on their part would endanger the blood supply to the brain (only a fraction of the body's present requirements) and so that they could be inter-changed and repaired without disturbing its functions. The brain thus guaranteed continuous awareness, is connected in the anterior of the case with its immediate sense organs, the eye and the ear—which will probably retain this connection for a long time. The eyes will look into a kind of optical box which will enable them alternatively to look into periscopes projecting from the case, telescopes, microscopes and a whole range of televisual apparatus. The ear would have the corresponding microphone attachments and would still be the chief organ for wireless reception. Smell and taste organs, on the other hand, would be prolonged into connections outside the case and would be changed into chemical testing organs, achieving a more conscious and less primitively emotional role than they have at present. It may perhaps be impossible to do this owing to the peculiarly close relation between the brain and olfactory organs, in which case the chemical sense would have to be indirect. The remaining sensory nerves, those of touch, temperature, muscular position and visceral functioning, would go to the corresponding part of the exterior machinery or to the blood supplying organs. Attached to the brain cylinder would be its immediate motor organs, corresponding to but much more complex than, our mouth, tongue and hands. This appendage system would probably be built up like that of a crustacean which uses the same general type of arm for antenna, jaw and limb; and they would range from delicate micro-manipulators to levers capable of exerting considerable forces, all controlled by the appropriate motor nerves. Closely associated with the brain-case would also be sound, colour and wireless producing organs. In addition to these there would be certain organs of a type we do not possess at present—the self-repairing organs—which under the control of the brain would be able to manipulate the other organs, particularly the visceral blood supply organs, and to keep them in effective working order. Serious derangements, such as those involving loss of consciousness, would still, of course, call for outside assistance, but with proper care these would be in the nature of rare accidents.

The remaining organs would have a more temporary connection with the brain-case. There would be locomotor apparatus of different kinds, which could be used alternatively for slow movement, equivalent to walking, for rapid transit and for flight. On the whole, however, the locomotor organs would not be much used because the extension of the sense organs would tend to take their place. Most of these would be mere mechanisms quite apart from the body; there would be the sending parts of the television apparatus, tele-acoustic and tele-chemical organs, and tele-sensory organs of the nature of touch for determining all forms of texture. Besides these there would be various tele-motor organs for manipulating materials at great distances from the controlling mind. These extended organs would only belong in a loose sense to any particular person, or rather, they would belong only temporarily to the person who was using them and could equivalently be operated by other people. This capacity for indefinite extension might in the end lead to the relative fixity of the different brains; and this would, in itself, be an advantage from the point of view of security and uniformity of conditions, only some of the more active considering it necessary to be on the spot to observe and do things.

The new man must appear to those who have not contemplated him before as a strange, monstrous and inhuman creature, but he is only the logical outcome of the type of humanity that exists at present. It may be argued that this tampering with bodily mechanism is as unnecessary as it is difficult, that all the increase of control needed may be obtained by extremely responsive mechanisms outside the unaltered human body. But though it is possible that in the early stages a surgically transformed man would actually be at a disadvantage in capacity of performance to a normal, healthy man, he would still be better off than a dead man. Although it is possible that man has far to go before his inherent physiological and psychological make-up becomes the limiting factor to his development, this must happen sooner or later, and it is then that the mechanized man will begin to show a definite advantage. Normal man is an evolutionary dead end; mechanical man, apparently a break in organic evolution, is actually more in the true tradition of a further evolution.

A much more fundamental break is implicit in the means of his development. If a method has been found of connecting a nerve ending in a brain directly with an electrical reactor, then the way is open for connecting it with a brain-cell or another person. Such a connection being, of course, essentially electrical, could be effected just as well through the ether as along wires. At first this would limit itself to the more perfect and economic transference of thought which would be necessary in the co-operative thinking of the future. But it cannot stop there. Connections between two or more minds would tend to become a more and more permanent condition until they functioned as dual or multiple organisms. The minds would always preserve a certain individuality, the network of cells inside a single brain being more dense than that existing between brains, each brain being chiefly occupied with its individual mental development and only communicating with the others for some common purpose. Once the more or less permanent compound brain came into existence two of the ineluctable limitations of present existence would be surmounted. In the first place death would take on a different and far less terrible aspect. Death would still exist for the mentally-directed mechanism we have just described; it would merely be postponed for three hundred or perhaps a thousand years, as long as the brain cells could be persuaded to live in the most favourable environment, but not for ever. But the multiple individual would be, barring cataclysmic accidents, immortal, the older components as they died being replaced by newer ones without losing the continuity of the self, the memories and feelings of the older member transferring themselves almost completely to the common stock before its death. And if this seems only a way of cheating death, we must realize that the individual brain will feel itself part of the whole in a way that completely transcends the devotion of the most fanatical adherent of a religious sect. It is admittedly difficult to imagine this state of affairs effectively. It would be a state of ecstasy in the literal sense, and this is the second great alteration that the compound mind makes possible. Whatever the intensity of our feeling, however, much we may strive to reach beyond ourselves or into another's mind, we are always barred by the limitations of our individuality. Here at least those barriers would be down: feeling would truly communicate itself, memories would be held in common, and yet in all this, identity and continuity of individual development would not be lost. It is possible, even probable, that the different individuals of a compound mind would not all have similar functions or even be of the same rank of importance. Division of labour would soon set in: to some minds might be delegated the task of ensuring the proper functioning of the others, some might specialize in sense reception and so on. Thus would grow up a hierarchy of minds that would be more truly a complex than a compound mind.

The complex minds could, with their lease of life, extend their perceptions and understandings and their actions far beyond those of the individual. Time senses could be altered: the events that moved with the slowness of geological ages would be apprehended as movement, and at the same time the most rapid vibrations of the physical world could be separated. As we have seen, sense organs would tend to be less and less attached to bodies, and the host of subsidiary, purely mechanical agents and perceptors would be capable of penetrating those regions where organic bodies cannot enter or hope to survive. The interior of the earth and the stars, the inmost cells of living things themselves, would be open to consciousness through these angels, and through these angels also the motions of the stars and living things could be directed.

This is perhaps far enough; beyond that the future must direct itself. Yet why should we stop until our imaginations are exhausted? Even beyond this there are foreseeable possibilities. Undoubtedly the nature of life processes themselves will be far more intensively studied. To make life itself will be only a preliminary stage, because in its simplest phases life can differ very little from the inorganic world. But the mere making of life would only be important if we intended to allow it to evolve of itself anew. This, as Mr. Whyte suggests in *Archimedes*, is necessarily a lengthy process,[14] but there is no need to wait for it. Instead, artificial life would undoubtedly be used as ancillary to human activity and not allowed to evolve freely

14 See pp. 92–94.

except for experimental purposes. Men will not be content to manufacture life: they will want to improve on it. For one material out of which nature has been forced to make life, man will have a thousand; living and organized material will be as much at the call of the mechanized or compound man as metals are to-day, and gradually this living material will come to substitute more and more for inferior functions of the brain as memory, reflex actions, etc., in the compound man himself; for bodies at this time would be left far behind. The brain itself would become more and more separated into different groups of cells or individual cells with complicated connections, and probably occupying considerable space. This would mean loss of motility which would not be a disadvantage owing to the extension of the sense faculties. Every part would now be accessible for replacing or repairing and this would in itself ensure a practical eternity of existence, for even the replacement of a previously organic brain-cell by a synthetic apparatus would not destroy the continuity of consciousness.

The new life would be more plastic, more directly controllable and at the same time more variable and more permanent than that produced by the triumphant opportunism of nature. Bit by bit the heritage in the direct line of mankind—the heritage of the original life emerging in the face of the world—would dwindle and in the end disappear effectively, being preserved perhaps as some curious relic, while new life which conserves none of the substance and all the spirit of the old would take its place and continue its development. Such a change would be as important as that in which life first appeared on the earth's surface and might be as gradual and imperceptible. Finally, consciousness itself may end or vanish in a humanity that has become completely etherialized, losing the close-knit organism, becoming masses of atoms in space communicating by radiation, and ultimately perhaps resolving itself entirely into light. That may be an end or a beginning but from here it is out of sight.

Chapter IV: The Devil

WHY do the first two lines of attack against the inorganic forces of the world and the organic structure of our bodies seem so doubtful, fanciful and Utopian? Because we can abandon the world and subdue the flesh only if we first expel the devil, and the devil, for all that he has lost individuality, is still as powerful as ever. The devil is the most difficult of all to deal with: he is inside ourselves, we cannot see him. Our capacities, our desires, our inner confusions are almost impossible to understand or cope with in the present, still less can we predict what will be the future of them. Psychology at the present day is hardly in a better state than physics in the time of Aristotle; it has acquired a vocabulary, the general movements and transformations of conscious and unconscious motives are described, but nothing more. Yet in the absence of scientific analysis something must be said, because all the changes I have predicted in the organic or inorganic world must, in the first place, start from some human psychological motive and effect themselves through the operation of human intellectual processes. We are obviously not in a position to predict the particular new orientations which a change in psychology would give to human development, beyond that which would result from the removal of what we know are inhibitory causes, so that here I will only attempt to estimate the effect of psychological forces in preventing or retarding the kind of processes outlined in the first two sections. The progress of the future depends no longer on physiological evolution but on the reaction of intelligence on a material universe. It will be hindered or stopped either by a failure in the capacity for maintaining creative intellectual thinking, or by the lack of desire to apply such thinking to the progress of humanity, or, of course, by both these causes together. Consider first the retarding factors that endanger the capacity for creative thinking. Some are apparent now. It is pretty clear that they are ineffective in stopping the course of thought at present, but they have not always been so in the past and we cannot be sure that they will not be so in the future. One of the most threatening retarding factors of the present is specialization, particularly as it is bound to increase with scientific knowledge itself. But it is doubtful whether specialization in itself is capable of bringing scientific thought to a standstill. It retards it so far as the specialist is ignorant of current thought in other fields, and the remedy for this is obviously an intelligently operated system of distribution and grading of knowledge so that each worker may have the amount that he requires outside his own field, in a form which can be absorbed with a minimum of mental effort. The problem is essentially that of communications to an army in action. After a rapid advance communications become disorganized, and there is a temporary halting until they are again in working order.

Such an organization of intellectual work for definite ends involves a fundamental change: it is analogous to change from a food-gathering to a food-producing society. The modern scientist is a primitive savage. If he is active and enterprising he tracks his prey down alone or in small parties; if he is industrious and thorough he gathers and piles up the natural products around him, but for his success he has to thank not only his own skill and the lore of his craft, but the richness of nature and the paucity of his companions. Good hunting will not last much longer, but the tilled ground is richer.

We shall be forced to attempt planned and directed researches employing hundreds of workers for many years, and this cannot be done without risking the loss of independence and originality. This is a serious and fundamental obstacle but it may be overcome in two ways. It should be possible so to improve educational methods, that mental activity, the capacity to form new associations, should not be incompatible with the performance of routine work: that is, every research worker should be potentially able to add to and modify the whole course of the research by his suggestions. At the same time it is certain that originality, organizing power and industriousness will continue as now to be very unevenly distributed; and it is an essentially social problem to make the best apportionment of functions, using for the more routine operations people who under present conditions would not be scientific workers at all, and using the organizers to translate into plans of action the incoherent ideas of the thinkers. Pedantry and bureaucracy—symptoms of an unintelligent respect for the past—are at present real dangers, but, once their genesis is understood, they can be made to vanish.

Specialization is brought about by the wideness of the field in which science operates, but as we go more deeply into nature the intrinsic complication of the phenomena increases and the modes of thinking used in ordinary life become more inadequate to deal with them. It is conceivable that the supply of minds capable of making any impression on these deeper problems may more and more fall behind the number required, and that all the efforts of education to produce ten genii where one grew before will be foiled by intrinsic difficulties in nature. It is impossible to know whether this will happen. One may guess, from experience of the past, that nature is never so complicated as it looks; that the value of theory and deductive thinking and the use of appropriate language and symbolism will reduce the difficulties in the measure that they are approached.

However they appear to the pessimist of the present day, it is not in specialization or complication that the chief danger to progress seems to lie: it is in something much more deep-seated and much more elusive. Bertrand Russell, in one of his "Sceptical Essays," predicting the approaching end of the scientific age, suggests that people will turn from physics to metaphysics because the hope that the former held out is seen to be vain except to new, half-cultivated peoples.[15] Perhaps after all it is hope that really determines whether an age is or is not creative. But the existence of hope in a society at any time itself depends on many unexplored psychological, economic and political causes. I do not think that the factors involved are of a mystical order, but that they require considerable disentangling.

There seem to be two psychological determinants in any culture: a crop of perverted individuals capable of more than average performance, and a mass of people effective not so much by their number as by their secure hold on tradition. In the normal state the perverse are dominated by the mass in two ways. Their mode of expression is dictated by the modes conceivable in the society; everywhere, even the most aberrant individual must conform to one of a small number of recognized types. The same type of mind that would now make a physicist would in the middle ages have made a scholastic theologian. Further, there is a process of selection in which the current tradition decides what is to be the relative value and effectiveness of each type. Thus, even if at all times types are always produced in the same abundance, only the selected are effective, as meditative ascetics in India or energetic salesmen in America. The mass of the people, or more properly the ruling class, pay the piper and they call the tune; genius is potent only when it fits the tendencies of the age. From this standpoint we are approaching the close of the period of respectable comfort which puritanism

15 *Sceptical Essays* was published in 1928. Bernal is referring to the essay 'Is Science Superstitious?,' in which Russell argues that 'As civilization becomes more diffused and more traditional, there is a tendency for the habits of mind of the appreciator to conquer those who might be creators, with the result that the civilization in question becomes Byzantine and retrospective. Something of this sort seems to be beginning to happen in science. The simple faith which upheld the pioneers is decaying at the centre [. . . .] the high priests [i.e. the leading scientists or philosophers of science like A. N. Whitehead] begin to weary of the worship to which they are officially dedicated' (London: Allen & Unwin, 1928), 35–44, 36.

demanded and mathematics and handicraft produced. But this period may not end in a regression to the mediæval state through the ultimate dissatisfaction with science; before that happens science, raised to power by industrialism, may in its turn become the directing tradition.

Political and social events must also be effective, but not in a very obvious fashion. Both political confusion and prolonged peace undoubtedly affect creative thought but whether they respectively hinder or help it is not at all certain. When one contrasts Athens, renaissance Italy and feudal China, on the one hand, with the Roman, the Spanish and the Chinese Empires on the other, war would seem positively to help mental activity. But as many examples could be found to the contrary. There may be something in the suggestion that wherever war appeared stimulating it was a war between approximate equals so that the disasters were seen to be due to human folly or perversity. In the case of the Empires, on the other hand, peace was achieved at the price of a submission to authority, bureaucratic or spiritual, which deprived men of their self-reliance and creative ability. However this may be, historical factors tend to have somewhat of a cyclic nature, and in the long run to cancel each other out, although it is always possible that one age will destroy, or cause to be forgotten, more than the previous age produced, and that a definite culmination may be reached in human progress. This may be closer than we think (if it is not already passed) and humanity may become static until it is destroyed by cosmic forces. Yet it seems more probable that we are on the point, owing to our mechanical achievements of reaching another order of cyclic changes, which may lead us to the stars.

Whether an age or an individual will express itself in creative thinking or in repetitive pedantry is more a matter of desire than of intellectual power, and it is probably more the nature of their desires than of their capacities that will determine whether or not humanity will develop further. Now it would seem that the present time is a very crucial one for the evolution of human desire. It is an age in which the nature of desire has been glimpsed at for the first time, and that glimpse enables us to see two very different possibilities. The intellectual life, both in its scientific and its æsthetic aspects, is seen no longer as the vocation of the rational mind but as a compensation, as a perversion of more primitive, unsatisfied desires. Now the question arises is this perversion in the line of evolution, or is it a merely temporary pathological process? If by a sounder psychology, a way of living more in accordance with nature, it should be found that the satisfaction of purely human—or, as we might almost say, purely mammalian—desires is capable of absorbing all the energy that suppression now forces into scientific or æsthetic channels, then the human race may well find itself statically employed in leading an idyllic, Melanesian[16] existence of eating, drinking, friendliness,

16 Melanesia is a subregion of Oceania in the southwestern Pacific Ocean (and northeast of Australia) stretching from New Guinea to the Fiji Islands. The lifestyle Bernal describes evokes Gaugin's Tahiti (in neighboring Polynesia). But Bernal is more probably thinking of the anthropologist Bronisław Malinowski's pioneering work on the Trobriand Islanders in *Argonauts of the Western Pacific* (1922); *Crime and Custom in Savage Society* (1926) and *Sex and Repression in Savage*

love-making, dancing and singing, and the golden age may settle permanently on the world. On the other hand it may be that though the desire, the necessity to escape life on the paths of intellectual or æsthetic creation may be weakened by the application of an intelligent psychology, yet a corresponding freedom from the internal conflicts which now hinder both these forms of expression may more than compensate for what is lost, and we may find the capacity to live at the same time more fully human and fully intellectual lives. The latter alternative is more in line with the recent developments in Freudian psychology which divide the psyche into the primitive id, the ego which is its expression of contact with reality, and the super ego which represents its aspirations and ideals. Rationalism strove to make the super ego the dominant partner; it never succeeded, not only because its standard was too high to allow any outlets for the primitive forces, but because it was itself too arbitrary, too tainted with distorted primitive wishes ever to be brought into correspondence with reality. Naturalism,[17] less definitely, aimed at giving the primitive wishes full play but equally failed because these wishes are too primitive, too infantile, too inconsistent with themselves to be satisfied even by the greatest licence. The aim of applied psychology is rather now to bring, by analysis or education, the ideals of the super ego in line with external reality, using and rendering innocuous the power of the id and leading to a life where a full adult sexuality would be balanced with objective activity. It is this alternative that makes the mechanical, biological progress that I have outlined not only possible but almost necessary, for a sound intellectual humanity will never be content with repeating itself in circles of metaphysical thinking like Shaw's Immortals, but will need a real externalization in the transforming of the universe and itself. Such a development could hardly leave unchanged the present types of human interests in art and science and religion.

It is here that prediction is most difficult and most fascinating. Under the influence of psychology it may well be that, just as all the branches of science itself are coalescing into a unified world picture, so human activities of art and attitudes of religion may be fused into one whole action-reaction pattern of man to reality. The recognition of the art that informs all pure science need not mean the abandonment for it of all present art, rather it will mean the completion of the transformation of art that has already begun. Art expressing itself on one side in a kind of generalized architecture, massive or molecular, gives form to the infinite possibilities of the application of science; on the other a generalized poetry expresses the ever-widening complexities of the understanding of the universe, while religion clarified by psychology remains as the expression of the desire that drives man through the universe in understanding and hope.

Society (1927), both published in Ogden's International Library; and *The Sexual Life of Savages in North-Western Melanesia* (1929).

17 The ethical view of 'Action arising from or based on natural instincts, without spiritual guidance'; *OED* sense 1; as in the writings of Jean-Jacques Rousseau or D. H. Lawrence.

It is not sufficient, however, to consider the absence or presence of desire for progress, because that desire itself will not make itself effective until it can overcome the quite real distaste and hatred which mechanization has already brought into being. The distaste is nothing to what the bulk of present humanity would feel about even the milder of the changes which are suggested here. The reader may have already felt that distaste especially in relation to the bodily changes: I have felt it myself in imagining them. The effectiveness of these conservative feelings is the balance of two opposing factors. The changes in question do not come all at once: envisaged in broad outline in the sequence given, their nature would suggest that they follow each other with increasing frequency, as the past has already shown. Now the more rapid the environmental changes the less will the individual mind be able to adapt itself to them and the more violent will be its emotional reactions. At the same time these changes give more and more power to those groups of men which are involved in them and are bringing them about, so that, up to the present in the war of the machines, the mechanists have always been the victors: but, of course, if the emotional reactions of the mass increased more rapidly than the power of the mechanists, the reverse would be the case. A severe crisis in mechanical civilization brought about by its inherent technical weakness or, as is much more likely, by its failure to arrange secondary social adjustments, is likely to be seized upon by the emotional factors hostile to all mechanism, and we may be closer to such a reversion than we suppose. Two recent books representing very divergent standpoints, the last works of Mr. Aldous Huxley[18] and Mr. D. H. Lawrence,[19] show at the same time the weakening desires and the imminent realization of futility on the part of the scientist, and a turning away from the whole mechanization on the part of the more humanely-minded. The same thought is echoed in the writings of Mr. Bertrand Russell They may be prophets predicting truly the doom of the new Babylon or merely lamenting over a past that is lost for ever. With these uncertainties before us, each must follow his own desires, accepting that his opponent may be as right as himself. The event will show which, but only after his own time.

There remains still another possibility: the most unexpected, but not necessarily the most improbable, the development of a di-morphism[20] in humanity in which the conflict between the humanizers and the mechanizers will be solved not by the victory of one or the other but by the splitting of the human race—the one section developing a fully-balanced humanity, the other groping unsteadily beyond it. But this possibility involves the consideration of mechanical and biological factors, the synthesis of which, with the psychological, will be attempted in my concluding pages.

18 *Point Counter Point* (1928) satirizes the reclusive scientist Lord Edward Tantamount, said to be based partly on Huxley's close friend J. B. S. Haldane, though also drawing on aspects of Huxley's father and biologist brother Julian; and possibly the arch-rationalist Bertrand Russell.

19 *Lady Chatterley's Lover* (1928), in which the paralyzed war veteran Clifford Chatterley wants to introduce new technology into his coal-mines.

20 A dividing into two distinct forms.

Chapter V: Synthesis

HAVING followed our main lines of change separately, it now remains for us to consider the interaction between the physical, physiological and psychological elements of future human evolution. It is very easy to see the relations of the first two: the colonization of space and the mechanization of the body are obviously complementary. The dissimilarity between the conditions of life in space and on the earth would in itself be sufficient to cause perfectly normal unassisted, evolutionary changes in human beings, but obviously spatial conditions would be more favourable to mechanized than to organic man. If he could get rid of the major part of his body and his necessity for a relatively large intake of oxygen and water-saturated food, the cellular nature of the celestial globes would cease to be necessary. This would give mechanized man an advantage similar to that which the relatively flexible and naked animal cell has over the rigidly demarcated plant. Besides, it is only in space that the potentialities of the more highly developed forms of complex minds would have an adequate field of functioning, particularly in their extended time relations.

It may be that we are approaching or will ultimately reach a conception of time that will make transit in time as easy as transit in space. But all our present knowledge, apart from our desires, suggests that it is improbable. Even if time and space were made equivalent, to gain a second of the future would be equivalent to travelling 180,000 miles. But even without a fundamental change in the conception of time the time faculties of mechanized man would be still very different from ours. Extension will be its chief character: already in the monkey stage the actual present of an animal embraces a short part of the past and future. Anticipation of movement, through muscular innervation and memory, by its retention of nerve impulse images, extend the present to the limit of a second or so. Every time we play tennis we are prophets without knowing it of the future position of the ball which is conceived of as present. In the human stage we have extended mostly backwards as memory, our immediate prevision being limited by lack of scientific knowledge. It is now rapidly increasing, but is not usually accepted as prevision because it is conscious and intellectual. However, prevision plainly tends to become more and more deductive, and, to the mechanized man, the immediately apprehended may include years or centuries of past and future.

One may picture then, these beings, nuclearly resident,[21] so to speak, in a relatively small set of mental units, each utilizing the bare minimum of energy, connected together by a complex of ethereal intercommunication, and spreading themselves over immense areas and periods of time by means of inert sense organs which, like the fields of their active operations, would be, in general at a great

21 Bernal may be coining the word 'nuclearly' here. This is the first recorded instance in the *OED*. The metaphor (like Bernal's scientific work) hovers between physics and biology—both atoms and cells have nuclei. He appears to be following the logic of his thought that, once mechanized to the point of no longer needing the atmosphere and nutrients supplied by the huge space-traveling spheres, the disembodied minds can be imagined as organized in a different kind of cellular structure; with

distance from themselves. As the scene of life would be more the cold emptiness of space than the warm, dense atmosphere of the planets, the advantage of containing no organic material at all, so as to be independent of both of these conditions, would be increasingly felt.

It is when we turn to the interaction on the psychological plane that the difficulties again occur. The physical and the psychological have a mutual influence which it is very difficult at the present moment to estimate. Undoubtedly, if modern tendencies have any elements of permanency in them, a great deal of the activity of the future will be devoted to the end of a greater understanding of the universe. Humanity, or its descendants, may well be much more occupied with purely scientific research and much less with the necessity of satisfying primary physiological and psychological needs than it is at present. This character may stamp the whole of future development, so that machinery will be organized not for production but for discovery. Indeed, the great necessity for production either of food or other articles of consumption will disappear rapidly with the progress of dehumanization. But such changes are small compared with those which would necessarily be involved by the physiological alterations which I have suggested.

The human mind has evolved always in the company of the human body, and of the animal body before it was human. The intricate connections of mind and body must exceed our imagination, as from our point of view we are peculiarly prevented from observing them. Altering in any perfectly sound physiological or surgical way the functionings of the body will certainly have secondary but far-reaching effects on the mind, and these secondary effects are unpredictable and probably will still be unpredictable at the time when the physiological changes take place. But it is thoroughly in accord with both human and natural evolution that secondary changes should not be taken into account when reacting to the primary desire or stimulus: in other words, the physiological steps will probably be taken without consideration of the psychological consequences, which may, of course, wreck the whole organism, or, on the other hand, lead to an unpredictably large increase in mental grasp and efficiency. It is on account of this delicate balance between physiological and psychological factors that the future, as well as the present, will be full of dangerous turning-points and pitfalls. We shall have very sane reactionaries at all periods warning us to remain in the natural and primitive state of humanity, which is usually the last stage but one in their cultural history. But the secondary consequences of what men have already done—the reactionaries as much as any—will carry them away then as now. Obviously certain considerable psychological displacements or perversions must occur to balance the physiological perversions. The sexual instincts in particular, which still find considerable direct gratification, would be unrecognizably changed. One may assume that there is some kind of principle of psychological conservation which will prevent them, as it has prevented them up to the present, being suppressed altogether. But what will

---

small numbers of them aggregated together like a cell nucleus, but surrounded by remote wireless sense-organs at great distances from them.

they be changed into? The solution may be an extension of sublimation, a process which is at present outside conscious control but which may not always remain so. A part of sexuality may go to research, and a much larger part must lead to æsthetic creation. The art of the future will, because of the very opportunities and materials it will have at its command, need an infinitely stronger formative impulse than it does now. The cardinal tendency of progress is the replacement of an indifferent chance environment by a deliberately created one. As time goes on, the acceptance, the appreciation, even the understanding of nature, will be less and less needed. In its place will come the need to determine the desirable form of the humanly-controlled universe which is nothing more nor less than art.

The psychology of a complex mind must differ almost as much from that of a simple, mechanized mind as its psychology would from ours; because something that must underlie and perhaps be even greater than sex is involved. By the intimate intercommunication of minds, the very existence of the ego would be impaired for the first time. Some kind of equilibrium will have to be found between each partial and corporate personality. This we can vaguely adumbrate when we think of the conflicts involved between ego and sexual impulse, the latter attempting always to break the isolation of the former and reach out to another individual or group. If it is once possible to achieve this reaching out of feeling, the results are bound to be enormous and perhaps overwhelming. Will the corporate personalities form greater and greater complexes until there is only one intelligence, or will there be a multiplication of separate and differently-evolving complexes with resulting conflicts? Spatial considerations seem on the whole to favour the latter view, but we must allow for enormous increases in communication and in the capacity for rational conduct.

Another even deeper psychological consideration arises at this point. What is to be the future of feeling? Is it to be perverted or superseded altogether? In other words, are the mechanical or corporate men of the future to be emotional or rational? Here we have very little to guide us; we are not certain whether the comparative coldness of modern intellectualism is the effect of a considerable development or of dangerous repression. Even if we did know the answer to this it would hardly help us, since our new beings would have a different physiological balance. This balance will not be, as in us, at the mercy of the uncontrolled interactions of individual and environment. Feeling, or at any rate, feeling-tones, will almost certainly be under conscious control: a feeling-tone will be induced in order to favour the performance of a particular kind of operation. Of course, it would be excessively dangerous for human beings in their present state to have this control of their feelings. A great majority would probably be content to remain in a state of more or less ecstatic happiness, but the man of the future will probably have discovered that happiness is not an end of life. This is as far as we may go even in guessing. The psychology of the completely mechanized organism must remain a mystery.

Viewed from the standpoint of the present the carrying out of such a programme of human development must seem a very pointless occupation; but it is doubtful whether the present civilization would appear to an educated Athenean as something worthy to mark the culmination of his efforts. We must not assume static psychology and a further static knowledge. The immediate future which is our own desire, we seek; in achieving it we become different; becoming different we desire something new, so there is no staleness except when development itself has stopped. Moreover, development, even in the most refined stages, will always be a very critical process; the dangers to the whole structure of humanity and its successors will not decrease as their wisdom increases, because knowing more and wanting more they will dare more, and in daring will risk their own destruction. But this daring, this experimentation, is really the essential quality of life.

Chapter VI: Possibility

BY now it should be possible to make a picture of the general scheme of development as a unified whole, and though each part may seem plausible in detail, yet in some obscure way the total result seems unbelievable. This disbelief may be well founded, for what is suggested is not so much a fulfilling as a transformation of humanity, a setting up of what is virtually a new species or several new species, and a mode of setting up which is in itself a departure from the time-hallowed methods of evolution. Now, I believe, that this scheme is more than a bare possibility, that it, or something like it, has about an even chance of occurring; but I must justify this belief not by hypothetics of the future but by the analysis of causes acting in the present. Perhaps the most fruitful way of beginning is to ask the question, "What is the effective purpose of the human race as it now is?" We can eliminate such satisfactory answers as "For the glory of God," because, however true, they do not differentiate humanity from other parts of creation. The answer one seeks is the historical and economic one. Human societies are recent products and, up to the present, can be essentially qualified as co-operative food-producing societies—or perhaps, to include comfort, as co-operative body-satisfying societies. They are distinguished in this way from insect societies which are essentially, as Wheeler[22] has pointed out, reproductive societies. True, in fulfilling the function of securing a brood, insect societies have gone far in becoming food-producing units, and the complementary process in man is shown by the increased care taken over education; but devotion to children has never been the mainspring of human activity. Hunger and sex still dominate the primitive mammalian side of human existence, but at the present time it looks as if humanity were within sight of their satisfaction. Permanent plenty, no longer a Utopian dream, awaits the arrival of permanent peace. Even now, through rationalized capitalism or Soviet state planning, the problem of the production and distribution of necessaries to the

22 William Morton Wheeler's *The Social Insects: Their Origin and Evolution* (London: Kegan Paul, 1928), was published in Ogden's International Library. Wheeler describes insect colonies as 'super-organisms'—a term which Haldane had used in his 1927 essay 'The Last Judgement,' and which underlies Bernal's thinking about the possible development of 'compound' and 'complex' minds. Wheeler had first used the term in 'The Ant-Colony as an Organism', *Journal of Morphology*, Vol. 22 (1911), 307–325.

primary satisfaction of all human beings, is being pushed forward with uniform and intelligent method. Stupidity and the perversity of separate interests may hold the consummation back for centuries, but it must come gradually and surely.

Now supposing this state achieved or approximating to achievement, what is to become of humanity? Is it, like the stabilized insect societies, to settle down to an eternity of methodical enjoyment, or is there appearing, by some unforeseen chance, a new objective, a new reason for existing beyond the calls of hunger and lust? The primates, and subsequently man, developed intelligence in order to satisfy their desire in a world that was getting more and more difficult to live in. They developed it as primitive plants develop the habit of eating, or fish that of breathing, and just as those plants became animals who lived to eat and those fish became animals that lived to breathe, so we may, in time, come to live to think instead of thinking to live. But this biological analogy carries a very suggestive element; more fish remain in the sea than ever came out of it. It is not the habit of evolutionary process to transform the whole of one state of living into another. Rather does nature pick some particularly happy development and allow it to expand in the place of and even at the expense of her earlier efforts. If man is to develop something new, the insistent question is, whether all humanity is going to develop or only a part of it? The biological analogy in favour of the latter would be overwhelming if man were an ordinary species, but it happens that at the moment, for the first time in his history, he consists virtually of one society, and we have no precedent for the development of any new types, particularly of solitary types, from the middle of a single society; but what, of course, could develop from a society would be another society, at first simply a part of it, but afterwards differentiating itself more and more clearly.

If we consider only those alternatives that lead to the development, leaving on one side the not impossible state in which mankind would be stabilized and live an oscillating existence for millennia, we have to consider, in the light of the present, the alternatives: whether mankind will progress as a whole or will divide definitely into a progressive and unprogressive part. Over and over again in history there has occurred the raising of a particular class or a particular culture to a point at which there seemed a permanent gulf between it and other cultures or classes. Yet the gulf was not permanent; the particular aristocracy fell or its advantages spread themselves so widely that they became common stock. The cause for this is not obscure: first, the aristocrats differed only superficially from the many, and secondly they were not progressing themselves in such a way as to increase their distance and leave humanity behind. The present aristocracy of western culture, at the very moment when it most clearly dominates the world, is being imitated rapidly and successfully in every eastern country. It is not on the lines of a cultural aristocracy or the formation of a class more able to lead the good life that the splitting of the human race is likely to occur; because such aristocracies are only reaching to a more complete humanity, and where they lead the race will follow. It is rather the aristocracy of scientific intelligence that may give rise to new developments. They have come down the earlier centuries, scattered singly or in small groups, but the mechanical revolution and its consequences have increased their number and at the same time their compactness. More and more, the world may be run by the scientific expert. The new nations, America, China and Russia, have begun to adapt to this idea consciously. Scientific bodies naturally are first conceived of as advisory and they will probably never become anything else; but, with every advance in the direction of more rational psychology, the power of advice will increase and that of force proportionally decrease. This development, coupled with the broadening of the idea of private interest to include, almost necessarily, some consideration of humanity, will tend to centre real sovereignty in advisory bodies. The scientists would then have a dual function: to keep the world going as an efficient food and comfort machine, and to worry out the secrets of nature for themselves. It may well be that the dreams of "Dædalus" and the doom of "Icarus" may both be fulfilled.[23] A happy prosperous humanity enjoying their bodies, exercising the arts, patronizing the religions, may well be content to leave the machine, by which their desires are satisfied, in other and more efficient hands. Psychological and physiological discoveries will give the ruling powers the means of directing the masses in harmless occupations and of maintaining a perfect docility under the appearance of perfect freedom. But this cannot happen unless the ruling powers are the scientists themselves. For a state in which the present rulers impose themselves in this way, the prospect of which so appals Mr. Bertrand Russell,[24] though possible, is essentially unstable and bound to lead to revolution, which would be brought about by the gradually increasing inefficiency of the rulers and the increasingly effective insurgence of the excluded intelligent. Even a scientific state could only maintain itself by perpetually increasing its power over the non-living and living environment. If it failed to do so, it would relapse into pedantry and become a perfectly ordinary aristocracy. In the earlier chapters I have given some ideas of one way in which this scientific development could take place by the colonization of the universe and the mechanization of the human body. Once this process had started, particularly on the physiological side, there would be an effective bar between the altered and the non-altered humanity. The separation of the scientists and those who thought like them—a class of technicians and experts who would perhaps form ten per cent. or so of the world's population—from the rest of humanity, would save the struggle and difficulty which would be bound to ensue if there were any attempt to change the whole bulk of the population, and would, to a certain extent, lessen the hostility that these fundamental changes would necessarily produce. Mankind as a whole given peace, plenty and freedom, might well be content to let alone the fanatical but useful people who chose to distort their bodies or blow themselves into space; and if, at some time, the magnitude of the changes made them aware that something important and terrifying had happened, it would then be too late for them to do anything about it. Even if a wave of primitive obscurantism then swept the world clear of the heresy of science, science would already be on its way to the stars.

23 The first two To-day and To-morrow volumes by J. B. S Haldane and Bertrand Russell, both included in this *Reader*.

24 See pp. 57–58.

In tracing this development, however, we have neglected other weighty considerations. Up to the present the cumulative edifice of science has been erected by assistance as much from the practical world as from the learned, and scientists themselves have never formed an hereditary or even a closed caste. In two ways the progress of science depends upon non-scientific humanity. As experimentation becomes more complex, the need for the co-operation in it of technical elements from outside becomes greater and the modern laboratory tends increasingly to resemble the factory and to employ in its service increasing numbers of purely routine workers. If development is to follow, even in the earliest stages, on the lines I have indicated above, this necessity for economic and technical assistance will be multiplied many times. More important still, the complexities of scientific—and particularly of theoretical scientific—thought, calls for an ever greater number of first-class intelligences, and the modern development of science can hardly be disconnected from the political and economic changes which make it possible to recruit the personnel of science from wider and wider circles. For until we can know from the inspection of an infant or an ovum that it will develop into a genius, or else can from any infant produce one by a suitable education we shall have to rely on the diffusion of a general education in order to ensure that all capable minds are utilized.

This recruiting of science is the surest way of preventing a permanent human di-morphism from arising, because it reinforces what is probably the strongest factor involved, the emotional conservatism of the scientists themselves. The mere observation of scientists should be sufficient at the present to show that any fear of immediate di-morphism is unfounded. In every respect, save their work, they resemble their non-scientific brothers, and no one would be more shocked than they at the suggestion that they were raising up a new species and abandoning the bulk of mankind. For whether they are inventing submarines or depth charges, they feel they are serving humanity. The consciousness of solidarity—and even more, the unconscious emotional, identification with the group—is a terrific force binding humanity together, and so long as individual scientists have it, di-morphism would be impossible.

But the scientists are not the masters of the destiny of science; the changes they bring about may, without their knowing it, force them into positions which they would never have chosen. Their curiosity and its effects may be stronger than their humanity.

These two obstacles to the separation of the scientists, though weighty, are of the kind that would lose force with time, while those favouring their separation tend to increase. The technical importance of the scientist is bound to give him the independent administration of large funds and ends the mendicant state in which he exists at present. Scientific corporations might well become almost independent states and be enabled to undertake their largest experiments without consulting the outside world—a world which would be less and less able to judge what the experiments were about. It is very probable that before the real independence of science could make itself felt, the organization of the world would have to pass through its present semi-capitalistic stage to complete proletarian dictatorship, because it is unlikely that a scientific corporation would, in an ordinary capitalistic state, be allowed to be so wealthy and powerful. In a Soviet state (not the state of the present, but one freed from the danger of capitalistic attack), the scientific institutions would in fact gradually become the government, and a further stage of the Marxian hierarchy of domination would be reached. Scientists in such a stage would tend very naturally to identify themselves emotionally rather with the progress of science itself than with that of a class, a nation or a humanity outside science, while the rest of the population would, by the diffusion of an education in which the highest values lay in a scientific rather than in a moral or a political direction, be much less likely to oppose effectively the development of science. Thus the balance which is now against the splitting of mankind might well turn, almost imperceptibly, in the opposite direction. The whole question is one largely of numbers, and would become entirely so as soon as the quantity and quality of population were controlled by authority. From one point of view the scientists would emerge as a new species and leave humanity behind; from another, humanity—the humanity that counts—might seem to change *en bloc*, leaving behind in a relatively primitive state those too stupid or stubborn to change. The latter view suggests another biological analogy: there may not be room for both types in the same world and the old mechanism of extinction will come into play. The better organized beings will be obliged in self-defence to reduce the numbers of the others, until they are no longer seriously inconvenienced by them. If, as we may well suppose, the colonization of space will have taken place or be taking place while these changes are occurring, it may offer a very convenient solution. Mankind—the old mankind—would be left in undisputed possession of the earth, to be regarded by the inhabitants of the celestial spheres with a curious reverence. The world might, in fact, be transformed into a human zoo, a zoo so intelligently managed that its inhabitants are not aware that they are there merely for the purpose of observation and experiment.

That prospect should please both sides: it should satisfy the scientists in their aspirations towards further knowledge and further experience, and the humanists in their looking for the good life on earth. But somehow it fails by the very virtue of its being a possible and probable solution on the lines of our own knowledge. We do not really expect or want the probable; all, even the least religious, retain in their minds when they think of the future, an idea of the *deus ex machina*,[25] or some transcendental, superhuman event which will, without their help, bring the universe to perfection or destruction. We want the future to be mysterious and full of supernatural power; and yet those very aspirations, so totally removed from the physical world, have built this material civilization and will go on building it into the future so long as there remains any relation between aspiration and action. But can we count on this? Or, rather, have we not here the criterion which will decide the direction of human development? We are on the point of being able to see the effects of our actions and their probable consequences in the future; we hold the future still timidly, but perceive it for the first time, as a function of our own action. Having seen it, are we to turn away from something that offends the very nature of our earliest desires, or is the recognition of our new powers sufficient to change those desires into the service of the future which they will have to bring about?

25 A supernatural being who appears in order to resolve human problems (especially in a play).

Excerpts

From: *Prometheus; or, Biology and the Advancement of Man*

H. S. Jennings (1925)

Headnote

Herbert Spencer Jennings (1868–1947) was an American zoologist and geneticist. He was Henry Walters Professor of Zoology and Director of the Zoological Laboratory, The Johns Hopkins University. Many inter-war thinkers believed Eugenics offered the prospect of transforming and improving humanity by controlling human breeding selectively, as humans have done with crops and animals. *Prometheus* offers one of the most powerful critiques of such claims. Jennings went on to expand the arguments here in his 1930 book *The Biological Basis of Human Nature.*

******

From the Preface

[. . . .]

The beginnings of knowledge often point to clear and simple conclusions that range themselves in the pattern of some obvious theory. Such conclusions are drawn with confidence: they are set down in the text-books; they are fixed in the mind; they acquire authority; they become part of science. Only details, illustrating and applying these principles, require, it seems, yet to be discovered. As the details are worked out, they are ranged as best may be under the theory; if some of them fit it awkwardly or not at all, these are but laid aside as indigestible. In time there comes to be a great store of facts that do not fit the accepted generalizations. But to revise the generalizations—hic labor, hoc opus est;[1] this is not to be lightly undertaken; it throws all into confusion; it discredits what science has pronounced; it is avoided. Theory and facts become hopelessly at variance, yet they continue to rest side by side in a single framework. Such has been the course of events in the science of genetics. The hen of Mendelian theory has hatched a brood of swans. She would fain claim them as her own, but they have lost all semblance of chicks. [6–7]

[. . . .]

1 Virgil's *Aeneid*, VI, line 129, has it the other way round: 'hoc opus, hic labor est': that is the task, that is the hard work.

DOI: 10.4324/9781032625522-10

From Chapter II. The Biological Background

Heredity and Environment

[. . . .]

Other known cases illustrate the effect of the environment in altering the totality of the organism, its entire personality, as it were. Many years ago there was discovered in Mexico a salamander that lives throughout its life in water, has a heavy, broad body, a tail flattened for swimming, and external gills. In this condition it becomes mature, lays its eggs in the water; produces young that inherit its characteristics, and finally dies. This continues for generation after generation. A number of these axolotls were kept for years in the Zoological Gardens at Paris; they showed the inherited characteristics above set forth. Breeding experiments on these animals would show these characteristics to be inherited in the usual Mendelian manner.

But after years in which these were the only inherited characteristics that they were known to possess, certain different environmental conditions were brought into action, and thereupon, to the astonishment of the observer, the axolotls developed a new set of inherited characteristics, a new and diverse personality. The external gills disappeared, the body became smaller, slender, and of a very different shape, the animals came out on the land and remained there, breathing air. They now became mature in this amblystoma[2] condition, laid eggs, and produced offspring—which again, under these conditions, developed into land animals of the same sort; and this, too, may continue for generation after generation. The inherited characteristics are now these land characters; these are, in detail, inherited in the typical Mendelian manner.

Here we have two extremely different sets of inherited characteristics; which one shall appear is determined by the environment under which the organism develops. If "hereditary" means dependent for development on peculiarities of the germinal constitution (the sense in which it is employed) both sets are hereditary. If "environmental" means dependent for development upon peculiarities of the environment, both sets are environmental. Any characteristic requires for its production both an adequate stock of chemicals and an environment adequate for its production through proper interaction of these chemicals with one another and with other things. In these senses all characteristics are hereditary and all are environmental, but no characteristic is exclusively hereditary or exclusively environmental.

Beyond all other organisms, man is distinguished by the possession of many sets of inherited characteristics; the decision as to which shall be produced depending on the environment. The axolotl may be compared to an uneducated man, the amblystoma to an educated one. The educated man has characteristics very diverse from those he would possess if uneducated. We say, when we think of this fact, that these are acquired characters, environmental characters, due to education. This is correct; but there is a tendency to go farther and say that these are not inherited

2 The genus of mole salamander is now normally given as 'ambystoma.'

characters. This is a mistake—unless we are prepared to cancel *in toto* the designation "inherited characters." The characteristics of the educated man are his native, inherited characters, just as truly as are any that he has. For all his characteristics depend on the conditions under which he develops, and would be diverse under different conditions, just as is true of the characteristics that develop under education. And the characters developed under education depend upon the hereditary materials derived from the parents, changing as these materials are altered, just as do all others. "Hereditary" has no consistent meaning other than this. The teachings of genetics may be summarized in Aristotle's saying—"the nature of man is not what he is born as, but what he is born for"—paraphrased perhaps into the form "the inheritance of man is not alone what he is born with, but what he can develop."

Why it seems paradoxical to call the characteristics developed under education inherited, while we make no difficulty in thus designating the colour of the eyes and the stature, lies in certain practical difficulties, not in any difference of principle. In the group of organisms to which man belongs there is an early period in which it is practically difficult to change effectively the conditions under which the organism develops, because it is enclosed within the mother's body, or within a resistant eggshell. So we have gotten accustomed to calling inherited those characteristics which are determined before it leaves its mother's body or the egg, while those determined later are called acquired characters. But this is an artificial distinction, based on practical considerations. In many organisms there is no such distinction into two periods; in them it is possible to alter the conditions at any period, even the earliest. And when this is done it is found that all the characters depend on the conditions; that such fundamental characters as "the number of eyes an animal has" or "the position of the eyes in the body" may be altered. In fish, for example, two eyes, one at each side of the middle line, form as distinctly an inherited characteristic as in man, yet fish can be subject so early to changed conditions (as Stockard and others show) that the animal has a single median eye instead of two lateral ones; and many other equally striking changes are producible by changes in the chemical environment. If the fish lived continuously in these conditions they would regularly inherit a single median eye; the two lateral eyes would be looked upon 'as a rare abnormality, produced by special conditions and not inherited. In truth, all characters are as certainly due to the conditions of development as to the materials of the germ cells.

If there were not practical difficulties in the way, similar fundamental changes of structure could be made in man or any of the higher animals. In these higher creatures, a time comes, before development stops, in which it is possible to change the conditions; that is, after what we call birth. And then it is found that changing the conditions does change the characteristics that later develop—exactly as the characteristics of the fish are changed by changing the conditions. We call this process education. If we could give the same education for many generations to a number of different human families, we should find that the characteristics resulting from education are inherited characteristics in the same sense as are colour of eyes and form of head. Under the same education they would differ in individuals of diverse stock; and in crossing they would follow Mendelian rules, as do physical characters.[3] Every creature has many inheritances; which one shall be realized depending on the conditions under which it develops; but man is the creature that has the greatest number of possible heritages. Or, more accurately, men and other organisms do not inherit their characteristics at all. What their parents leave them are certain packets of chemicals which under one set of conditions produce one set of characters, under other conditions produce other sets. In man, the number of diverse sets that may thus be produced is very great; although it is, of course, not unlimited. But what the limitations are cannot be stated from general biological principles or from what we know of any other organisms; they can be discovered only by concrete studies of man himself.

Adequate recognition of these facts and principles, which appear fully established by the advance of genetics, would greatly alter some of the current discussions and attitudes on the relation of biological science to human affairs. The biologist is pained to find that the medical man resists the introduction of the concept of heredity into the domain of disease. This is because of the current fallacy that what is hereditary is certain, fixed, unchangeable. Very properly the medical man rejects that, in its application to disease. But with the recognition that to assert that a thing is hereditary signifies merely that the organism has received such a constitution as to produce it under given conditions, all such ground of objection vanishes. This does not deprive of significance recognition of the part played by heredity in medicine; which is but recognition of constitutional differences that have origin in diversities of the germ cells. The individual who may produce an inherited defect under certain conditions need not produce it under others. Some individuals receive a constitution which resists disease under conditions in which others succumb to it. Some respond in one way to particular therapeutical agents, others in another way, depending on their hereditary genes. It is only against what Davenport has characterized as purely impersonal medicine that the implications of genetic science lie.

The same fallacy reappears in discussions of racial problems. The recent immigrants into the United States show certain proportions of defective and diseased persons; and we are informed that "these deficiencies are unchangeable and heredity will pass them on to a future generation." There is no warrant in the science of genetics for such a statement; under new conditions they may not appear. It is particularly in connection with racial questions in man that there has been a great throwing about of false biology. Heredity is stressed as all-powerful; environment as almost powerless: a vicious fallacy, not supported by the results of investigation. We are warned not to admit to America certain peoples now differing from ourselves, on the basis of the resounding assertion that biology informs us that the environment can bring out nothing whatever but the hereditary characters. Such an assertion is perfectly empty and idle; if true, it is merely by definition: anything that the environment brings out *is* hereditary, if the word hereditary has any meaning.

3 [Author's note] Obviously, nothing in the text relates to the effect of education on the descendants of the educated person; that is, to "the inheritance of acquired characters." The propositions set forth are equally valid whether there is or is not such effect on the descendants.

But from this we learn nothing whatever as to what a new environment will bring out. It may bring out characteristics that have never before appeared in that race. What the race will show under the new environment cannot be deduced from general biological principles. Only study of the race itself and its manner of reaction to diverse environments can give us light on this matter.

All characteristics then are hereditary in that they depend upon the germinal material and change as that material is altered. All characteristics are likewise environmental, in that they depend upon the conditions under which development occurs, and change as those conditions are altered. The concepts of the hereditary and the environmental cannot properly be employed in the absolute way now practised, as distinctive of particular characteristics. But their respective roles can nevertheless be disentangled with precision if they are applied, not to characteristics in themselves, but to the diversities between particular concrete cases. [53–67]

[. . . .]

III. The Advancement of Man

SUCH being the biological background, what of the advancement of man? Through his manifold alternative inheritances, what a man shall become is determined by the conditions under which he develops to a greater extent than in any other organism. The lot of given individuals and societies is enormously alterable and improvable by change of conditions, by inventions, by increase and dissemination of knowledge. The tremendous changes in the state of the Germanic races in the last 20,000 years have been brought about, for better or worse, by this means; equally great changes can, therefore, be similarly brought about in this or other races. These are facts of human biology that cannot be altered by anything learned of the biology of other organisms. All groups of organisms have some common features; but every group has, too, its distinctive features. When the biologist, from his knowledge of other organisms, is tempted to dogmatize concerning the possibilities of human development, let him first ask himself: How correctly could I predict the behaviour and social organization of ants from a knowledge of the natural history of the oyster? Man differs from other organisms in these respects certainly as much as the ant does from the oyster; for these distinctive aspects of his biology only the study of man himself is relevant. There is nothing in the general principles of biology which says that increase and diffusion of knowledge shall or shall not do away with warfare, solve the problem of over-population, and otherwise bring to realization the dreams of reformers. These are questions of the special biology of *Homo sapiens*, on which knowledge of Amoeba, the oyster and the ant have no bearing.

But environmental changes have another class of effects: through their selective action they alter the constitution of later generations. As conditions become easier, combinations of genes that would have been eliminated survive, develop and propagate, so that the later population contains less resistant individuals than did the earlier; saving the tubercular results in a population less resistant to tuberculosis.

This effect of improved conditions has given rise to a pessimism of unrelieved blackness, thorough-going, irremediable. Every improvement in conditions, it holds, inevitably weakens the stock. By a dreadful paradox, present amelioration means later deterioration. The more rapid the upward movement in conditions, the more headlong the downward plunge into degeneration.

Every important cultural improvement certainly does shift the incidence of selective elimination. Nothing in the history of the race has so preserved and multiplied the weak and (previously) unfit, as did the Promethean bringing of fire into the service of man. The invention of clothing, of tools, of social organization, of writing, have had the same degenerative effects. Compared with the disastrous results of these, vaccination for smallpox and typhoid, abolition of conditions that bring plague, cholera and yellow fever, though tending to fill the population with non-resistant types, can have been only moderately degenerative. Does biological science support and demonstrate this paradox?

If it is certain that the downward racial tendency induced by the changed selective action is more potent than the immediate amelioration resulting directly from the improved conditions, then, indeed, is man headed for destruction; for it is idle to attempt to so alter his nature that he will not improve his living conditions. No one will propose to go back to the days of no fire and no clothing. But nothing demonstrates that the softening due to the changed selective action is more potent than the directly beneficial effects of such inventions. No one knows that the vigour of later generations is not increased by the greater well-being of the bodies in which the genes live and develop. There is no certainty that the invention of fire, clothing, social organization, and vaccination have not augmented the well-being and staying-power of the race.

On the other hand, there are a few social practices that recognizably and directly work toward racial degeneration; these should obviously be stopped. Such are the freedom and encouragement of reproduction among the feeble-minded, the criminal, the insane. But it is a far cry from these to the inventions that have improved the lot of all mankind.

But such negative and palliative eugenic measures, directed against the obviously pathological, leave untouched the futility, incompetence and unreason that have of late brought gigantic disaster. There is a positive eugenic programme, designed to meet this problem. What are its opportunities?

Under the usual present conditions, omitting the clearly pathological, some combinations of genes produce desirable individuals, others undesirable ones. It is to the prevalence of the latter that the troubles of humanity are due. Shall we not bring it about that in future generations only the individuals shall develop that have the better assortments of genes?

Insight into the difficulties and possibilities of this programme demands the taking of a wide and long view of the organism that we wish to improve. Man, or any of the complex animals, is to be looked upon as a single organism, temporarily divided into pieces or individuals. These pieces reunite at intervals and again separate. The great organism thus forms a network. The enduring parts of the network are the genes. They form the continuous strands of the net, living through from generation to generation, and taking part successively in the formation of the bodies of many different individuals. The number of strands or genes in the entire network is very great; much greater than the large number present in any individual. A different combination of strands is present in every individual.

As the generations pass the combination of strands in a given individual separates repeatedly into two halves; each half unites with a corresponding half from another individual to make a new combination for the next generation. From each new combination of the strands an outgrowth appears; this is what we call an individual. But this outgrowth is not permanent. After a time it decays and drops off, while the combination of genes that formed it separates and passes into new combinations, yielding new individuals.

Thus upon this great immortal vine the individuals hang like grapes from a pergola, each individual constituted of all the strands that the netted vine happens to have at that point. What the fruit becomes depends, under given conditions, upon which of the strands go into it; size, form, colour, structure, vigour, activity, type of behaviour, senses, power of learning, power of reasoning—all depend, other things being equal, on the strands that are present.

Thus we find that, in all organisms, some of these combinations produce fruits that are better fitted to the prevailing conditions of life than are others. Some in man are vigorous, sensible, efficient, valuable to themselves and others. Others are defective, weak, foolish, worthless, injurious to themselves and to others. What the eugenecist desires to do is to save, propagate and multiply the good combinations, letting the poor ones disappear.

In the lower animals and plants there is what might be called a special device for carrying out precisely and regularly this eugenic problem. Any one of the fruits on the vine—the individuals—can multiply without forming new combinations of genes. It simply divides frequently, or gives off small pieces, producing thus a great number of individuals like itself—all with the same identical combination of genes. An individual of vigour thus brings forth thousands like itself; in time almost or quite the entire population may consist of this single type, while the weak individuals wither and disappear. This happens on a grand scale in the lower organisms under natural conditions.

By the intervention of man, supplying special conditions, almost any type of individual may thus be induced to multiply till an entire population consists of that single kind. This is what has happened in most of the cultivated fruits. Once a good combination of genes is obtained, it is preserved and propagated unchanged.

This is what the eugenicist desires to do in man. But in man and other higher animals no combination is permanent. None ever lasts beyond the life time of the single individual. No individual can be multiplied in such a way as to retain the same combination of genes. No matter how vigorous, how well adapted to the conditions, how valuable from every point of view, every combination must disintegrate and a differently constituted one must appear for the next generation. The new one may be less vigorous, less valuable in' every way, but it must appear in place of the old one. Indeed, the outstanding individuals, the geniuses, are always rare special combinations, occurring, according to the laws of permutations, only once in thousands of instances. When they are taken apart, the new combinations made are almost certain to be the commoner types, less valuable than their parents.

What occurs in such cases is seen when one of the valuable fruits—a fine variety of apple or orange—is allowed to reproduce by seed, forming thus new combinations of genes. Among the offspring are many types, mostly inferior ones, thorny, irregular, weak plants with worthless fruits. Almost never is one produced that equals the parent. This is the sort of thing that occurs regularly in man. But the same is true for the poor combinations. They, too, must disintegrate and pass into new groupings; and now the offspring may be better than the parents; certainly they will be diverse. And from the large population of commonplace types appear continually, as the generations pass, a few rare ones—for genius or for inferiority—then after a generation these drop back again into the great reservoir.

So, though in lower organisms the valuable types may be directly saved and propagated, this cannot be done in man. The genes must continually change their combinations, now producing a superior individual, now a commonplace one, now an inferior one:

> Only while turns this wheel invisible
> No pause, no peace, no stopping place can be;
> Who mounts will fall, who falls may mount—the spokes
> Go round unceasingly.[4]

So far as the genes go, reincarnation is a literal fact; one gene in its time plays many parts in persons of the most diverse kinds:

> Who toiled a slave may come anew a prince—
> Who ruled a king may wander earth in rags.[5]

The disappearance into the melting pot of the rare and excellent combinations has long been a ground for criticism of the way the universe is run. The author of Ecclesiastes hated all the labour that he had taken under the sun because he should leave it to the man that should come after him, "and who knoweth whether he shall be a wise man or a fool; yet he shall have rule over all my labour wherein I have shown myself wise under the sun." The Eugenicist agrees with these animadversions, and wishes to change the system which justifies them; he desires to continue the combinations that have proved valuable, and to suppress the others.

If uniparental reproduction could be brought about in man, as it is in apples and oranges, this could be done. And the process that underlies this—the division of a single combination so as to produce identical individuals—does at times occur in man, though in a way that is not under control, and that does not help in attaining the eugenic ideal. In the production of identical twins one of the fruits of the vine divides into two that have the same combination of genes. But this occurs but once, and at so early a stage in development that there is no way of knowing whether the

4 Sir Edwin Arnold's narrative poem on the life of the Buddha, *The Light of Asia* (1879), Book the Eighth.

5 Ibid.

combination duplicated is a good one. What the eugenic plan requires is that the adult, after he has shown his value, should be multiplied without change of genetic combination. If this could be done, man would have his fate in his own hands. He could multiply the desirable combination until the entire population consisted of that type.

Of course, he would then come squarely against the question as to what kind of individual should thus be allowed to multiply and inherit the earth. There is little doubt that, as Mr. Bertrand Russell[6] suggests, the ruling class, whatever its character, would be the one allowed to prevail. As they now attempt to suppress the opinions that do not agree with their own, so with this efficient weapon in hand they would radically wipe out types holding other ideals. With the capitalist class in complete possession, the world would become uniform; doubtless more comfortable; certainly less entertaining as a spectacle. The unexpected, the exciting, the amusing, the melodramatic, would disappear. With this would go poetry, art, literature. The human orchard would consist of well-trimmed fruit trees in set rows, each bearing a crop of succulent fruit.

Unfortunately or fortunately (whichever one's philosophy inclines him to hold), this cannot be done in mankind. If an ingenious inventor were set to devise a system for the purpose of heading off completely anything of this sort, he could hardly produce one so effective as the one found in nature. This might rather seem devised to the end of giving the greatest possible variety; of yielding the extremes of diversity at any one time; of inducing most thorough-going and continuous changes as generations pass. Personified nature might well be held to abhor uniformity and constancy, for the higher organisms.

Is there no way by which the Eugenicist can circumvent this propensity of nature? Is there no way to substitute uniform excellence for extreme and heterogeneous variety?

Quite aside from the difficulty of getting agreement as to the human type that is to prevail, the enterprise is extremely difficult even from a merely technical standpoint. But a certain amount of effect could be produced by operating on statistical principles. While it is not certain that given superior parents will produce superior offspring, nor particular inferior parents inferior offspring, it is true in the long run that a million superior parents will produce, on the whole, offspring of higher average grade than those produced by a million inferior parents. If, therefore, on a grand scale, the superior parents could be induced to produce more offspring, the inferior ones fewer, this might on the whole raise very slowly the level of human kind. The uncertain element here is the great reservoir of genes held by the undistinguished individuals, constituting nine-tenths or more of humanity. From these come, through kaleidoscopic recombination of genes, a large proportion of the unusual individuals that occur—both those that are superior and those that are inferior. If all these unusual individuals, above and below the commonplace, could

6 [Author's note] Russell, *Icarus* (Kegan Paul and Co., Ltd., London; E. P. Dutton and Co., New York). [See 56–57 above.]

be cut off as parents, they would be replaced in at least large measure from the unusual progeny of the large commonplace group.

But whatever eugenic measures are attempted, so long as biparental inheritance is kept up, the variety, the surprises, the perplexities, the melodrama, that now present themselves among the fruits of the human vine will continue. Capitalists will continue to produce artists, poets, socialists, and labourers; labouring men will give birth to capitalists, to philosophers, to men of science; fools will produce wise men and wise men will produce fools; who mounts will fall, who falls will mount; and all the kinds of problems presented to society by the turns of the invisible wheel will remain. They may be softened by improvement of conditions, by increase of knowledge, and perhaps by selective eugenics; but they will not disappear so long as biparental reproduction continues. [73–94]

From: *Automaton; or, the Future of the Mechanical Man*

H. Stafford Hatfield (1928)

Headnote

Henry Stafford Hatfield (1880–1966) was an inventor and a prolific writer and broadcaster about science and invention. His *The Inventor and his World* (London: Kegan Paul, Trench, Trübner & Co., 1933) was reissued as a Pelican book in 1948. *Inventions; and their Uses in Science Today* was published by C. K. Ogden's Orthological Institute (1939) and reissued by Pelican in 1940. Joseph Agassi, *Science in Flux* (Dordrecht: D. Reidel, 1975), p. 282, refers to Hatfield as: 'the most important writer on technology, perhaps.' He also wrote *The Conquest of Thought by Invention in the Mechanical State of the Future* for another of Ogden's series, the Psyche Miniatures (London: Kegan Paul, Trench, Trübner & Co., 1929).

The title and subtitle of *Automaton* might lead you to expect a science fantasy about robots. Hatfield certainly discusses robotics and automation; but he places it within the much larger context of the long history of humanity's relation to technology, as well as the early twentieth-century's response to mass production and factory assembly lines.

In outlining the challenges to future developments of automata, Hatfield effectively maps out some of the key technological fields of robotics, cybernetics, computing, image and voice recognition, machine learning, and drone guidance. He also advocates developing inventions we now take for granted, such as answering machines and central heating thermostats.

******

[. . . .]

The Constitution of an Automaton.—An automaton, by analogy with the human model, should consist of three parts: limbs to work with, senses to perceive what it is working with, or what result it is producing, and a brain to regulate the action of its limbs in accordance with the perceptions of its senses. Needless to say, we are striving to create, not a Frankenstein's monster, a Robot, a mechanical servant which can be set to any simple task, but thousands of different automata each specialized for a certain task. In our machines we have already developed *limbs* of a power and precision exceeding our own many many thousand-fold.

DOI: 10.4324/9781032625522-11

In our instruments, we have developed *senses* exceeding our own, in many cases, a million-fold in sensitivity. Indeed, they are capable of receiving impressions, such as magnetism, which are qualitatively imperceptible to our natural senses. What we have still to develop is the mechanical *brain*, the link between instrument and tool. At present, we make use in most cases of a human brain. The ship is driven, and its rudder moved, by machinery; the compass senses the direction; but a man still stands at the wheel with his eyes on the compass. Our instruments are means for "amplifying" or translating their subtile perceptions into indications perceptible to the coarse senses of our human agents. The man at the wheel cannot sense directly the magnetic field of the earth, but the compass does so, and translates the result into a movement perceptible to his eyes. His function is purely mechanical. It could be, and indeed, already has been in some cases superseded by a mechanism.

The Human Automata of Modern Industry.—This case is typical of the function of a large number of workers in modern industry. Their task is often of the simplest, most monotonous description. They watch a dial or two, and move a lever or two accordingly. They copy figures into a book, and add them up. They count or weigh material into packages, rejecting bad stuff.

The steady development of machinery of all kinds, and of highly elaborate processes requiring careful attention to the indications of instruments in order to regulate them closely, calls more and more for a type of employee completely inured to boredom, capable of sustained vigilance under conditions devoid of all incident to rivet attention. From the point of view of industry, it should be the task of the "social reformer" to see that the supply of such semi-automata shall keep pace with the demand, and that they shall be clean, healthy, simple minded, undisturbed by rude passions, whether animal, political, or spiritual; regular in their habits, fertile to a sufficient degree but not beyond it, and dependable in the matter of attention to their task.

It need hardly be said that the individual social reformer is usually completely unconscious of his real function. It is sufficient for him to see people dirty, drunken, quarrelsome, lustful, envious, insubordinate, and to imagine how miserable he would be if he were compelled to be one of them. He regards the mainspring of his activity as the desire to reduce the sum of human misery; a view undoubtedly correct in its way, though his remedy appears to involve emptying the human baby out with its bath of misery. [14–17]

[. . . .]

The Present State of the Technology of Automata.—This field of work is at present in a most primitive stage of organization, compared with that of other branches of technical science. The initial work is largely done by inventors of the most "classical" type, striving to solve a problem because it happens to fascinate them. Often they are men without the most elementary training in physical and chemical principles. The development into practicability, also, is carried on by men equally ignorant of fundamental principles, for these fundamental principles have never been systematically studied. While the design of prime movers, electric generators, machine tools, and so forth has been elucidated and written about theoretically and practically, there is hardly any literature available concerning the design of such matters as electric relays, contacts, complicated mechanical motions. Nothing at all is published concerning the relative merits and dependability of all the various alternative devices which may be used to solve an automaton problem. Each problem is attacked by a costly process of trial and error usually devoted to one system only; and a successful result may well be superseded by a solution based upon a totally different system. Technical alternatives will, of course, always exist, but in more developed branches of engineering the relative merits and demerits of alternatives can be estimated from theory and experience before work is begun. No such possibility exists at present in the design of automatic controls, and the establishment of a research and teaching centre to elucidate these problems would be a great step forward. The multiplicity of methods, both in principle and in detail, would be reduced; sound ideas which have remained untested would be brought to light and tested out. The fundamental problems peculiar to small mechanisms would be investigated, whereas no one but the practical constructor seems at present to be aware that such peculiar problems exist.

The Future of the Automaton.—The purpose of this essay is to consider how far we may reasonably expect this development of an automatic link between instrument and machine to go in the future. We have first considered the matter from the point of view of the human worker, and we have concluded that he will not improve in dependability. On the other hand, it is certain that the automaton can be largely improved and cheapened. Also, that the processes of the future will be more exacting from the point of view of control than those of the present. Hence it is likely that the only limits which will be set to the development of automata will be those inherent in the limitations of our science.

Recording Instruments and Alarms.—We ought to notice, before passing on to technical details, a modern development which represents another method of coping with human frailty. Every kind of instrument can now be made self-recording, and it is thus possible for a works-manager to receive an automatic report each day of the exact manner in which the plant has been worked in the past twenty-four hours. Jointly with this development, we have instruments with giant dials, and others which sound alarms or exhibit coloured lights when all is not well. The moral effect of these devices on the men in charge is very great, and ingenious means are being developed for translating into terms of bonuses and fines the goodness or badness of each record. But instruments that can do all these things may, in many cases, equally well set machinery going which dispenses with human control. We may still, however, retain our records and alarms. The latter then correspond to the pain mechanism of our automaton. It cries to us in distress when all is not well with its internal economy.

The Mechanical Brain

Examples of Simple Automata.—We have already defined the automaton as consisting of three parts, corresponding to the senses, brain, and limbs of the human body. It is thus distinguished from an automatic tool, such as a lathe, which is devoid of senses, and is thus incapable of adapting itself to varying conditions of material or working. Perhaps the simplest everyday example of a true automaton is the gramophone motor. Its function is to turn the disc at a constant speed, in spite of varying friction, strength of spring, and other causes. In it we find a very simple instrument, consisting of a pair of spring governor balls, the divergence of which is a measure of the speed of the motor. This device "perceives" any change in speed, from whatever cause. It acts upon a brake applied to the motor, presses this on when the speed increases, and takes it off when it decreases. This is the brain of the arrangement, which controls the limbs, the motor.

Another very common device is the thermostat, which regulates the temperature of an oven, furnace, cold-storage chamber, or even an ordinary room. Here we have some suitable form of thermometer (very many forms are in use) which keeps the temperature regulated. It is linked, by mechanisms of the most varied kind, with the source of heat or cold, in such a way that whenever the temperature which is to be controlled rises or falls slightly, suitable automatic action, such as increase or decrease of fuel supply, takes place. This ancient device is being fitted, as a great novelty, to modern domestic gas ovens. It was quite well known when the first gas ovens were put on the market.

The Corrector Automaton.—These two devices are typical of one fundamental kind of automaton. It is a kind which may be said to wait for slight trouble, and then immediately take steps to correct it. The diagram illustrates the principle in an electrical form of quite general application (Figure 7.1). This form is applied in the automatic steering of ships. Here the instrument is the compass, in practice the gyrostatic compass,[1] though the system might easily have been applied long before the invention of the latter. A minute swing of the compass from its correct set course makes an electric contact, which sends a weak current to a magnetically operated switch capable of sending a current sufficiently strong to cause a motor to turn the rudder so as to correct the deviation from course.

The Prescient Automaton.—The system just described serves very well for large ships, which are not noticeably deflected from their course by single waves or gusts of wind, as are small vessels. The helmsman of the latter, however, does not wait for these deflections to take effect, but watches for waves and anticipates their effect by a turn of the rudder. It would obviously be a very difficult matter to make an automatic steersman which would act in this case in a manner equivalent to that of the helmsman. But in many other cases it is possible to construct automata which do not wait for trouble to occur, but, instead, anticipate it. Such automata perceive a change in conditions, such as size or composition of material, and immediately adjust the tool so as to meet correctly the changed conditions. As an example we may take the case of softening water by the addition of chemicals which throw down the lime and magnesium salts to which the hardness is due. The dose of chemical added must be exactly proportioned to the

1 This includes a gyroscope, which maintains the stability of its axis of rotation as it spins.

SENSE. INSTRUMENT
(such as pyrometer, pressure gauge, speed indicator, compass) measuring whatever the automaton is to keep constant (temperature, pressure, speed, direction).

BRAIN. TWO FIXED CONTACTS
close together, one on either side of instrument pointer. They are set so that when pointer is touching neither, its reading is the correct constant value desired. The two contacts are connected to a battery and to

TWO RELAY SWITCHES
one of these receives a weak current from the battery, whenever instrument pointer touches one of the contacts: that is, whenever that which should be constant varies in either direction. It then closes and holds on a switch which sets going one of

LIMBS. TWO MOTORS
which then moves

REGULATING LEVER
in correct direction to rectify error.

As soon as error is rectified, instrument pointer ceases to touch contact, relay switch ceases to receive current, and lets motor switch go. Motor stops.

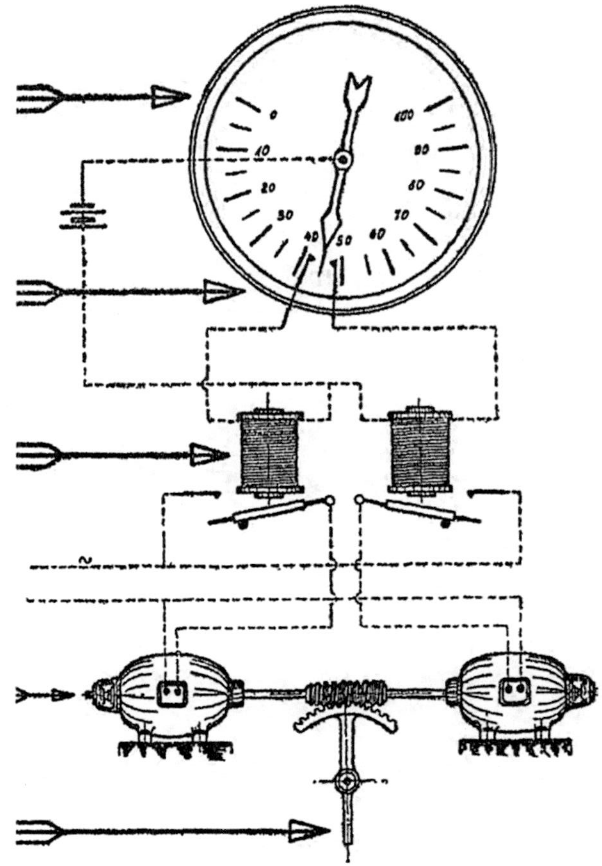

Figure 7.1 Diagram of corrector automaton.

hardness of the water. Now the process of softening and settling takes two hours to complete, so if we were to try and work on the "corrector" principle, and have our automaton readjust the dose of chemical when it found that the treated water was too hard or overdosed, we should always be two hours late, and the device would work with very great inaccuracy. Instead, we use an instrument (invented by the writer) which analyses the raw water as it enters the plant, and at once adjusts the valves to give the correct dose of chemical.[2] The automatic analyst exactly imitates the usual method by which liquids are analysed by hand in the laboratory, and can thus be used for many other purposes besides the analysis of water. Innumerable technical processes carried out on natural materials, such as ores, wood fibre, vegetables (sugar beet), and so forth, require at present continual control by analysis, since the raw material varies in a way which cannot be foreseen. These processes will no doubt be rendered largely automatic in the future .

The second diagram illustrates this "prescient" type of automaton (Figure 7.2). The electrical scheme is used merely because it is simple, to exhibit the principle. It is evident that this second type of automaton has inherent advantages which make it of wide application. Broadly speaking, whenever we can construct an instrument to measure the quality to which our tool or plant has to be adjusted we can construct an automaton of this type.

Sorting.—Perhaps one of the simplest cases of this kind is that of sorting. The scope for future development is here very great, and the problems to be solved are very various. It is of importance to consider it at this stage, because it illustrates very well a matter of fundamental importance to our subject. The problem is clearly one for the "prescient" type of automaton, which should perceive the size, weight, or other quality according to which sorting is to take place, and direct the objects to be sorted into their appropriate receptacles.

Now in the matter of size, the problem is solved in the most complete fashion by one of the oldest devices known to man, namely, the sieve. A series of closely graded sieves placed one above the other in order of fineness will, when set vibrating, separate a pulverized mass thrown upon the uppermost into as many classes of fineness as we may desire. The various products may be passed on to machines each adjusted to treat material of the particular size.

The treatment of ores in mining makes use of this and numerous other classifying devices. The actual ore-dressing machines, which separate the valuable from the worthless material are also closely allied in function. These devices, classifiers, and dressing plant are automata of the most ideal prescient description. But they can hardly be said to be built upon the human tripartite model. They perform their tasks at a vastly greater speed and lower cost than could devices measuring each individual grain, and actuating mechanism in accordance

2 The dust-jacket of the first UK edition said: 'Dr Hatfield is the inventor of the "Chemical Robot" which performed such marvels at recent demonstrations in London.' He had filed a US patent for it in 1927.

SENSE. INSTRUMENT

Measuring property of material to be worked upon by automaton (chemical composition, size, electric conductivity).

BRAIN. TWO MOVEABLE CONTACTS

Are on either side of instrument pointer. They are fixed to a rod attached to the regulating lever (see below) and move over the scale of the instrument as the lever moves. The lever is so geared to the apparatus under control that the latter is correctly regulated, in accordance with the indication of the instrument, when the contacts (and hence the lever) are in such a position that the instrument pointer touches neither contact. When it touches one or the other, a current is sent

from a battery to one of TWO RELAY SWITCHES
which then switches in one of

LIMBS. TWO MOTORS

The contacts are moved in the correct direction until the one in contact with instrument pointer breaks away from it, whereupon the relay switch ceases to receive current and lets motor switch go. Then motor stops.

which moves the

REGULATING LEVER
and with it the contacts on the instrument.

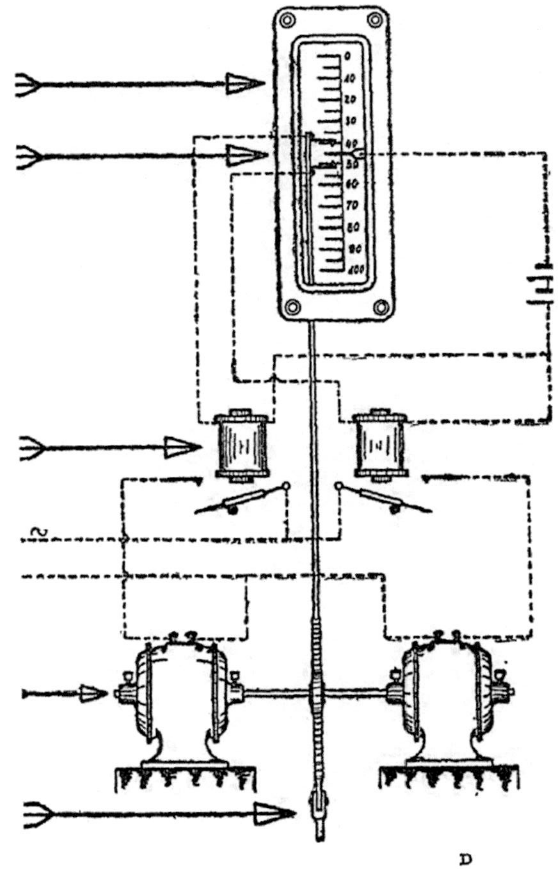

Figure 7.2 Diagram of prescient automaton.

with the result to send that grain to its appropriate destination. They warn us that there may be a way round the anthropomorphic solution of any automaton problem.

Hand-Picking.—Yet in this very field there is a problem which has hitherto resisted all attempts at solution. In treating minerals, it is often very desirable for many reasons to avoid having to crush up the rock containing the mineral to the point at which all the constituents are set free from one another. Often the valuable material occurs in large lumps sparsely distributed through barren rock, and to crush the whole rock would mean heavy expense and great waste. So instead, what is called "hand-picking" is resorted to. The rock is broken into large lumps, which pass on a travelling belt between rows of workers who pick out by eye the lumps containing valuable material in visibly large quantity. The same process is, of course, resorted to in numerous other cases where good material is to be sorted from bad.

Leaving aside for a moment the question of whether we shall ever succeed in making an instrumental eye to work actually in the same way as the human worker does, it is evident that solutions of the problem are possible in cases where we can satisfactorily perceive what is to be perceived by means other than eye. For instance it should not be difficult to make an automaton to sort lumps of rock or other objects in accordance with their magnetic qualities, weight, or electric conductivity.

It would be impossible to go into details here, but there can be no doubt that a very great development will take place in this field. An immense amount of sorting and counting is done by hand which could be done automatically. Each problem, however, is quite individual. We will therefore content ourselves with considering quite generally the possibilities of automatic sorting or recognition by means analogous to the human eye.

The Seeing Automaton.—Ever since the discovery of the sensitivity of the element selenium to light, which causes it to change its electrical resistance very greatly, inventors have promised themselves the most marvellous results by an application of its properties. It was early proposed to switch street lighting lamps on and off automatically as required by means of a selenium cell which would "perceive" the amount of natural light, and operate the switch accordingly. A recent invention is Mr. Fournier d'Albe's apparatus by which the blind are enabled to read ordinary print.[3] Of recent years other forms of cell sensitive to light have been constructed, which have been recently written about a great deal in connection with television. We need not therefore enter into detail here, but content ourselves with noting that all these devices can be made part of an electric circuit, the current in which varies in intensity according to the amount of light falling on the

3 E. E. Fournier D'Albe was a physicist who wrote two excellent volumes for To-day and To-morrow: *Quo vadimus? Glimpses of the Future* (excerpted here) and *Hephaestus; or the Soul of the Machine* (both 1925). The apparatus mentioned is the optophone: "An instrument designed to enable a blind person to read, in which a photoelectric cell is used to scan a text and produce electric signals that are converted into audible ones corresponding to the different characters" (*OED*).

light-sensitive cell. Nowadays we have extraordinarily simple and sensitive means for detecting and amplifying small variations in electric currents.[4] They are known to every user of wireless receivers. We may safely prophesy that the day is at last at hand when a perfectly satisfactory and highly sensitive "light-relay" will be available; an apparatus, that is to say, that will respond to light falling upon it by good strong variations in an electric current, which can then be set to work any kind of apparatus.

Possibilities of the Seeing Automaton.—The writer is, indeed, puzzled to know why such an apparatus is not already on the market. Its applications would be innumerable. The old scheme of turning artificial light on and off automatically according to need could at once be put into practice. Races could be automatically timed by sending a beam of light across the track into a light-sensitive relay, which would start or stop a chronograph[5] when horse, man, car, greyhound, or other competitor interrupted the beam for a moment. Objects of any sort passing a given point could be counted by similar means. The advent of vehicles at cross-roads could be signalled in advance.

Many of these things could be done, by the way, by apparatus much less delicate than a light-relay, since the beam of light could be powerful enough to actuate a relay sensitive to its heating effect.

But the applications to the construction of automata of a highly sensitive relay would be innumerable. They would present no difficulty wherever the human hand is guided by colour, as in hand-picking, sorting of fruit and other natural products, testing eggs by transmitted light, controlling chemical processes where colour is a guide, dyeing, and colour printing.

Limitations to the Powers of the Electric Eye.—Let us now consider the limitations of our electric eye. We may assume that there is no limit to its sensitiveness, and that it could be made to react with certainty to the minutest variations in light of any selected colour. The limitations we are about to consider are of a different description, and raise one of the most fundamental questions of our subject. Very low forms of life possess a faculty of responding to visual impressions which depends, not upon an extreme sensitiveness of the eye, but upon the comparison of the impression received at the moment with past impressions stored in the memory, the latter word being used in the widest sense, of a "conditioned reflex".[6] Once an animal has experienced what food, or an enemy, looks like, it afterwards responds to the visual impression in a reflex, that is, a purely automatic fashion. Now this faculty has a further peculiarity, namely, that it is not limited to recognition of an

4 That is, diodes and triode valves.

5 "An instrument for recording time with extreme exactness; also, a watch or clock to which various mechanical devices are attached for the same purpose. It is used in astronomical and other observations, in the timing of races, etc." (*OED*)

6 Term associated with the work of Russian physiologist, psychologist, and physician Ivan Petrovich Pavlov (1849–1936), training animals (especially dogs) to respond with a reflex action (such as salivating) in response to a usually unconnected stimulus (such as the sound of a bell), through repeated combinations (as by ringing the bell during mealtimes).

impression exactly like that which originated the reflex, but is equally well exercised by similar impressions. An animal responds by instant flight to an impression of human beings of the most varied sizes, shapes, and colours. It appears to have a general idea of what men are like, and is able to compare a visual impression with this general idea (not always correctly, of course) and judge whether it has been received from a human being. It need hardly be said that I am not suggesting any kind of conscious comparison; the whole process takes place, not only in animals but in ourselves, in a purely reflex fashion. Even fish are said to recognize individual human beings who feed them regularly, and they certainly respond instantaneously to the flight of a bird enemy over the water.

How far can we ever hope to imitate this in our automata? Could we ever, for instance, devise an apparatus which would signal "sail ho!" at sea?

Let us proceed step by step. Fournier d'Albe has given us, in his reading machine for the blind, one hint of a possible solution. A tiny camera serves as its eye, taking in one letter at a time, and projecting an image of that letter upon a screen. The screen, like our own retina, is a network of separate light-sensitive spots. In his machine, it is true, the impression on the artificial retina is converted into sound in a telephone applied to the ear of the blind reader. Each spot on the retina plays a certain note when it is excited by light, and the reader thus learns to distinguish the particular mixture of musical sound which corresponds to each letter. We could, however, equally well arrange that each sensitive spot, when excited, should transmit the effect to a set of indicators on a board corresponding to the retina. Say the image of a white cross fell upon the instrument. Then, let us say, studs lifted by little magnets would rise up on the board to form a sort of image of the cross. How is our automaton to be made to "know" that this is a cross, and to do whatever we want it to do when it receives the image of a cross? Obviously, if all we ask is that it should respond to a particular size of cross set the same way every time, the matter is quite simple. The same set of studs would rise every time, and we have only to arrange that that particular set shall complete an electric circuit. But if we demand a response to any cross, no matter what size, but to no other shape; or worse still, to a cross of any size in any position, the problem will need some hard thinking. We will return to it later, and see whether we can imagine some solution.

We may safely say, however, that we could devise a machine which would typewrite, or set type, from a typewritten or printed page presented to it. It would be highly complicated and expensive, but it could be done. It would be in the nature of an adaptation of the blind reader.

The Automaton which Reads Manuscript.—Suppose, however, the ingenious constructor were now asked to adapt his machine to read manuscript, or even type of different size and "face". The difficulty would be stupendous; I should say insuperable. Our automaton has to "recognize", not a given thing of fixed size and shape, but the class to which a vast variety of similar sizes and shapes belongs.

Automatic Traffic Control.—Another example is the problem of automatic traffic control. A good light-sensitive relay would, as I have already remarked, readily enable a signal to be sent that *something* was approaching a cross road. It would be next door to impossible, however, to signal "man," "car," "horse and cart," let alone "policeman," "Ford," "Rolls-Royce." It is not, impossible, however, that we could imitate the sensitiveness of men and animals to movement in the field of vision. We might even be able to signal "object moving rapidly", and like matter, but it would be very difficult.

It appears to be worth considering whether this problem of the automatic signalling of traffic might not be approached from quite a different side. All the vehicles the advent of which requires to be signalled are made of magnetic material, and it would be a perfectly easy matter to arrange that the passage of such a vehicle at a point some distance from the cross roads should cause a signal to be exhibited there, which would be washed out again at the moment when the vehicle reached the cross road. The difficulty would be to avoid the complete disorganization of the system by vehicles not behaving in a regular fashion.

The Ear of the Automaton.—We pass naturally from sight to sound. A brilliant achievement in this field was the development of sound ranging during the war.[7] As with the eye, so with the ear; our device will respond to a certain sound by sending or varying an electric current. Given such a response, and we can do almost anything with it by modern methods. Recent inventors have been busy with devices which respond selectively to certain notes of the musical scale. Others make use of sound waves in water and other media. As sound is propagated at quite a slow rate it is possible to measure distances automatically by sending out sound waves and noting the time they take to reach their destination, or to be reflected from it and return to the sender. Very successful devices of this kind enable ships to take soundings to any depth.

It is extremely likely that sound-perceiving instruments will, in future, play a part in the construction of automata. They may well serve as means for detecting the nature of media, whether solid, liquid, or gaseous. For instance, the pitch of the note of a whistle depends upon the nature of the gas used to blow it. This fact was used by Haber[8] to construct an extremely simple means for indicating the presence of dangerous gases in the air of mines. He used two whistles, which were tuned so as to be of exactly the same pitch when blown by the same gas. They were taken down the mine, and one was blown by the air of the mine, while the other was blown by pure air from a cylinder of compressed air. A very small amount of fire-damp[9] sufficed to cause a slight change in the note of the former whistle, which then "beat" against the one blown with pure air. Our ear is very sensitive to beats, that is, those most unpleasant oscillations produced when two notes very close in pitch are sounded together. A sound-relay sensitive to beats could, of course, readily be constructed. The field is worth the attention of inventors, and should afford them a great deal of amusement. Not profit, of course. Pioneer invention is

7 Method of locating enemy artillery using the sounds of its firing.

8 Fritz Haber (1868–1934) was awarded the Nobel Prize in Chemistry for 1918.

9 Methane gas, common in coal mines and hazardous because it forms an explosive mixture with air.

the least financially profitable pastime that exists; or rather, the most disastrously unprofitable.

The Automaton that Types to Dictation.—Every now and then we hear from the inventor who plans a typewriter that will write to dictation. This may appear to the non-technical no more fabulous than a forecast of the achievements of wireless would have appeared twenty years ago. But the cases are not on all fours. I venture to say that if any technical problem can be flatly termed insoluble, this is one. We are brought back at the start to the same type of difficulty as that discussed in connection with the electric eye, namely, of the "recognition" by our automaton of similarity, and not identity. The voice and pronunciation of every person differs very greatly. Were we to succeed, however, in making an electric ear which would respond in the same way every time to the same sound, it would still respond only to syllables, and not to letters. Its only conceivable response would be a sort of pattern corresponding to each syllable, which would then have to be sorted automatically amongst not 26, but thousands of possibilities. Finally, of course, the matter of spelling would present entirely insuperable difficulties, since the human typist can only decide by the sense whether to write "wood" or "would," and often decides wrongly.

This matter, however, takes on a somewhat different aspect when considered in connection with the endeavours, so persistent and yet hitherto so unsuccessful, towards the formation of a universal language. The latest suggestion of the kind, however, appears to me to have far more prospect of success, and, curiously enough, to reopen the question of the typist automaton. I mean the plan for the simplification of English proposed by Mr C. K. Ogden. He hopes that when his present researches are completed it will be possible, by a new and fundamental analysis of the mechanism of language, to produce a modified English with a basic vocabulary of about 500 words in which almost everything can be expressed.[10] It is certainly within the bounds of possibility that an automaton might be devised to take dictation in this language from a practised speaker, provided only that no ambiguities of sound occur in the final vocabulary.

The recording of sound upon the photographic film, now in commercial use in the talking film,[11] should have a future in connection with office work. It would enable a record, secret if desired, to be made of important conferences or interviews. It would be a much more capacious and handy apparatus than the

10 [Author's note] A specimen translation of pages 44–50 of the present work appears as an Appendix. The vocabulary is constructed from an ingenious system of panoptic conjugates. [Editor's note] Ogden's 'Editorial' in *Psyche*, Vol. 8, No. 3 (January 1928), 1–2, had announced the project in the year *Automaton* was published. The 'Editorial' for *Psyche*, Vol. 9, No. 3 (January 1929), 1–9, gave the first publication of what had become the 850-word list, then under the name of 'The Universal Language', of what he would name BASIC – for 'British, American, Scientific, International, Commercial' (*Psyche*, Vol. 9, No. 5 [10, 1] (July 1929), 1–30). BASIC is a regularised form of English designed for international use as an auxiliary language. It became Ogden's main project thenceforth.

11 See *Wireless Possibilities*, 160, n. 10.

dictaphone.[12] It could, of course, be used to avoid typing altogether by sending records of spoken messages by post; an early dream of the future, which was common when the phonograph was first invented. It should also find a wide application in the construction of automatic announcing devices, such as would be very useful to railways.

The Recognition Problem Again.—It is not at all impossible that light might be thrown upon some of the fundamental problems in psychology by close study of recognition. We can only go so far here as to discuss visual recognition a little more in detail, and leave the reader to consider it further for himself. A lens throws an image of a shape upon a screen made up of light-sensitive spots. Can we arrange matters so that when and only when, a certain shape appears on the screen a certain response is given? The answer is certainly, "Yes," if the shape is the same size and occupies the same place every time. Can we further provide the apparatus with a repertoire of shapes, so that it gives a separate signal for each shape in its repertory? The answer is "Yes," if the same shape always occupies the same part of the screen. Suppose it does not? It is quite conceivable that the apparatus might then be made to centre itself automatically upon, say, the centre of figure of the shape, whatever it was. It would get a general disturbance, "figure in the field," and then proceed to move until it was "looking" at the centre of the figure. The image of the latter might then, however, be set at the wrong angle for recognition, so that the next step of the automaton would be to turn the screen right round the compass. If recognition still failed, it might be because the figure was too large or too small. It would then be necessary to arrange that the automaton would change the size of the image on the screen (by optical means) in minute gradations, and test out each gradation all round the compass. It would no doubt require a lifetime of work and a fortune in money to construct a workable apparatus of this kind, but it would not be impossible.

When, however, we tackle the question of the recognition by the automaton of a number of similar but not identical forms, the only solution we can think of is one which simply enlarges the repertory of the automaton by a number of *definite* shapes. It is, however, pretty certain that this is not Nature's way. The mental process is no doubt a sort of working to limit gauge.[13] The same shape is recognized as the same until distortion has proceeded in various directions up to a certain point. But in the mental process, the shape appears in some way to be taken out of the three dimensions of space. The mirror image of a completely unsymmetrical shape is at once recognized as the same, or nearly the same, as the original.

Colour.—Colour is an entirely different matter. The recognition of colour by a sensitive relay would be quite a simple matter, for light can be filtered by screens so as to pick out any desired part of the spectrum, or, if that be not fine enough,

12 See *Wireless Possibilities*, 62, n. 8.

13 A limit gauge is a precision measuring tool used to check if manufactured parts fall within the desired size range.

a prism may be used. The applications to industry will be numerous. Wherever the human hand is guided by colour, the automaton could take its place. Thus in hand-picking and sorting by colour of natural objects, such as minerals, fruits, eggs, and so on, automata can be safely predicted. Mention has already been made in the press of an automatic device for picking out cigars of bad colour. Chemical processes are often controlled by observations of colour, turbidity, or other optical properties of the substances, and all these controls could well be made automatic.

Colour as a Label for Automata.—We can readily imagine that colour might be chosen as a convenient label to attach to objects of different sorts which are to be dealt with automatically. Cards could readily be sorted in accordance with their colours, and this might find application in automatic systems of accounting and book-keeping, instead of, as at present, the use of a code of holes punched in the cards. A standard-sized envelope with a certain repertory of colour schemes might be used to enable the automatic sorting of letters. One imagines that the labour-saving despatch of letters in the future will involve automatic correct franking according to weight, by the insertion of a coin, or, in the case of business houses, by the use of a franking counter of the kind recently introduced. The letters would then be automatically sorted by colour, it being, of course, a matter for the sender to use an envelope of the correct colour, appropriate to the destination. The old-fashioned style of things could be kept up alongside the new, with appropriate excess of cost to those who preferred to send their letters in envelopes of their own choice.

It is obvious that the light-relay could equally well be made to respond to patterns printed in black on the white envelope, each destination having a particular key pattern.

Even at present, it is surprising that the post-offices of the world do not combine to bring pressure to bear in the direction of uniformity, of stationery, address, and so on. Hand-sorting would be greatly facilitated. [21–50]

[. . .]

Counting.—We may now consider another case in which the call upon the human intellect is of the simplest description, and hence the field for automata is a promising one, namely, the operation of counting. The application of automatic methods is here making rapid progress, but there are many directions in which applications of mechanical counting could still be made. For instance, the annual disturbance of stock-taking might be largely avoided. Very simple and cheap counters could be made to record, on a central board, the taking in and taking out of stock as it proceeds. The counter on the board would indicate the stock held at the moment. Instead of this, work is often at a standstill while time is spent counting thousands of articles of trifling value, for the purpose of the insertion of that value into a balance sheet. As, however, the other items on the balance sheet are estimated, very often, with a margin of accuracy of hundreds or thousands of pounds, the estimation of the stock to the nearest penny is an unscientific procedure.

Only quite recently was the proposal to introduce electrical indicators of the number and position of seats vacant in a theatre or cinema carried into practice in America. The system could readily be extended. One might be enabled to select a vacant seat and pay for it by the insertion of a coin in a machine, receiving a check which would serve as a key to enable that seat to be tipped down for use.

Accounting.—There is no need to remind the reader of all that has been said concerning the life at the ledger. Until the development of mass production, it stood uniquely for all that is most monotonous, mechanical, and dreary in human bread-winning.

Machinery is already in use by which a very great deal of the purely mechanical work can be eliminated. It depends fundamentally upon the same principle as the well-known Jacquard loom.[14] Whatever is to be recorded is translated into a code of holes punched in a card. This card is then inserted into a machine, which is supplied with fingers or electric contacts which come up and look, as it were, for the holes. According to the combination of holes found, certain records are made, or the card may be passed on to another machine to be dealt with. We may imagine an order coming in for so many size so and so things at so much to be delivered in such and such a time. All this may be translated into the hole code by a single clerk. The subsequent operations of booking the order, notifying the correct department, checking the delivery time, totalling by classes all orders received, deducting goods when taken from stock, and invoicing them, could be performed mechanically. Moreover, the hole code can always be made to translate itself back to legible type, though the reverse operation would fall under one of those very difficult cases which we dealt with earlier.

The still simpler operations of banking afford a large field for such developments. The German Postscheckamt,[15] an institution which enables small sums to be paid by the simple process of dropping a card in a post-box, has installed a system of this kind. Perhaps, after a time-lag similar to that which the C.O.D.[16] system required, we shall be blessed with this convenience elsewhere. Bankers, however, seem to possess a passion for dreary architecture of a very expensive kind, rather than labour-saving systems.

The Automatic Steering of Ships and Airplanes.—We have already mentioned the matter of automatic steering of ships, and we may now consider it in somewhat greater detail, as it is one of considerable interest to the general public.

14 The Jacquard loom system, controlled by punched cards, was adopted by Charles Babbage and Ada Lovelace in their plans for the Analytical Engine, never fully realized, but now seen as a forerunner of the computer. See Lovelace's landmark Notes to the 'Sketch of The Analytical Engine Invented by Charles Babbage', by Luigi Menabrea, *Scientific Memoirs*, ed. Richard Taylor, vol. 3 (London: Richard and John Taylor, 1843), 666–731.

15 Post Office Girobank.

16 Cash (or collect) on delivery. Customers order without needing to pay in advance.

Such mechanism is always of the corrector type of automaton. We have referred to the use of the compass for this purpose. For some not very obvious reason, it is only quite lately that this possibility has been realized in practice, the type of compass being the gyrostatic, a very recent invention. The ordinary magnetic compass could, however, be adapted to the same purpose.

The automatic steering of an airplane is, however, a somewhat different matter. Here any kind of compass behaves very badly, and in any case, automatic steering by the compass would be of little use. In practical navigation, the airman uses what is called a turn-indicator, which consists of a simple spinning gyrostat hung in bearings. This is not to be confused with the gyrostatic compass, which tells the true direction of the earth's meridian. The turn-indicator, on the other hand, simply swings to one side or the other whenever the plane deviates from its course in one direction or the other. The airman is thus enabled to keep a dead straight course for time long enough to enable his troublesome compass to settle down and tell him what the course is. He can then correct it if necessary.

This simple apparatus may obviously be made to act upon the rudder in the way we have grown familiar with, so as to cause any deflection from the course to be automatically corrected.

This is the pilotless plane in its simplest form. In all the principal countries, the greatest efforts are being made to devise apparatus of much greater elaboration. The problem is, of course, to steer a plane correctly by automatic means to any desired objective, in order that it may deposit there as much high explosive or poison gas as it can carry. Merely setting the plane on a straight course will not do, for the variation in the direction and strength of the wind would inevitably destroy any accuracy of aim. Efforts are being made to control from the ground or the air such pilotless planes, but at present this involves two disadvantages that make the prospect of doubtful value. In the first place, it would be only possible to observe the pilotless plane in daylight, and it would then be equally visible to the enemy, and could easily be destroyed. Secondly, the use of wireless radiation as the means of control would be easily rendered futile by the enemy, unless it were so constructed as to be safe from jamming. However, work of this kind is so much more attractive to the inventor than work in conjunction with financiers out for quick profits, that it is sure to occupy some of the very best brains available. Methods of fantastic originality and difficulty can be tried out, regardless of expense. The moral aspect of the matter will play very little part. After all, the work of the inventor of automatons is in any case to abolish unnecessary human beings.

The Universal Automatic Pilot.—We have already said that the difficulty of automatic navigation of airplanes lies in the fact that the effect of changes in the strength and direction of the wind cannot be taken account of. The compass, or similar instruments, tells the navigator in what direction the head of his craft is pointed, but it does not tell him in what direction, relatively to the earth's surface, he is really moving. Similarly, a log or speed indicator on a ship or plane tells the navigator his speed relatively to the water or air around him, but not relatively to the earth. The physical principle of relativity asserts that it is impossible to construct an apparatus that will do this, unless it is linked in some way with the earth. There must be some physical connection of some kind. When the navigator can see the earth, he can easily ascertain his speed and true direction by observing the rate at which objects on the earth appear to move across the field of a telescope. But we have not yet devised means of seeing the earth which clouds will not interfere with.

There, is however, a possible solution which is so interesting that it may be discussed here. Relativity denies the possibility of the pilot's ascertaining at any instant the true direction and amount of his velocity, relatively to earth, by means independent of connection with the latter. But this does not mean that he cannot measure the amount of any *change* in that velocity, whether produced by changes in the power of his engine, or changes in direction or speed of the air currents in which he is immersed. On the contrary, as all travellers know, we may be unconscious of the speed at which we are travelling, so long as it is uniform, but any change in it is at once perceptible to our physical sensations. Not only so, but loose objects in the vehicle in which we are travelling are also affected. A piece of luggage falls off the rack, or a bottle upsets. In fact, a heavy object in the carriage behaves, when the vehicle accelerates or slows down, as if it were subjected to a push, opposite in direction to that in which acceleration is taking place. It is thus quite easy to measure in a closed carriage the acceleration; a simple pendulum suffices. Hang a heavy object by a string from the rack next time you travel by train. When the train is at rest or travelling at a uniform speed, the string will be vertical. When the train accelerates, the suspended object is left behind, as it were; the string points in a direction opposite to that in which the train is travelling. When the train slows down, the object carries on, and the string slopes in the same direction as that in which the train is travelling. The deflection of the string from its original vertical position is a measure of the acceleration or deceleration of the train.

The next point is not quite so easy to grasp. Take the case of the train starting from rest. The pendulum deflects by an amount which indicates the rate at which the train is increasing its speed. If you then observe for how long the pendulum is deflected until the speed becomes uniform and the deflection ceases, you are able at once to calculate the speed which the train has attained. The deflection of the pendulum indicates that the train is increasing its speed, in one second, by so many feet per second or miles per hour. Thus if we find that the train is increasing its speed by one mile per hour every second, then it is evident that if the pendulum indicates this rate of increase of speed for a period of a minute, at the end of that time the train will be moving at the rate of sixty miles per hour. It is possible to devise a mechanism to do this sum mechanically, that is, to sum up every deflection of the pendulum multiplied by the time during which it has persisted. The mechanism would, of course, take account of slowing down as well as speeding up, and present at any instant the sum total of positive and negative effect. This sum would be the true speed of the vehicle at the moment.

Furthermore, it would be the true speed relatively to the starting point. Let us imagine ourselves in an airship anchored by the nose to a mast. Let there be a strong wind blowing. We set the propellers in motion until the power they develop is just sufficient to overcome the wind, so that we can cast the nose of the airship adrift and she remains stationary relatively to the earth. If, however, those on board could not see the earth they would imagine their ship to be moving through the air at a certain speed which would be exactly equal, and opposite in direction to that of the wind. It is obvious, however, that a pendulum on board would not have been deflected, and hence its adding device would indicate zero, the true speed of the ship relatively to the earth. If the wind suddenly drops to nothing, the ship will bound forward, and we, and our instruments will certainly be conscious of that bound. The ship will now, provided it continues to work its propellers at the same speed, attain a speed relatively to the earth equal but opposite to the original speed of the wind, and this will be truly indicated by our instrument.

We can thus see that the problem is fundamentally soluble on these lines. The practical difficulties are, however, very great. For while the accelerations due to the ship's or plane's propelling force are considerable, the accelerations or retardations due to air or water currents are very small in amount, but may continue for very long periods. Thus the plane may get up to full speed in a minute or two, while a wind may spring up against it, and take an hour or two to attain a speed, say half as great as that of the plane. The speed of the plane relatively to the earth is now half what it was at the start, but the retardation due to the wind has never been more than a small fraction of the original acceleration of the plane. This means that our accelerometer must be of extreme sensitiveness over a very long range, and this is the most difficult condition that any instrument can be required to fill. Nearly every instrument has a limit of accuracy expressible as a fraction (say 1/1,000) of its total range. Thus, when it is indicating at the top of its range, it is doing so with an accuracy which may be expressed as one in a thousand. When, however, it is working say at a hundredth only of its full range, its accuracy is only one in ten.

The problem presents other and even greater difficulties, but its fascination and importance must not lead us too far in discussion.

Seeing the Earth by means other than Light.—Another hopeful line of development consists in the possibility of finding some means of physical connection with earth other than light rays. Other kinds of ray are hardly practicable, though infra-red rays have some dim chance of utilization. The difficulty is, that we must needs form an image in an optical system of objects on the earth. Obviously, when there are no objects, even seeing the earth is no use to us, as over the sea. The use of the earth's magnetic field has been suggested. A wire stretched across the plane at right-angles to its direction of motion is cutting the magnetic lines of the earth's field at a rate proportional to the true speed of the plane. Hence, according to the principles of electro-magnetism, a minute electrical voltage is developed between the ends of the wire, just as if it were a wire of the armature of a dynamo. We might hope to measure this if we could lead it back to our voltmeter. Unfortunately, the wires we should need to use would also cut the earth's lines, and hence neutralize the effect on the instrument. This is an interesting problem for those who know something of electricity and magnetism to puzzle over. If it could be solved, we should at once be able to tell both the true direction and the true speed of the plane, if we knew the strength and direction of the earth's field at the place of observation. We should thus have to know the position of the plane pretty accurately.

Another possible plan is as follows. If you move a magnet near a conductor, the magnet experiences a slight drag, due to the eddy currents induced by its field in the conductor. The earth being a conductor, a magnet on the plane experiences this drag. Unfortunately, a magnet is a heavy thing compared with the drag it experiences, and the earth is a bad, and worse still, a variable conductor. The drag on any conceivable magnet at any reasonable height would be excessively minute. But of recent years we have grown accustomed to measuring the incredibly minute.

The Railways.—Let us pass from considering the most recent means of transport to the question of one which is now regarded by many as doomed to gradual extinction; I mean the railway. One may truly say that the running of most of our railways is almost medieval from the standpoint of modern technical possibilities. Miniature railways are run in many towns for postal purposes entirely automatically. Very nearly the same thing could be done with a full-sized railway transporting passengers and goods, provided of course, it was electrically driven. The tube railways have already shown the way. There is no reason at all why one should not purchase a ticket to Aberdeen at King's Cross in exactly the same way that one purchases a ticket at Liverpool Street for Dover Street, from the man at the barrier. There is no reason why the Scotch express should not be a train in charge of one man, with automatic doors. There is no reason at all why it should be greeted at intermediate stations by hosts of officials. There is no sense in the present elaborate system of dealing with luggage. The taximan should dump it onto an automatic weighing machine which would discharge a printed ticket. A porter would then put it on the train in the same coach as the passenger, who would pay for it at the barrier as he took his ticket, on the base of the check from the weighing machine.

It is impossible to foresee whether the capital will be found to modernize our railways. It is possible that they will revert to a primitive simplicity. Fifty years hence, railway men may be as unmodern and picturesque a class as bargees to-day, sleeping with their wives and families on their trains, and carting heavy goods slowly about the country, by means of steam locomotives, in a fashion even more primitive than that of to-day. The alternative is complete electrification, with a service of trains at very frequent intervals but moderate speeds, each manned with but one official. The driving could be made purely automatic; the word is a misnomer in connection with a really modern railway such as the Tube. The official in charge would press the starting button; the journey of the train to its stop at the next station would then be purely automatic, and safe against any mishap short of a landslide onto the line. If it arrived at a block in which another train still lingered, it would be stopped automatically. All this is possible by means already well tried. I am inclined to think that it will come sooner than we expect. Train speeds are quite un-necessarily high; comfort and absence of overcrowding would compensate for their reduction, while rapid travel in case of urgent need is provided for by the airplane. The capital cost of modernizing would be high, but the main item would be electrification, which is already being carried out rapidly on the Continent. The automata part of the business would be a comparatively small item of cost, provided that previous experiment had enabled a standardized equipment to be produced by methods of mass-production. It is high time that the experimental development of these modern possibilities be put in hand in a small closed area such as the Isle of Wight, in charge of competent engineers.

At the time of writing, an omnibus is being put on the road, provided with sleeping berths, which will do the journey from London to Newcastle during the night at less than third-class fare. It is an evident folly to transport passengers over a rough, hilly winding road, in the dark, with their safety dependent upon the nerve and skill of the driver, when they could be transported with much less expenditure of power over a smooth steel road requiring no driving in any real sense, and watched over by automatic signalling in the surest possible fashion.

The Automaton in Power Generation.—One of the greatest advantages of the electrification of railways is the possibility of introducing the automaton principle in the generation of the power by which they are run. In few fields of modern technology is it more widely applied than in power generation. One can buy quite cheaply small electric generating plants which are complete automata; they are widely used in country places. In Denmark (but in no other country as far as I am aware) one meets almost everywhere small automatic plants driven by wind power. These all depend upon the principle of keeping a battery of accumulators charged. As soon as an instrument indicates a drop in the charge, the dynamo is automatically set in motion to remedy the defect.

In large power stations such a use of accumulators would be out of the question, but the automatic regulation of the voltage produced by the dynamos has reached a state of the highest perfection. The periodicity, in the case of alternating current, is also regulated with the greatest exactness, and a by-product of this accuracy will in the future be of great importance in connection with automata.

A New Kind of Clock.—Modern current supply is almost exclusively alternating that is, the current changes its direction many times (usually 50) per second. Now alternating current may be used to drive motors of many different types, one of which is the so-called synchronous motor. This type of motor works by exactly keeping time in its rotation with the alternations of the current; the number of revolutions it makes thus bears some simple proportion to the number of alternations of the current. It has either to work at this speed and in this way, or not at all. Hence, if the rate of the current-alternations is timed very exactly, the motor which it drives also revolves with the same constant accuracy of speed. Such a motor may then be used to turn a clock train in place of spring and pendulum or balance. So accurate is the alternation of modern current, that a clock run in this way keeps good time. Furthermore, it is only necessary to have a clock of this kind at the central station, and observe its timing, in order to make any slight correction in the rate of alternation of the current. All other clocks connected to the network will then be corrected at the same time. The immense advantage of such a clock, apart from its simplicity, is the unlimited power behind it. Its rate does not depend upon the work it is called upon to do. So long as this is not great enough to stop the motor altogether, the clock runs true to time.

Inventors and Clocks.—From the earliest times inventors have had visions of doing marvellous things automatically by the use of clocks. They have always been hampered by the fact that you cannot get a clock to go accurately if you try and make it do work. Hence, it is necessary to use clocks so large and powerful that the work they are called upon to do is negligible in comparison with the power of the clock. All this trouble can now be banished. It is possible to make a simple and robust mechanism which will do anything in reason at any time you like to set it for.

The Time-Switch.—Before long, it should be possible to buy at a low price a clock which will switch on or off such currents as are used in everyday work at any predetermined time. Such "Time-switches" have, of course, been on the market for very many years, and have been widely used for controlling the lighting of streets, and for other purposes. On the whole, however, they have been found to be more bother than they are worth, for to be good they had to be fairly expensive. The synchronous-motor switch could be made, in large numbers of course, for a few shillings.

Automata in the Home.—We are thus led to the question of the place automata will occupy in the home of the future, assuming that such a thing as a home in the old-fashioned sense will exist at all. A few years ago, it was a favourite pastime of rich Americans to construct automatic homes. The door opens automatically when a member of the family, but not a stranger, approaches. The door-mat automatically brushes your boots. The thermostatic oven has been switched on at the correct time in anticipation of your arrival, and a touch of a switch causes it to discharge its contents onto a table, which then travels to the dining-room. When the dinner is eaten, another switch dismisses the remains of the feast to the kitchen. The romance of science used to go on like this for pages, but I suspect that the staff of skilled mechanics and electricians necessary to keep the whole business going proved more expensive than normal service, even in America.

Nevertheless, the time-switch should find many uses in the home of the future, especially in conjunction with electric cooking and heating. It has already been found capable of heating water and making tea, and waking the sleeper when these are ready for him. Whether it would be necessary to have it go to the length of turning the bed upside down, would depend upon the sleeper.

Answering the Telephone Automatically.—Another great need for the modern home is an automatic apparatus to take telephone messages. It would be somewhat costly, but would certainly find a large number of users, if satisfactory. It is one of those inventions which has been announced again and again, but it is to be supposed that a workable apparatus has not yet been produced, or at any rate, not put on the market. So many new developments which would seem to facilitate it have been made recently, that one may expect that a new effort will be made to achieve success. It is to be hoped that the introduction of automatic telephones will not render the application of a recorder any more difficult. One can imagine a device which would automatically raise the receiver from the hook when the bell had rung in vain a certain number of times, and apply it to the sound-recording device. The latter might well be photographic or magnetic, rather than the usual wax cylinder. A certain signal would be given to the caller, and he would then have a certain time in which to give his message; a most salutary arrangement. At the end of this time the receiver would sink back onto the hook. The sound recorder might be serviceable as a dictaphone for office use as well.

The Control of Domestic Heating.—Thermostatic control of domestic heating is long overdue. Here again the time switch could be used to cut the heating in and out at predetermined times. This would be easiest, of course, with electric heating, but would not present any great difficulty in connection with steam or gas, if an electric supply was available to work the necessary valves.

The satisfaction of going to bed, knowing that one would be called at a precise moment to a hot bath, tea, and a warm room, would be very great. Perhaps we might feel a little doubtful about the return in the evening, for it would be annoying, if unexpectedly detained, to know that all the automatic preparations for one's return were taking place with the inevitability of Fate. [58–79]

[. . . .]

Discovery and Invention.—The distinction between discovery and invention is well-known in Patent Law. It is less well-known to the public. The two are often confused. We discover the laws of nature, the facts about the world we live in. It is the task of the inventor to apply the knowledge thus gained to the solution of practical problems. If matter is the enemy, pure science discovers its habits, and invention enslaves it.

For a long time invention was well ahead of discovery. The steam-engine was invented long before the laws of thermodynamics were discovered. At the present day, discovery is far ahead of invention. The most modern and subtle of inventions, such as the triode valve or the photoelectric cell,[17] only make use of a minute part of recent discoveries in chemistry and physics. Pure research attracts to its service an increasing proportion of the most gifted of the scientifically educated. Great endowments have made a life devoted to pure research possible and even moderately affluent. Those possessing a practical inventive gift find a full outlet for it in the problems presented by pure experimental research. Dewar[18] invents the vacuum flask, Einthoven[19] the string galvanometer, merely by the way, in the course of

17 See 73 n. 16 on triode valves. A solar cell or photovoltaic cell is a device that converts solar energy into electricity by the photovoltaic effect.

18 Sir James Dewar (1842–1923), chemist and physicist.

19 Willem Einthoven (1860–1927) was a Dutch doctor and physiologist. His string galvanometer the first practical electrocardiograph (ECG) in 1895. In 1924 received the Nobel Prize in Physiology or Medicine.

the experimental pursuit of pure knowledge. The incentive of financial gain by monopoly is lacking, but to compensate for that, the pure research worker has the inducement of academic and public honour, a large and comparatively free life as his own master, and the satisfaction of feeling that his work is of permanent value, whereas that of the inventor may be entirely superseded and forgotten.

The Mastery of Matter.—Shall we realize the ideal of our civilization? When we know all we can ever find out about the world around us, living and dead right down to the last detail of its atomic structure, will invention serve us to turn all this knowledge into mastery? Will great automata feed, clothe, house, warm, light, and amuse a population from which all but the brains that can understand, and the skilled hands that can tend, have been eliminated? Will the dreams of past centuries be realized?

The first need is enthusiasm, incentive, and material support for invention. Enthusiasm for science in general will no doubt continue, at least for several generations, but for the reasons I have indicated we may well tend to a condition of affairs when scientists will be a kind of caste apart, pursuing knowledge for its own sake, in directions determined solely by intrinsic interest. Pure mathematics has, indeed, almost reached that stage to-day. Because all Universities are now teaching engineering, it is popularly supposed that science and practice are coming nearer together, whereas in truth all that is happening is that engineers are being trained in these institutions without a thorough scientific foundation, in the formulae of existing practice, and then sent out in the world without the equipment for applying to their practical work the discoveries of pure science.

The Humanitarian's Opportunity.—It is a curious fact that the humanitarian has hitherto done little or nothing to do away with the dreary, monotonous, and disgusting labour inseparable from industry as carried on to-day. Instead, as we saw at the outset, his efforts to alleviate the lot of the masses actually compete against the efforts of science and invention to abolish manual labour. The supersession of hand-labour is decided almost entirely on financial grounds by the management of industry, and the certainty that cheap and contented labour will always be available, calling for no capital investment, tells heavily against the automaton. If the humanitarian really feels that the type of population resulting from his best efforts is a worthy and desirable product, there is nothing one can oppose to him that has not been said by the advocates of Mediævalism, craftsmanship, and the peasant life. But if there are humanitarians who have their doubts, one might suggest that they should seriously consider the employment of their resources on behalf of, and not against, the automaton.

Hitherto, comparatively trifling prizes have been offered, from humanitarian motives, for inventions designed to abolish particularly poisonous or dangerous occupations conducted by hand-labour. If, however, the whole weight of humanitarian influence were directed towards the substitution of automata for all forms of monotonous hand-labour, great progress might be made. The lot of the workers in the field should be made as attractive as that of the pure research worker. The idea, inevitable at a time when the problem of employment is so serious, that the first function of industry is to find employment for the masses, must be abolished in favour of the endeavour to make it supply our needs with the minimum of labour.

The supreme desire of our civilization must be realized, if its old age is to be honourable. The masses of human automata of which it makes use to-day must not be found among our descendants. Not the manual labour of teeming populations, but the power of great automata, must supply its physical needs; and, over and above these, resources for yet undreamed-of efforts to conquer matter, space, and time. [88–92]

From: *Eos; or, the Wider Aspects of Cosmogony*

J. H. Jeans (1929)

Headnote

Sir James Hopwood Jeans OM, FRS (1877–1946) was an English physicist, astronomer, and mathematician. He was an important contributor to quantum theory, the theory of radiation, and stellar evolution. He and the other great science popularizer of the period, Arthur Eddington, were the leading figures in British cosmology.

Jeans was a Fellow of Trinity College, Cambridge from 1901; professor of applied mathematics at Princeton from 1904; then returned to Cambridge in 1910. He was Secretary of the Royal Society. He was well known for books such as *The Universe Around Us* (1929) and *The Mysterious Universe* (1930).

******

Foreword

This little book contains the Trueman Wood Lecture delivered before the Royal Society of Arts on 7th March, 1928, expanded to more than double its original length by the addition of what I would have liked to say, but had not time to, at the lecture, and by the inclusion of a small part of an earlier lecture "Recent Developments of Cosmical Physics" delivered before the University of London on 9th November, 1926.

My thanks are offered to the Royal Society of Arts, and also to the Editor of "Nature" in whose columns the reports of these lectures first appeared, for permission to use them in preparing the present book. While the "To-day and To-morrow" series has been the medium of publication of many brilliant flights of imagination, the present book is limited to scientific facts and such inferences as I think may properly be drawn from them. Happily astronomy is a science in which exact truth is ever stranger than fiction, in which the imagination ever labours panting and breathless behind the reality, and about which one could hardly be prosaic if one tried.

J. H. Jeans. Dorking, 12th May, 1928.

Interest in scientific cosmogony is a recent, and still a very tender growth. Anthropologists and geologists tell us that man has existed on earth for something like 300,000 years; we must go this far back to meet our ape-like ancestry. Between

DOI: 10.4324/9781032625522-12

them and us some 10,000 generations of men have walked the earth, most of whom have probably given some thought, in varying degrees, to the significance of their existence and the plan of the universe.

Of these 10,000 generations of men, the first 9990 unhesitatingly regarded the earth as the centre, and terrestrial life as the central fact, of the universe. As was suited to its majesty and dignity as the abode of man, the earth stood still while the celestial sphere spun round it, covering in the earth much as a telescope-dome covers in the telescope; and this dome was spangled with stars, which had been thoughtfully added so as not to leave the central earth unillumined at night. Ten generations at most have been able to view the problem of their existence in anything like its proper astronomical perspective.

The Position of Man in the Universe

The total age of the earth far exceeds the 300,000 years or so of man's existence. The evidence of geology, and of radio-activity in rocks in particular, shows that it must be something like 2000 million years, which is several thousand times the age of the human race. Old Mother Earth must regard man as a very recent apparition indeed; he has just appeared to burrow into her, burn her forests, put her waterfalls into pipes, and generally mar the beauty of her features. If he has done so much in the first few moments of his existence, she may well wonder what is in store for her in the long future ages in which he is destined to labour on her surface. For in all probability the life in front of the human race must enormously exceed the short life behind it. A million million years hence, so far as we can foresee, the sun will probably still be much as now, and the earth will be revolving round it much as now. The year will be a little longer, and the climate quite a lot colder, while the rich accumulated stores of coal, oil, and forest will have long been burnt up; but there is no reason why our descendants should not still people the earth. Perhaps it may be unable to support so large a population as now, and perhaps fewer will desire to live on it. On the other hand, mankind, being three million times as old as now, may—if the conjecture does not distress our pessimists too much—be three million times as wise.

Looked at on the astronomical time-scale, humanity is at the very beginning of its existence—a new-born babe, with all the unexplored potentialities of baby-hood; and until the last few moments its interest has been centred, absolutely and exclusively, on its cradle and feeding-bottle. It has just become conscious of the vast world existing outside itself and its cradle; it is learning to focus its eyes on distant objects, and its awakening brain is beginning to wonder, in a vague, dreamy way, what they are and what purpose they serve. Its interest in this external world is not much developed yet, so that the main part of its faculties is still engrossed with the cradle and feeding-bottle, but a little corner of its brain is beginning to wonder.

Taking a very gloomy view of the future of the human race, let us suppose that it can only expect to survive for two thousand million years longer, a period about equal to the past age of the earth. Then, regarded as a being destined to live for three-score years and ten, humanity, although it has been born in a house seventy years old, is itself only three days old. But only in the last few minutes has it become conscious that the whole world does not centre round its cradle and its trappings, and only in the last few ticks of the clock has any adequate conception of the size of the external world dawned upon it. For our clock does not tick seconds, but years; its minutes are the lives of men. A minute and a half ago the distance of a star was first measured and provided a measuring-rod for the universe. Less than a quarter of a minute has elapsed since Professor Hertzsprung of Leiden and Dr. Shapley, now Director of Harvard Observatory, showed how the peculiar stars known as Cepheid variables provide a longer measuring-rod, and taught us to think in distances so great that light takes hundreds of thousands of years to traverse them. With the very last tick of the clock, Dr. Hubble, of Mount Wilson Observatory, using the same measuring-rod, has found that the most remote objects visible in the biggest telescope on earth are so distant that light, travelling 186,000 miles a second, takes about 140 million years to come from them to us.

Not only is our vision of the universe continually expanding, but also it is expanding at an ever-increasing rate. Is this expansion destined to go on for ever? So far as we can at present see, no; for a general guiding principle, that of generalised relativity, fixes a limit, which we are fast approaching. According to this theory, space cannot extend for ever; it has no limit, but is nevertheless finite like the surface of the earth. Without exploring and surveying the whole of the earth's surface, we can make a fair estimate of its total area by measuring its radius, which we can do by measuring its curvature at any one point. In the same way the total volume of space is fixed by a quantity, the curvature of space, which can be determined by measuring the density of distribution of matter in space. Space which contained no matter would go on for ever, but the parts of space we can survey with our telescopes contain enough matter to show that we already see an appreciable fraction of the whole of space. It is as though our baby, watching ships coming from over the horizon, concluded that the earth's surface was curved, and formed a general rough conception of its size by imagining the observed curvature continuing until the earth's surface rounded back on itself. Exact figures are impossible, but Dr. Hubble has calculated that space is not likely to extend to more than about a thousand times as far as the farthest nebula visible in the biggest telescope. Nothing prevents our going on and on in space beyond this distance, but, if we do, we merely come back to ourselves. The possessor of a sufficiently sensitive wireless apparatus may emit signals and pick them up a seventh of a second later after they have travelled round the world. In the same way a not inconceivable increase in the size of our telescopes would take us round the whole of space, and we should see the stars surrounding our sun by light which had travelled round the universe, not of course as they now are, but as they were 100,000 million years ago.

Such considerations make it improbable that the expansion of the universe can continue at its present rate for much longer. Having grasped that the world is round, the infant speedily forms a fair idea of its size. Our particular infant, mankind, has made the great discovery of the existence of the outer world, has formed some conception of its size, and adjusted his ideas, not by a process of slow revelation, but by a brain-flash of the last few seconds. In his mature years and his staid old age he is no doubt destined to make many sensational discoveries, but he can never again live through the immortal moment at which he first grasped the immensity of the outer world. We only live through a few ticks of his clock, and fate might have ordained that they should be anywhere in the three days that the child has already lived, or in the seventy long, and possibly tedious, years yet to come. The wonderful thing is that she has selected for us what is, perhaps, in some ways the most sensational moment of all in the life of our race.

The child sets its newly awakened mind to work to adjust and co-ordinate a new array of facts. If the world was not made to surround its cradle, what purpose can it serve? If the lights of the great ships in the harbour were not designed to light its nursery at night, what can they possibly be for? And, most interesting problem of all, if the world is such a big affair, can there be other cradles and other babies?

These remarks will have served their purpose if they suggest that the cosmogony of 1928 should not be judged as a finished science or the solution of a problem; it is rather the first confused gropings of the infant mind trying to understand the world outside its cradle. And if the impression produced by its first inexperienced glance at the outer world had to be described in a single word, it would probably select the word "immensity".

The Immensity of Space

The immensity of space is measured by the figures already mentioned. Light and wireless signals travel at the same rate because, of course, they are essentially the same thing; and this thing takes a seventh of a second to travel round the world, and probably something like 100,000 million years to travel round the universe. The ratio of these times (2×10^{19})[1] measures the dimensions of the universe in terms of the familiar dimensions of the world; incidentally, it also measures the expansion of our spatial ideas since Copernicus. The disparity of size is too great to be easily visualised. Suppose: the size of our earth represented by a single atom, whose diameter is about a hundred-millionth part of an inch. Then the range of vision of the biggest telescope is about represented by the whole earth, and the size of the whole universe, according to the theory of relativity, is represented by a stack of a thousand million earths.

Scarcely less bewildering than the immense extent of space is the immense amount and variety of matter it contains. The sun, which is a million times as big as the earth and 300,000 times as massive, proves to be something less than a grain of sand on the seashore. It forms one of a family whose number must certainly be counted in thousands of millions; Dr. Seares has estimated it at thirty thousand millions. This is not the only family of stars in space. Each of the great spiral and

1 [Author's note] Here, as elsewhere, 2×10^{19} is an abbreviation for 2 followed by 19 zeros, or 20,000,000,000,000,000,000. A million is 10^6, a million million is 10^{12}, and so on. The largest number which occurs in the present book is $10^{420,000,000,000}$. Failing our convenient mathematical shorthand, this number would be expressed by a 1 followed by six million volumes similar to the present, all full of 0's.

other extragalactic nebulae [. . .] is either a family of stars, or consists of stars in the making, or of matter which is in all probability destined ultimately to form stars.

We can estimate the masses of these great nebulae by gravitational means, and each is found to contain enough matter to make a thousand million suns. This of itself will give some conception of the vast size of these nebulae, but to tell the whole story, it must be added that their colossal masses are so tenuous that each millionth part of an ounce is, on the average, as big as the Matterhorn. Think of a body which is bigger than the Matterhorn by as much as a thousand million suns is heavier than a millionth part of an ounce, and we have the size of any one of these great nebula. Any one of the photographs reproduced in this book would have to be enlarged so as to cover at least the whole of Asia before a body of the size of the earth became visible in it at all, even under the most powerful of microscopes.

Dr. Hubble estimates that about two million such nebulae are visible in the great 100-inch telescope at Mount Wilson, and that the whole universe is about a thousand million times as big as the part of space which is visible in this telescope. Let us now multiply 1000 million by 2 million, and the product by 1000 million. The answer (2×10^{24}) gives some indication of the probable number of stars in the universe; the same number of grains of sand spread over England would make a layer hundreds of yards in depth. Let us reflect that our earth is one millionth part of one such grain of sand, and our mundane affairs, our troubles and our achievements, begin to appear in their correct proportion to the universe as a whole.

The Variety of the Stars

While the stars may be compared to grains of sand in number, they differ too much *inter se* for the comparison to be carried further. They used to be regarded as mere points of light which differed from one another in glory but were too remote for any investigation as to their nature and condition to be practicable. Astronomy can now take their temperatures, by methods similar to those used with factory furnaces, can measure their sizes with a specially-designed stellar interferometer, and can calculate their weights from the gravitational pull they exert on companion stars, just as we calculate the weight of the earth from the pull it exerts on the moon to keep it in its orbit. And the measurements disclose an immense variety of big and little stars, of bright and faint stars, and of hot, hotter and still hotter stars.

The faintest star known, Wolf 359, emits only a fifty-thousandth part of the light of the sun. If this star were suddenly substituted for the sun, the seas would immediately freeze into solid ice, and the atmosphere condense into liquid air; people at the Equator would receive rather less light and heat from their new sun than one receives from a coal fire a mile away, and we should meet an icy death. At the other extreme, the most luminous of known stars, S Doradus, emits 300,000 times as much light and heat as the sun. If this star were to replace our sun, our temperatures would run up to about 7,000 degrees Centigrade, which is about twice the temperature of the hottest part of the electric arc, and the solid earth, with its cloud capped towers and gorgeous palaces and ourselves, would immediately dissolve into vapour.

There is almost as much variety in the sizes of the stars. The smallest known star, Van Maanen's star, is about the size of the earth; a million such stars could be packed inside the sun and leave room to spare. The largest known star, Betelgeux, is so large that 25 million suns could be packed inside it. If we represent the sun by a golf-ball, Van Maanen's star becomes something smaller than the dot of an 'i' on this page, while Betelgeux must be represented by a good-sized house. If Van Maanen's star were to replace our sun, it would appear smaller than Jupiter or Saturn; if Betelgeux were to replace our sun, we should find ourselves inside it, its radius being greater than that of the earth's orbit.

As compared with their enormous ranges of brightness and size, ranges greater than those between a searchlight and a glow-worm, or between balloons and bird-shot, the stars shew a restful uniformity in their weights. No star exists whose weight is known or suspected to be less than about a tenth of that of the sun, while few exist which have as much as ten times the sun's weight—probably about one in a hundred thousand, the other 99,999 all having weights between a tenth and ten times the weight of the sun.

It follows that the different sizes of the stars do not result to any great extent from the different amounts of matter the stars contain; they result rather from the varying closeness or looseness with which the matter is packed. An average ton of matter in the sun occupies about as much space as an ordinary ton of coal in a coal-cellar, but the same amount of matter in Betelgeux takes up as much room as the Albert Hall, while the average ton of matter in Van Maanen's star takes up only as much room as a pea in a pea-pod; a hundred tons of it could be packed into a pocket book with ease. Judged by the standards of solidity which prevail on Van Maanen's star, everything on earth is of the very filmiest of gossamer. [7–25]

[. . . .]

The need for caution in our interpretation of the universe is borne in upon us with overwhelming force when we reflect that the new world in which astronomy moves to-day is all a discovery of the present century. It is not merely that our present concepts of the extent of the universe in space and of its duration in time are new revelations to us; our understanding of its fundamental mechanism is equally new. The conversion of matter into radiation, which appears to be the primary physical process of the universe, did not come within our terrestrial purview at all until 1904. The primary matter of the universe appears to consist of elements whose existence we are only just beginning to suspect, and to exist in the state of almost completely broken-up atoms, a state of matter which, again, was not contemplated before 1917. The primary radiation of the universe is not visible light, but short-wave radiation of a hardness which would have seemed incredible at the beginning of the present century. Indeed, our whole knowledge of the really fundamental physical conditions of the universe in which we live is a growth of the last quarter of a century.

The simple explanation of this situation is to be found in the fact that life, naturally enough, begins its exploration of Nature by studying the conditions which immediately surround it; the study of the general conditions of the universe as a whole is a far more difficult task which life on this planet is only now approaching.

Now the physical conditions under which life is possible form only a tiny fraction of the range of physical conditions which prevail in the universe as a whole. The very concept of life implies duration in time; there can be no life where the atoms change their make-up millions of times a second and no pair of atoms can ever become joined together. It also implies a certain mobility in space, and these two implications restrict life to the small range of physical conditions in which the liquid state is possible. Our survey of the universe has shown how small this range is in comparison with the range of the whole universe. Primeval matter must go on transforming itself into radiation for millions of millions of years to produce an infinitesimal amount of the inert ash on which life can exist. Even then, this residue of ash must not be too hot or too cold, or life will be impossible, and endless other conditions, of which we know nothing, must in all probability be satisfied before life comes into being. We cannot be too suspicious of the interpretation which our minds, trammelled with long brooding over, and experience limited to, one tiny corner, puts on the greater universe we are just beginning to discover. [71–74]

[. . . . .]

The Birth of Planets

After their birth stars do not live entirely uneventful lives. They may meet with a variety of accidents and these result in different observed astronomical formations. A star may rotate too fast for safety, just as a flywheel may; when this happens it breaks into two, and the two stars so formed revolve endlessly about one another as a binary system. Two stars may run into one another, although this is very rare. A more common occurrence is for two stars to escape running into one another by a narrow shave. When this happens, huge tides are raised on the two stars involved, and these may take the form of long streamers of gas, which ultimately condense into 'drops' just as did the gas in the outlying regions of the spiral nebula, It seems reasonably certain that the planets were formed in this way. [. . . .]

The birth of the solar system, then, resulted from the close approach of two stars; if a second star had not happened to come close to our sun, there would have been no solar system. It may be thought that with a life of millions of millions of years behind it, one star or another would have been certain to come near enough at some time to tear planets out of the body of our sun. Calculation shows the reverse; even after their long lives of millions of millions of years, only about one star in 100,000 can be surrounded by planets born in this way. A quite unusual accident is necessary to produce planets, and our sun with its family of attendant planets is rather of the nature of an astronomical freak.

In the thousand million stars surrounding our sun there are, at a moderate computation, not more than ten thousand planetary systems, because there has not been time for more than this number to be born. They are of course still coming into existence; calculation suggests a birth-rate of about one per thousand million years. Thus we should have to visit thousands of millions of stars before finding a planetary system of as recent creation as our own, and, even if life similar to our own exists on other planets, we should have to visit millions of millions of stars before finding a planet on which civilization, and interest in the outer universe, were as recent a growth as are our own. Utterly inexperienced beings, we are standing at the first flush of the dawn of civilization. Each instant the vision before us changes as the rosy- fingered goddess paints a new and ever more wonderful picture in the sky, while on earth the rolling back of the morning mists discloses new, mysterious and unsuspected vistas to our bewildered gaze. We call it living in an age of progress.

In time the glory of the morning must fade into the light of common day, and this in some far distant age will give place to evening twilight presaging the final eternal night. But we children of the dawn need give but little thought to the far-off sunset.

It may be suggested that the creation of planetary systems is also only at its beginning, and that in time every star will be surrounded, like our sun, by a family of planets. But no; the stars will have dissolved into radiation or disappeared into darkness before there is time for this to happen. So far as we can judge, our part of the universe has lived the more eventful part of its life already; what we are witnessing is less the rising of the curtain before the play than the burning out of candle-ends on an empty stage on which the drama is already over. There is not time for many more planets to be born.

Life and the Universe

The planets are the only places we know where life can exist. The stars are too hot; even their atoms are broken up by the intense heat. Nebulae are in every way unsuitable; even if cool solid bodies exist in them, they would probably be so drenched with highly penetrating radiation as to render life impossible. We have already noticed how life demands a special type of matter, such as does not produce intense light and heat by transforming itself into radiation. We find it only in the surfaces of the stars, which are too hot for life, and in the planets which have been pulled out of these surfaces.

On any scheme of cosmogony, life must be limited to an exceedingly small corner of the universe. To our baby's wonderings whether other cradles and other babies exist, the answer appears to be that there can at best be very few cradles, and there is no conceivable means of knowing whether they are tenanted by babies or not. We look out and see a universe consisting primarily of matter which is transforming itself into radiation, and producing so much heat, light, and highly penetrating radiation as to make life impossible. In rare instances, special accidents may produce bodies such as our earth, formed of a special cool ash which no longer produces radiation, and here life may be possible. But it does not at present look as though Nature had designed the universe primarily for life; the normal star and the normal nebula have nothing to do with life except making it impossible. Life is the end of a chain of by-products; it seems to be the accident, and torrential deluges of life-destroying radiation the essential.

There is a temptation to base wide-reaching inferences on the fact that the universe as a whole is apparently antagonistic to life. Other quite different inferences might be based on the fact of our earth being singularly well-adapted to life.

We shall, I think, do well to avoid both. Each oak in a forest produces many thousands of acorns, of which only one succeeds in germinating and becoming an oak. The successful acorn, contemplating myriads of acorns lying crushed, rotten, or dead on the ground, might argue that the forest must be inimical to the growth of oaks, or might reason that nothing but the intervention of a special providence could account for its own success in the face of so many failures. We must beware of both types of hasty inference.

In any case, our three-days-old infant cannot be very confident of any interpretation it puts on a universe which it only discovered a minute or two ago. We have said it has seventy years of life before it, but in truth its expectation of life would seem to be nearer to 70,000 years. It may be puzzled, distressed, and often irritated at the apparent meaninglessness and incomprehensibility of the world to which it has suddenly wakened up. But it is still very young; it might travel half the world over before finding another baby as young and inexperienced as itself. It has before it time enough and to spare in which it may understand everything. Sooner or later the pieces of the puzzle must begin to fit together, although it may reasonably be doubted whether the whole picture can ever be comprehensible to one small, and apparently quite insignificant, part of the picture. And ever the old question obtrudes itself as to whether the infant has any means of knowing that it is not dreaming all the time. The picture it sees may be merely a creation of its own mind, in which nothing really exists except itself; the universe which we study with such care may be a dream, and we brain-cells in the mind of the dreamer. [81–88]

Part II

Society, Politics, and Gender

Complete Texts

Hypatia; or, Woman and Knowledge[1]

Dora Russell (1925)
(Mrs. Bertrand Russell)

Headnote

Dora Black (1894–1986) had been a member of the Heretics Society from 1909, when she was an undergraduate at Cambridge, where she had won a modern languages scholarship to Girton College. She was elected to a fellowship in Girton from 1918, but resigned it to accompany Bertrand Russell to China and Russia in 1920. She became his second wife in 1921. She supported the Bolshevik revolution, and birth control for women, helping to form the Workers' Birth Control Group in 1924. In 1929 Dora Russell organized the World League for Sexual Reform's successful Congress in London with the Australian sexologist Norman Haire (who contributed *Hymen; or, the Future of Marriage* to To-day and To-morrow in 1927). The Congress is discussed in Vera Brittain's *Halcyon.*

The Russells put their progressive educational ideas into practice, founding Beacon Hill School in 1927. She worked for the Ministry of Information during the Second World War, and was, like Bertrand Russell, a founder member of the Campaign for Nuclear Disarmament in 1947.

******

To My Daughter Kate Russell

Preface

Hypatia[2] was a University lecturer denounced by Church dignitaries and torn to pieces by Christians. Such will probably be the fate of this book: therefore

1 It may strike today's readers as odd that a feminist essay arguing that (as the original dust-jacket had it) 'Woman must be free to handle the problems of love and maternity in her own way' identifies its author on the title-page as a married woman. It may have been a defensive move on the publisher's part, to reassure readers and reviewers that, if Dora Russell's explicit celebration of sexual fulfillment shocked them, it came from someone respectably married. It may also have been promotional; an attempt to appeal to readers of Bertrand Russell, who was already a celebrity.

2 Hypatia of Alexandria (c. 350/370–415 ACE) was a Greek or Egyptian mathematician, philosopher and astronomer. She followed the Neoplatonist philosopher Plotinus, encouraging logical and

DOI: 10.4324/9781032625522-15

Figure 9.1 The Unascetic Sage (see p. 213).

it bears her name. What I have written here I believe, and shall not retract or change for similar episcopal denunciations.

—*January*, 1925

I. Jason and Medea[3]

Is There a Sex War?

A FEATURE of modern life is that matrimonial quarrels, like modern war, are carried on on a large scale, involving not individuals, nor even small groups of individuals, but both sexes and whole classes of society. In the past, Jason and Medea, neither of them quite an exemplary character, measured their strength against one another as individuals; and, though each voiced the wrongs and the naked brutality

mathematical studies instead of religion. She lived in Roman Egypt, and was killed by a Coptic Christian mob. Her story was well-known via Charles Kingsley's 1853 novel *Hypatia*.

3 In Greek mythology, Medea fell in love with Jason when he came to Colchis to claim the Golden Fleece. She agreed to help him get it if he would marry her and take her away. But later he abandons her in Corinth for Glauce, the King of Corinth's daughter. In the best-known version of the story, Euripides' tragedy *Medea*, she kills her own children by Jason. Russell plays inventively on To-day and To-morrow's predominantly classical names for titles, choosing different mythological figures for each section, who bear on the situation of women. Medea features as the type of the woman enraged by male injustice.

of their sex, it did not occur to either to seek in politics or in social reform a solution or a compromise. Jason, indeed, as the reactionary face to face with a turbulent and insurgent female, called to his aid the powers of kingship and the State—to suppress and exile, but not to remedy. Medea, driven mad—like so many able and remarkable women—by the contempt and ingratitude of men as individuals or in the mass, and aware that the law was a mockery where she was concerned, expressed herself in savage protest after the manner of a militant suffragette. While I can open my newspaper to-day and read of mothers desperate with hunger, misery, or rage drowning themselves and their children, I cannot bring myself to look upon Medea as some elemental being from a dark and outrageous past. As for Jason, he never did appear to anybody as other than an ordinary male.

During the last twenty or twenty-five years, when women were struggling for their right as citizens to a vote and to a decent education,[4] began what has been called the sex war. No woman would deny that we began it, in the sense that we were rebels against a system of masculine repression which had lasted almost unbroken since the beginning of history. In a similar sense, the proletarian to-day begins the class war. Those who remember the heroic battles of suffrage days know that the sequence of events was as follows: We made our just demands and were met with ridicule. We followed with abuse—all the pent-up anger, misery, and despair of centuries of thwarted instinct and intelligence. Man retaliated with rotten eggs. We replied with smashed windows; he with prison and torture. People forget so readily, that it is well to remember that this was in the immediate past; it is not a nightmare picture of one of those future sex wars with which our modern Jasons delight to terrify the timorous of both sexes.

Is there a sex war? There has been. It was a disgraceful exhibition, and would not have come to a truce so soon, but that it was eclipsed by the still more disgraceful exhibition of the European War. In 1918 they bestowed the vote, just as they dropped about a few Dames and M.B.E.'s,[5] as a reward for our services in helping the destruction of our offspring. Had we done it after the fashion of Medea, the logical male would have been angry. They gave the vote to the older women, who were deemed less rebellious. Such is the discipline of patriotism and marriage, as it is understood by most women, that the mother will sacrifice her son with a more

4 Women in the UK were enfranchised in two stages. The Representation of the People Act of 1918 granted the vote to women only if they were over the age of 30. That was the situation when *Hypatia* was written. The age threshold was only lowered to 21 (the same as for men) in 1928. Women could take university examinations, but were not allowed to be awarded degrees until 1878 at the University of London, 1920 at Oxford, and 1948 at Cambridge.

5 Part of the UK's Honours system, 'The Most Excellent Order of the British Empire' was established on 4 June 1917 by George V. It includes five classes; in decreasing order of seniority:

1 Knight Grand Cross (GBE) or Dame Grand Cross (GBE)
2 Knight Commander (KBE) or Dame Commander (DBE)
3 Commander (CBE)
4 Officer (OBE)
5 Member (MBE).

resigned devotion than the younger woman brings to the loss of her lover. There may be more in this than discipline. If honesty of thought, speech, and action were made possible for women, it might transpire that on the average a woman's love for her mate is more compelling than love for her offspring. Maternal instinct—genuine, not simulated—is rarer, but, when found, more enduring.

There was a promise, as yet unredeemed by any political party—for the politician has yet to be found who will realize that the sex problem is as fundamental in politics as the class war, and more fundamental than foreign trade and imperial expansion—to extend this franchise on equal terms with men. "Good fellowship" between the sexes as between classes was the key-note of the war. It was held that women had proved their mettle and that mutual aid was to be the basis of all future activities, public and private. The sex question was deemed settled, and everyone was led to suppose that all inequalities would be gradually eliminated. On this partial victory and this promise feminists called a truce, and abandoned the tactics of militarism.

But you never know where you have Jason. He was a soldier, mark you, and a gentleman. Forbidden open warfare, he takes to sniping. He snipes the married women out of those posts for which they are peculiarly fitted—as teachers or maternity doctors—although it is against the law to bar women from any public activity on the ground of marriage. He cheats unemployed women out of their unemployment insurance more craftily and brutally than he cheats his fellow-men. Instead of realizing that the competition of women in industry and the professions is a competition of population pressure rather than of sex, he seeks by every means in his power to drive woman back to matrimonial dependence and an existence on less than half a miserably inadequate income; and then he mocks at her when she claims the right to stem the inevitable torrent of children[6] whose advent will but aggravate man's difficulties as well as her own. But worse than all the sniping is the smokescreen of propaganda. While feminists have, in a large measure, stayed their hand, anybody who has anything abusive to say of women, whether ancient or modern, can command a vast public in the popular press and a ready agreement from the average publisher.

It is a very insidious propaganda. Thus the fashion-papers tell us that grandmamma's ways are coming back into their own; elsewhere we are flattered for the frank honesty of the modern girl and then warned not to ask for equal pay or opportunity.[7] Again, we hear that woman, like the Labour Party when in office,[8] has done nothing with the opportunities given her by the vote; or that the country rejected the women candidates wholesale. This, regardless of the fact that the steady increase of the Labour poll has been due in great part to the votes and more to the organization and propaganda of large numbers of intelligent working women who know not only what they want, but how to get it. They are backed now by

6 That is, through birth control.

7 [Author's note] Lovat Fraser in a cunning article in *The Sunday Pictorial*, January 4th, 1925.

8 When Russell was writing, Labour had only been in power once, and only with a minority government led by Ramsay MacDonald, which only lasted for nine months in 1924.

many of the middle-class women who were young enough to be revolted by war politics in 1914, and are old enough to claim their citizenship in 1924. Hundreds of thousands of others, now between twenty and thirty, mothers, professional and working women, will make themselves heard before long. To them, the principle of feminine equality is as natural as drawing breath—they are neither oppressed by tradition nor worn by rebellion. I venturc to think that, had the Labour Party machine been less dominated by the masculine perspective, to which the equal franchise Bill was a matter of secondary importance, they would not have lost so heavily in the 1924 election.[9] Votes for women at 21 would have greatly increased the poll of many Labour candidates. I have seen young mothers almost sobbing outside the polling-station on polling-day because they had no vote to cast for the future of themselves and their children. As for the defeat of women candidates, everybody, including the leader-writers who spread the adverse propaganda, knew perfectly well that the majority of them stood in constituencies where even a man of their party would not have had the ghost of a chance. Here again Jason at head-quarters displayed his well-known chivalry.

It is no part of my thesis to maintain that women display their feminism in proportion as they vote for any particular political party—Labour, for example. But I do suggest that it is the progressive working woman rather than the woman of the middle class who will in the future make the most important contribution to the thought of feminism and to a solution of our practical difficulties. One of the most inveterate anti-feminists, the author of *Lysistrata*,[10] as an avowed anti-democrat, has based his thesis and his strictures on observations that do not go beyond the bounds of upper- and middle-class people, barely even beyond the bounds of the night club or the suburban dance-hall. In his eyes we are to blame for everything. Our worst crime is to "blaspheme life and man"; our next, not to have prevented food being put into tins; our next, to have adhered faithfully to that ascetic view of life and sex so firmly instilled into us by medieval monks and bullying Puritan fathers and brothers. We are to blame for the industrial revolution in that we let weaving, spinning, milling, and baking go out of our hands. We are to blame for the iniquities of doctors in that we did not maintain our position as the dispensers of healing potions and simples. We are to blame in that we have not learned to bring forth our children without pain, those children whose brows bear the marks of obstetric instruments that were used to spare their mothers, and whose lips have not been laid to those inhuman mothers' breasts. (There are no scars of war, O Jason!) Where is salvation for us, and how shall we rid us of the burden of our iniquity? We who have waxed so arrogant that we have even aspired to let science build our

9 On the 29 October.

10 Anthony M. Ludovici's *Lysistrata; or, women's future and future woman* had appeared in the series in 1924. Russell clearly intended *Hypatia* as a refutation. Ludovici (1882–1971) was a British philosopher and social critic. He started as secretary to the sculptor Auguste Rodin. He is known as a Nietzschean proponent of aristocracy, anti-egalitarianism, eugenics and racial purity. He met Hitler in 1936, and wrote admiringly about him.

children outside their mothers' bodies[11] must humble ourselves once more and take upon us the whole duty of woman. We must use our votes to restore aristocracy[12] and take the food out of tins; spin and weave, no doubt, the while we nurse and bear our yearly child, delivering it over to infanticide when necessary, since birth-control is artificial and displeasing to the male. In our leisure moments—of which, doubtless, we shall find many under this humane *régime*—we are to discover by what means of diet, or exercise it may be, we can fulfil our maternal functions with pleasure instead of suffering.

A joke, you say? No, no, my poor Medea, it is a man called Rousseau, risen from the dead. Not long ago he preached this sort of thing to women who pinched their waists and wore a dozen petticoats. They were not educated enough to follow Voltaire,[13] so they listened to what Rousseau called the Voice of Nature.[14] Soon thereafter, they found they were being abused for being less civilized, more ape-like than the male, irrational and unsuited to take a part in public life. So they tried again, poor things, and then there was an awful thing called the Industrial Revolution, and the food got into tins. They may be pardoned, as may all of us, if at this point they became a little bewildered. Some people blamed science, some civilization, some the meat-trusts and the millers, but the true culprit, as ever, was Woman. A thousand voices cried her down—she hadn't enough children; she had too many; she was an ape; she was a dressed-up doll; she was a Puritan; she was an immoral minx; she was uneducated; they had taught her too much. Her pinched waist was formerly abused—now it was her slim and boyish body. Eminent surgeons[15] committed themselves to the view that the boyish figure with its pliable rubber bust-bodices and corselets would be the ruin of the race, that race which had been superbly undegenerate through four centuries of armour-plate corset and eighteen-inch waists, that race which, then or now, can hardly compete in toughness with the Chinese, among whom the boyish figure has been for centuries the ideal, and whose women cannot conceivably be accused of shirking any of the responsibilities of maternity. Others told us that the woman-doctor has no nerve to tend confinements, and conveniently forgot that, since the world began, and until quite modern times, it is women who have ministered to one another in that agony

11 *Lysistrata*, 92, objects to Haldane's concept of ectogenesis.

12 [Author's note] An ingenious method of accomplishing this suggests itself. Since women do not sit in the House of Lords, suppose that all Peers' wives, following the example of the Duchess of Atholl, stand for Parliament where their husbands have estates. This would obviate the necessity, now felt by Conservatives, of restoring the veto of the House of Lords.

13 François-Marie Arouet (1694–1778), known by his pseudonym, Voltaire, was a French Enlightenment writer, essayist, and philosopher known for his wit, and defense of civil liberties, including freedom of religion.

14 Jean-Jacques Rousseau (1712–1778), philosopher and autobiographer. Though his political philosophy influenced the French Revolution, he believed in the subordination of women. In *A Discourse on Political Economy* (1755) he argued that 'the voice of nature is the best counselor to which a father can listen in the discharge of his duty.'

15 [Author's note] Sir Arbuthnot Lane, for whom I have hitherto entertained an entirely unqualified admiration, in a recent article. *Vide The Weekly Dispatch*, December 28th, 1924.

which now as in the past is the lot of every mother. Is there truth in the words of Jason? Is there truth or justice in the passion of Medea? Let us not ask the protagonists, but let us summon the inquiring intelligence of Hypatia to find us a way out of the intolerable tangle in which their quarrelling has landed us.

II. Artemis[16]

The Early Struggles of Feminism

WHEN the feminist struggle began during the last century, ignorance and beauty were the two qualities most admired in women. It is necessary to remind our masculine critics what was the soil from which the feminist movement sprang and what the current morality which influenced its direction. It was customary in those days to make fun of old or ugly women and to scorn those who showed any signs of intelligence. A man chose a young, beautiful, and blushing creature for his bride, and transformed her by one year of marriage and one childbirth into a gentle and submissive matron. Ugly or intelligent women, for the most part, paid a heavy price. Not only were they rejected in youth, and starved of all their natural joys, but as "old maids" they were the object of general scorn and derision. Small wonder that women adopted artificial aids to beauty and artificial hindrances to their native intelligence. Strongest of all the taboos laid by masculine custom and religion on feminine minds was that regarding sex-knowledge. Their purity was to be preserved only by ignorance, and even as matrons and mothers it was scarcely decent for them to refer to any of the physical changes of their bodies. It is impossible to over-estimate the strength of this tradition, or the harm which has been worked by it to the cause of women.

The feminists were, and are still, howled down by men on the pretence that they invented chastity and scorn of bodily values. History disproves such a ridiculous assertion. The early feminists were what history and tradition made them, and could not at the time of their rebellion have been otherwise. The origin of the stupid ideal of womanhood against which men as well as women to-day are still fighting was the asceticism of the Christian religion; and, unless St. Paul was a woman in disguise, I fail to see how woman is to be blamed for a conception of her place and duty from which she has suffered more than anybody else. Before the conversion of the West to Christianity, barbarian women of the North enjoyed a certain rough equality with their husbands. They stride through the sagas, these fierce women, brides of heroes, glad to reward the warrior with their favours, quick to avenge an

16 'Artemis, in Greek religion, the goddess of wild animals, the hunt, and vegetation and of chastity and childbirth; she was identified by the Romans with Diana. Artemis was the daughter of Zeus and Leto and the twin sister of Apollo. Among the rural populace, Artemis was the favorite goddess. Her character and function varied greatly from place to place, but, apparently, behind all forms lay the goddess of wild nature, who danced, usually accompanied by nymphs, in mountains, forests, and marshes. Artemis embodied the sportsman's ideal, so besides killing game she also protected it, especially the young.' (*Britannica*) In Russell's schema for *Hypatia*, Artemis is the early feminist advocating chastity; and Athena, who enters later in this section, women's education.

insult or a wrong. They had no need to stoop to cajolery. Savage and untamed, they were the fit and equal mates of savage men.

Then came the monks, and the white wimples and courtly dresses and chivalry, chants and cathedrals, and meek and reverent casting up and casting down of eyes. The savage breast that had swelled and throbbed untrammelled in love or anger learnt to flutter and to sigh. Quenched were the fires of Brunhilde, her sunlit rock deserted.[17] Agnes and Mary,[18] tamed and pious, sat cooing in the shade. But for meekness and maternity, the early days of asceticism might have seen a crusade to destroy that temptress—woman. Barely allowed a soul, she slipped through a life of oblivion, praying that it might be a pretty crown with which Heaven would reward her patience and submission at the last. Then came the Puritans and denied her even that, substituting ugliness in this life as well as the negation of body, and a heaven of people in starched nightshirts, rendered oblivious to the horrid spectacle of their figures by the still more horrid chanting of their nasal psalms.

A breath of rationalism—brief, soon choked, a breath of "nature"—and so to crinolines, pantalettes,[19] and a life still lived in terror of hell-fire, terror of parents, dread of husband, horror of the least breath of adverse public opinion. Anyone who reads the *Fairchild Family*[20] must marvel that from such nerve-destroying parental tyranny and the intolerable weight of prejudice and religious superstition the nineteenth-century woman ever found the courage to rebel.

Was it astonishing that the revolt had in it something frenzied and ascetic—that it seemed to express the anger of the spinster thwarted and despised in the current scheme of values? I do not think the pioneers were so much Puritan as votaries, hanging the tablet of each achievement in the temple of Athene[21] or of Artemis, pressing on, breathless, swift of foot, sure of aim, in dread of the fate of Atalanta[22] whom the Golden Apples lured to destruction and the marital embrace. "Chaste as the icicle, that hangs on Dian's temple."[23] They had need to be, perhaps, who, in

17 In the *Prose Edda* of Snorri Sturluson (c. 1220), Brunhild is a Valkyrie who lives on a mountain, where she is surrounded by a wall of flame. As Brünnhilde, she is a central character in Wagner's *Ring* cycle.

18 Presumably St Agnes of Rome, virgin martyr, and Mary, virgin mother.

19 Crinolines: stiff or structured petticoats designed to hold out skirts. Pantalettes: undergarments to cover women's and girls' legs.

20 *The History of the Fairchild Family; or, the Child's Manual* was an early nineteenth-century didactic children's book series by Mary Martha Sherwood (1775–1851), published in 1818, 1842 and 1847. It focuses on the children's realization of original sin and their need of redemption, and also provides worldly lessons on topics including etiquette and virtuous consumerism.

21 The ancient Greek goddess associated with wisdom, warfare, and handicraft (Roman equivalent, Minerva).

22 Atalanta was unwilling to marry, due to the memory of her dead lover Meleager. Under paternal pressure, she only agreed to marry a man who could beat her in a running race. If she won, the man would die. Many dead suitors later, Hippomenes prayed to Aphrodite, the goddess of love, and was given three golden apples. By dropping them one at a time he was able to distract Atalanta, win the race, and marry her.

23 Shakespeare, *Coriolanus*, V.iii.75–77: 'The moon of Rome, chaste as the icicle/That's curdied by the frost from purest snow/And hangs on Dian's temple!—Dear Valeria.'

an atmosphere of swoons and ringlets, won for us schools and colleges, free limbs, health and the open air; unlocked for us the classics, science, medicine, the history of our world; drew us from our paltry, ladylike accomplishments; wrote upon our schoolbooks: "Knowledge is now no more a fountain sealed,"[24] and flung wide the gate into the world.

They, these pioneers, childless, unwed, created and bore thousands of women, made them anew, body and soul, for lives of mental and physical activity unknown in the past to any but the very exceptional few. Just like the new learning of the Renaissance to men's minds in Europe was the opening of high school and university to the feminine mind of to-day. Thousands of women of the last generation and this, who would otherwise have passed their existence in genteel poverty and vacancy of mind, have found their happiness in teaching, in medicine, or in some other profession. Thousands of mothers have watched with delight the unfolding of their children's minds, and enjoyed co-operating over "lessons" and arguing politics with the adolescent.

We, who in a sense are the children of the feminist pioneers, whose thoughts embrace the universe, whose lives are one long round of mental and physical delight, at times intense to ecstasy—we at least will pay our tribute to those who lit the sacred fires, before we take up pen and paper to criticize.

When one reads the lamentations of would-be intelligent men about the iniquities of modern young people, chiefly those of the female sex, one cannot but laugh at their method of approach. It would seem according to them that our modern women just happen like that: no one had a part in forming their bodies or in training their minds. In so far as these people consider education or early training at all, it is to blaspheme at the sex-hating feminists who have trained modern women to dispense with their birth-right—the love of man. How this squares with the wail of the Bishops against the sexual immorality of the younger generation we will leave Jason or the eloquent author of *Lysistrata* to decide. Our business is not to condemn woman, past or present, but to chronicle faithfully the forces that have made her, and the aspirations which will mould her future. For she, and she alone, shall be the arbiter of her fate, and neither man nor creed stand between her and the realization of her ideals. Men have blasphemed woman and life too long, and it will not be until the issues are clearer, the battle more advanced, that the basis for co-operation between man and woman can be finally established. There is too much evidence at present that man, professing friendship and concern, is still ready to snatch from us what little we have won.

To those elderly gentlemen, then, who watch with horror the upper- and middle-class woman perpetrating similar follies to those of upper- and middle-class men, the first question we would put is: "What education did they give their daughters, and what was taught to their mothers before them? What were the current ideas about feminine destiny which encircled them in their impressionable years?" Many would answer, still far too many, that their daughters were given the education of

24 From Canto II of Tennyson's *The Princess: A Medley* (1847).

a gentlewoman and fitted to become the wives of gentlemen. This we know of old. The lady eats, drinks, digests, wears clothes, tinkles the piano, dances, sings, handles a golf-club, submits to sex, bears a child without the smallest notion of anatomy, turns from the whole thing disgusted, and probably bears no more. Whose fault? Not hers. They do not teach mothercraft or physiology in finishing schools for gentlemen's daughters, and it is no part of the duty of gentlemen's wives to reproduce their kind. Perhaps there is comfort in that.

A great many parents, however, would tell us that they gave their daughters a good and liberal education in such schools as were available, good ordinary boarding and day-schools which have sprung up during the last fifty years in response to the feminist propaganda. Then we have the working woman, who has shared with her brothers in what education is permitted to trickle through the elementary schools. It must not be forgotten that this ends at fourteen.

Is there something wrong with this education of women, and if so, what? I think we must judge that there is. The reason lies in the sense of inferiority bred in women by so much oppression, and the natural result that their chief aim as they struggled upwards was to prove that in all respects they were just as good as men. The second aim was to prove that they could jolly well do without them. In exactly the same way the worker, rising in the social scale, seeks to prove himself a *bourgeois*. Both efforts are mistaken. Each class and sex has that to give to the common stock of achievement, knowledge, thought, which it alone can give, and robs itself and the community by inferior imitation. The feminist movement, like one dissentient voice in an excited public meeting, was querulous, hysterical, uncertain of itself. It dared not cry out that women had bodies. Its one hope of success was to prove that women had minds. And it was right in this, that the primary fact about men and women is not that they are two sexes apart, but that they are human beings and as such should share all knowledge in the world and make it the basis of their partnership and the rearing of their children.

Many an ardent feminist spinster in a girls' secondary school has sighed over the state of public opinion which forced her to drive her girls' minds along channels for which they were not always suited, that they might do well at college and show that women could surpass the men. Many another, well drilled by a mother or tradition in ideals of feminine virtue, gloried in the sexless learned women she was creating and in the thought that one day they would force those savage, lustful men to conform to the ideals which they set up for women. Why blame her? Lay the blame where it is due. It will be but a just retribution for that lustful male and his ideal of feminine virtue if one day, in a world full of prohibitions, he finds himself forced to kneel before the Mumbo-Jumbo[25] he himself built up to terrify his wives and daughters to submission.

Feminist ideals of education, then, had the defect that they did in a certain measure deny sex, or ignore it. The feminists had a pathetic hope that by so doing they

25 [Author's note] Mumbo-Jumbo was an idol set up by the men in Nigeria to terrify erring women. The men, but not the women, knew him to be a fake. See *Mungo Park's Travels*.

would convince the dominant male that a woman might be learned and yet remain a lady. But I wish to emphasize the fact that this feature has belonged to all education of women, especially of ladies, from time immemorial, and it is, therefore, unbecoming in a male, whether young or old, to use this as a cause for reproach to our sex. We went as far as we dared with an eye to male hostility. Young feminists to-day would be the first to admit that it would probably have paid us to go further. There never has been a period when education has trained women for the possibility of motherhood, and it is time that such training was begun. There never was a period when the education of women was completely honest, and it is time that that training was begun. What knowledge is of more vital importance to women than anatomy and physiology? They were allowed it if they were to be doctors, and then only with caution. Turning casually the pages of a book on anatomy in a girls' secondary school library, I found the diagrams connected with sex and maternity carefully stuck fast. What is more calculated to inspire prying and prurience? We have no right to blame young women for shirking marriage, sex, or motherhood, or for moulding their figures on boyish lines, when we carefully treat them as boys and withhold from them as long as we can all knowledge of the difference of their physique and possibly of their destiny. I have no wish to go back on the great achievements of feminism, or to drive women from the professions in which they have a just right to be employed. I want to break down the last barriers. Artemis is slim and bold; Athene is stately. We have done well to worship at their shrines.

But the call of Demeter the Fruitful is insistent.[26] If we would add to the achievements of those who came before us, let us freely admit that we have but been playing mock modesty, and that to us the body is no mere box to hold the mind, but a temple of delight and ecstasy: a temple to hold the future if we will. To me the important task of modern feminism is to accept and proclaim sex; to bury for ever the lie that has too long corrupted our society—the lie that the body is a hindrance to the mind, and sex a necessary evil to be endured for the perpetuation of our race. To understand sex—to bring to it dignity and beauty and knowledge born of science, in place of brute instinct and squalor—that is the bridge that will span the breach between Jason and Medea.

III. Aspasia[27]

The Younger Feminists

WHILE we have admitted that the first aim of the feminist movement was to open to women the stores of learning, to develop their minds and to teach them to think,

26 The ancient Greek goddess of the grain and fertility, and the health-giving cycle of life and death; the preserver of marriage and the sacred law. Roman equivalent: Ceres.

27 Aspasia was a Milesian woman who was famous for her involvement with the Athenian statesman, Pericles. She is mentioned in the writings of Plato, Aristophanes, Xenophon and others. In the comedy of the period she was portrayed as a courtesan. For Russell, she represents the sexually liberated modern woman.

and that no attempt was made to handle the problem of sex, it is not quite fair to say that even early feminism has consistently denied or despised the body. The schools and colleges made it their business to give to women opportunities for physical development, for open-air exercise, swimming, tennis, hockey, lacrosse. The Victorian young woman learnt gradually to be ashamed of her tiny waist and fat hips. She learnt that a healthy appetite[28] as well became a young woman as a young man, gave up her snacks in private and did justice to good meals taken at proper intervals. Quietly, and without mention of the fatal word "sex," the spinster feminists, by emphasis on health and vigour, built up a generation of young women who were to be frank about other desires besides eating and drinking.

I cannot see what is the matter with our figures. Steel rods and rubber are more modern materials than oak beams and pink plaster. Neither we nor our modern lovers admire the opulent Venuses, indolent and rose-embowered, who adorned the ceilings of old-fashioned ballrooms. They were stupid, self-indulgent creatures, not even good mothers, whatever the sentimental elderly gentlemen in their top-hats and whiskers may have to say. What is a good mother we will discuss in a later chapter, but for the present it is enough to say that more dangerous childbirths are due to narrow pelves caused by rickets than to hips contracted by the corsets of vanity. Let the doctors turn socialist and feed the poor, instead of spending their time lamenting the inadequacy for childbirth of a few fashionable women who don't very much matter. Middle- and upper-class girls nowadays—and most working-class girls, too—go corsetless up to maturity. They do gymnastic exercises, and dances that give suppleness to the body. They swim and they play out of doors. Those who are rich enough to be adequately fed are graceful and active as kittens, and as healthy. By the time adolescence brings, as it always does, a few years of intensive sex-vanity, the corset can do very little harm. The muscular little body does not tolerate it stiff, or very tight, and the bones are well grown. The mystery of feminine dress helps the appearance of slimness. There are few clothes, and no lumpy gathers. Beneath that boyish outline are firm little breasts, clean arching hips, abdominal curves and thighs, lovely as anything the Venus of Milo has to show.

Artemis fashioned this modern woman. That is admitted. Has Artemis her vows?

I'm afraid for once we have to admit that the Bishops are right. In spite of everything the Church can do, in spite of an education committed, so far as the authorities can control it, to sour or religious spinsters, the modern young woman is not very moral. It is a pathetic picture which the author of *Lysistrata* has drawn for us, of sexless beings going to and fro in tube and bus like shuttles in a machine to dull work robbed of all joy, earning their livelihood and turning their backs on man in response to feminist propaganda. Man, the enemy—to be defeated in his own professions, to be repelled in every onslaught upon feminine virtue: I wonder? I would

28 [Author's note] There may be a biological cause of the alleged smallness of feminine appetite. Watching a raven and his consort with fresh meat, I observed that she obtained only a minute portion beneath the contempt of the male. Can it be that, in the savage state, only those females survived who could exist on the little the male allowed them? Is this a case of sex-linked heredity?

hazard a guess that, relatively to the population, fewer women retain their virginity till death than in the Victorian period or the Middle Ages. In all probability it is sex, not sexlessness, which makes women cling so tenaciously to the right to earn their living. Marriage brings a jealous intolerant husband, children, prying and impertinent neighbours—degraded and humiliating slavery for the vast majority of women. Thirty shillings a week and typing or working in a shop, a still tongue, or a toss of the head and the assertion that independence is the best; and, in the background a lover with whom somehow evenings are spent—a lover who has no claim and cannot tyrannize. A lover, perhaps, who pleads to become a husband, but has no chance unless his income is good or secure. Marriage would change him: Aspasia knows it. Marriage would also rob her of that thirty shillings a week, which alone stands between her and the abyss of primeval submission. Or else Aspasia teaches in a school or college. She is a skilled teacher, devoted to her work and pupils. She may be a Research Fellow in some difficult branch of learning which is to her the very breath of life. She may be a doctor in the public service, tending and advising mothers and children. She is lovely, vital, creative. Man approaches. There are holidays of delight and secret dread of the scandal which will end the work Aspasia loves—or marriage and the certainty of that end at once. "Choose," say the Bishops and the school-managers (often the same thing); "choose," say the public authorities who support the Church and rather wish women would get out of this indelicate profession of surgery and medicine; "choose between love and duty to the male and service to the community." This is not feminism—feminists have fought it persistently—it is medieval Christianity. It presents a choice between physical pleasure and service to the mind or soul; it upholds the time-honoured theory that renunciation of the world, the flesh, and the devil is the path to duty and salvation. I am fully aware of all the arguments about economic pressure, the primary right of married men to work, the awful situation of their dependent children and their wives. None of this is fundamental, and the jealous male knows it. "Divide to conquer" is the principle in dealing with trade unions; it works equally well in the feminist struggle. Persuade the single women that the married woman is an unfair competitor,[29] terrify them so far as you can into believing that to succumb to sex is something unbecoming and disgraceful and punishable with misery everlasting, whether in marriage or outside it, and you can prevent the women combining against you.

But not if Aspasia will speak. If she but would, and put an end to this lie for ever. She could tell us how, especially during the years of war, young women took the last step towards feminine emancipation by admitting to themselves and their lovers the mutual nature of sex-love between man and woman. It sounds a platitude, but is, in fact, a revolution. Strange to say, the nearness of death from enemy bombs or enemy fire did not intensify the thought of holiness and heaven. It made the little footrules to measure morality look absurd; it mocked the emptiness of female

29 This scheme no longer works, as is evidenced by the attitude of the National Union of Women Teachers this year (1925). Intelligent women are more appreciated than they were, and teachers know they may all want to marry someday.

virtue. While poverty and parents forbade the certainty of marriage, with nothing but instability and death around them, our modern Aspasias took the love of man and gave the love of woman, and found this union, free and full on either side, the most priceless gift the immortal gods can bestow. There is nothing new in this, the moralist will say—it is just wickedness. Yes, there is this that is new: that, though these younger women may be driven from fear of starvation to the outward acceptance of old codes and conventions, inwardly they know they have done no wrong and will not admit a conviction of sin. Sex, even without children and without marriage, is to them a thing of dignity, beauty, and delight. All Puritans—and most males so long as they can remember—have tried to persuade women that their part in sex is pregnancy and childbirth, and not momentary delight. As well tell a man his part is the hunting and skinning of animals for food and clothing. To enjoy and admit we enjoy, without terror or regret, is an achievement in honesty. We will go further and say that polygamy, proffered by the male as a solution to our sexless lives, is no solution at all when we are polyandrous.[30] It is useless to go on pretending, as both sexes do, about this question. The plain truth is that there are as many types of lover among women of all classes as among men, and that nothing but honesty and freedom will make instinctive satisfaction possible for all. Grant each man and woman the right to seek his or her own solution without fear of public censure. Moral questions of this kind cannot be decided by some abstract rule. It would not be wrong for a man to have six wives, provided he and they all found mutual happiness in that arrangement; nor for a woman to have six husbands and a child by each, if she and they found such a life satisfactory. The wrong lies in rules that are barriers between human beings who would otherwise reach a fuller and more intense understanding of one another. And any man or woman of intelligence and vitality can testify that to have known each other as lovers is to have completed mental and spiritual, as well as physical, understanding, and to have permanently enriched each other's lives, capacities, energies, imaginations. There is no need to make these divisions into mind and body. There is no difference. A way of walking, laughter, thoughts spoken or written, gestures of love or anger, colour and light of eyes or hair—these are the human being, man or woman. It is thus that modern individuals think of one another. When we think so, it seems absurd to argue whether or no love between man and woman should stop short of a certain kind of physical expression. It is useless to say that a mental exchange is sufficient. On the contrary, lovers know that it is through sexual understanding they best apprehend the quality of each other's minds. It is equally futile to argue that woman is cheated of her full rights if children do not result. That is not true.

It is said that modern human beings, by dint of not valuing the body, are physically degenerate and lose the finest ecstasies of love. Their digestions are poor, we are told, their breath foul, their teeth bad. Was love more delightful, then, in the old days when baths were unknown, when "sweet breath" in a woman was so rare as to be sung by poets, and the reek of stale sweat was barely stifled by a strong

30 Polygamy: having more than one spouse; polyandry: having more than one husband.

perfume? John Donne wrote verses to the flea he saw nestling in his lady's bosom.[31] There is scarcely a fine gentleman to-day who could face the prospect of making love to one of the fine ladies of the past six or seven hundred years in Europe, if she could be presented to him just as she was to her contemporary lovers. It is true that neither vermin, filth, nor squalor—being equal for both—can stay the passion of sex whether now or in the past, but I do not believe in the theory that the rougher our physique the more intense our bodily delights. Health, to be sure, is essential; but health is to be secured in the modern world, not by a return to savagery, but by the use of intelligence. I believe the bodies of young people of to-day to whom fair opportunities have been given are more healthy within and without than they were in past times. And I believe that the disappearance of religious and moral dualism between mind and matter—not by an oppressive victory of either, neither by rational and moral control, nor by abandonment to sensual materialism, but by a better understanding of psychology and physiology based on the discoveries of physical science—is bringing to the whole of life, but especially to sex-love, maternity, the rearing and education of children, joy and rapture and promise surpassing anything known to the purely instinctive life of the past. Of course we are bewildered. Civilization without decay is at last a possibility. Let us have knowledge and patience: blaspheming and violence will ruin all.

It is for modern women and for men who can understand the problem to make an end to secrecy, shame, and starvation where sex is concerned. There has been a good deal of freedom in action, but less boldness in speech, because of the heavy penalties involved. For some women speech is impossible; those who are secure must fight their battle. How old and proper people love a vigorous and god-like young male! How they look askance upon, brow-beat, and bully his equivalent in the opposite sex! Here is a community for ever starving and choking its finest women, stifling their voices in education and public life; then turning and rending the submissive residue for being what years of intimidation have made them. Let them marry, you say, and make a success of that and their children. That would be well enough but for the taboos and disabilities with which marriage is surrounded. Feminism led women away from the home that they might return armed and unsubdued to make marriage tolerable. Women who have been free remember the horror of the approach to marriage: a barrier for most of us to free public activity; a life-long contract only to be broken in disgrace and public disgust; aunts, uncles, social duties that exasperate and are totally unnecessary; the common view that henceforward husband and wife are one and indivisible, and the wife for ever to be burdened with her husband's duties and affairs; looks of surprise and reproach if we enjoy other male society; constraint in the manner of men formerly our friends; income, if we have any or can still earn, taxed as a part of our husband's; children, which, had we had them illegitimate, would have been our own but now are our

31 John Donne (1572–1631), poet, scholar, soldier, and eventually Dean of St Paul's Cathedral in London. His poem 'The Flea' (addressed to a lover, about the flea) was probably written in the 1590s.

husband's; worst of all, the looks and smiles from silly women broken in to slavery, congratulating us on having done well and made ourselves secure for life.

Let no one think this is petulant abuse. It is the accumulation of these details and the pressure of public opinion which gradually destroy the nerve and independent judgment of married women who, in their free state, have been brilliant and remarkable. It is the fact that, by marriage, we conform and place ourselves in a line with millions of others whose view of what we have done is entirely foreign to our own. As a Labour Minister is corrupted by Court dress, so is a free woman by the marriage-contract. Nothing but our desire for children would make us endure it. We, to whom the mutual nature of sex-love is sacred, to whom a partnership involving children is of equal dignity on both sides, to whom the surrender of our whole being in love is a free gift—the highest we can bestow; who would neither bind ourselves nor others where love is non-existent; we must submit to a contract based on rights of property and possession, buying and selling of our bodies; a law whose conception of conjugal wrongs is sin, punishment, and just revenge; and a Church whose utmost concession is to bid us "serve" instead of "obey" our husbands. Build, O Aspasia, a trade union of lovers to conquer the world, and cry aloud that feminism is nowhere so much needed as in the home.

IV. Hecuba[32]

Feminist Mothers

So far I have refrained from any detailed discussion of modern women and maternity because it is still necessary to make it clear that a full life of activity for women is perfectly possible and permissible without it. I am quite aware that certain religious people assert as a moral principle that the purpose of sex-love is not mutual enjoyment but the perpetuation of the race. I am also aware that militarists enjoin on women the necessity of marriage and large families as a patriotic duty. Further, certain doctors have gone out of their way to try to prove that the use of contraceptives is contrary to health and nature. These same people, we may note, have no aversion from the wearing by women of internal remedial rubber supports for months on end nor to patching up with silver, pâpier maché, and other foreign materials, the insides and outsides of human beings mutilated in the natural and healthy pursuit called war. I am not concerned with the morals of convention or superstition, but with the morals of experience. It is the experience of modern women that sex is an instinctive need to them as it is to men, and further that the prevention of conception brings to them no loss of poise, health, or happiness. On the contrary, when once they embark on the task of maternity, contraception is a blessed safeguard to health and recovery in the periods of rest between pregnancies. I am not going to deny that the most perfect delight known to human beings

32 In Greek mythology, Hecuba was the wife of King Priam of Troy and had 19 children with him. She laments the death of her son, Hector, in Book 24 of Homer's *Illiad*. Russell uses her to discuss the future of maternity.

is the completely reckless, mutually adoring union of two people of vitality and intelligence who hope to create another human being as a constant reminder of the beauty of that moment. But many considerations, which we shall discuss, forbid a yearly child. I read recently in an article by G. K. Chesterton, that sex without gestation and parturition is like blowing the trumpets and waving the flags without doing any of the fighting.[33] From a woman such words, though displaying inexperience, might come with dignity; from a man they are an unforgivable, intolerable insult. What is man's part in sex but a perpetual waving of flags and blowing of trumpets and avoidance of the fighting? The vast majority of men are not even tender or kindly to their pregnant or nursing wives, nor will they give help or consideration to the care of their young children.

A revolt against motherhood under present conditions is not surprising, nor is it entirely regrettable. There are quite a number of women whose minds and bodies are not fitted or have not been fitted by their upbringing and education to produce and care for children. This is a source of distress to many people, who, as was suggested earlier, did not think of it at the right moment, when the education of women in public and private schools was being developed. Even now these same people stand in the way of the surest remedy: which is to teach science, physiology, and the beauty of sex and maternity to boys and girls at an early age. The London County Council, many of whom are certainly distressed beyond measure at the falling birth-rate and the discontent and irresponsibility of modern young people, have just, in consultation with suitably selected moral headmasters and mistresses, turned down the suggestion of sex-teaching in elementary and secondary schools. We are always told that boys and girls of all classes nowadays acquire this knowledge easily for themselves, but the mere knowledge is not the only thing to the adolescent mind. Things not spoken of by parents or teachers, things dealt with in hushed voices by moral and spiritual leaders, surrounded by cant and humbug and false sentiment, are bound to be thought nasty by mild young people and to provide ribald laughter for the obstreperous.

This is not to say that sex-information should be given in a spirit of evangelical solemnity and exhortation, nor even of soft sentimentality. All that is needed is lessons in physiology, taught as a matter of course, as botany or nature-study are often taught; and then explanations to boys of the working of their bodies, how to keep them in health, how not to dissipate and destroy their energies too soon. Further, they should be told that woman is neither a chattel nor a servant, nor even an inferior, but a partner in joy as in the business of life; that there is no question or difficulty, public or private, which cannot be brought to her for discussion and judgment; and that she has a right to share in all decisions affecting a joint life, children, money, and the conduct of affairs of State. To girls in the same way could be explained the physical changes of puberty, marriage, and maternity, how the child grows, what food and care the mother, and afterwards the baby, will need. There is nothing in this too difficult or shocking to young or adolescent minds. So many of

33 The writer G. K. Chesterton (1874–1936) had become a Roman Catholic in 1922.

us can remember the secret conclaves with our friends when we puzzled out and pieced together what scraps of information we could glean, awakening instinct darkly supplementing this knowledge. Some of us can remember, perhaps, having noticed obscene writing on school-walls, instantly reported by shocked prefects, instantly effaced by school-mistresses with an awful and portentous gravity which made us feel we had stumbled on the brink of a secret of incredible wickedness and horror. One straightforward lecture of concise information could have dispelled the lurking mystery once and for all and imparted a sense of magic and wonder and ambition. Some of the more fortunate of us, through study in libraries and dreaming over poems, created for ourselves a finer attitude. With no teaching other than that we might find someone who would marry us some day, and that marriage was an excellent destiny even for educated women, and with no belief in any of the moral taboos current around us, some of us can none the less remember the pride of caring for the body, safeguarding health and looks, avoiding excess, severe strain, and overwork, because we cherished our dreams of the children that our bodies were to make—not ordinary children, of course not: Promethean creatures, endowed with every gift that mortal man could steal from the jealous gods, strong, beautiful, intelligent, and bold—kings and conquerors, not of their fellow-creatures but of nature and the mystery of the world.[34] There is not a woman, unless completely warped by early training, in whom such dreams and visions will not stir if we try to wake them. If not, then let her pass: we do not need her to perpetuate the race. And do not trick her into motherhood by false sentiment and information, or by withholding from her the means to protect herself if she is not fully resolved upon bearing a child.

We want better reasons for having children than not knowing how to prevent them. Nor should we represent motherhood as something so common and easy that everyone can go through it without harm or suffering and rear her children competently and well. Without arousing dread or horror, we should tell young women frankly the pain and agony of childbirth, and the anxiety and griefs which are the fate of every woman who is a mother by choice and therefore loves her children. Nothing whatever is to be gained by driving the timorous and weak by lies or compulsion into pain which they will resent and responsibility which they will evade. Everything is to be gained by training a woman in knowledge, courage, and physical strength, and leaving it then to her own instinct and her mind to tell her that to create new human beings is worth the discomfort and the suffering which she must necessarily undergo. Those in whom the courage to create survives when choice is free and all the facts are known are those best fitted to bring children into the world, and breed in them eagerness and intrepidity. The others will only pass

34 In Greek mythology Prometheus was a Titan, known for his intelligence. He is most notorious for stealing fire from Zeus and, subsequently, giving it to mortals for their use. H. S. Jennings' *Prometheus; or, Biology and the Advancement of Man* (1925) was published in the series around the same time as *Hypatia*, so Russell had probably not read it yet; though she may have heard of it from Ogden, who was a close friend of hers and Bertrand Russell's. See pp. 139–48.

on fear and distaste for life from which individuals and the community suffer far too much already.[35]

I do not mean by this that we should, scorning the aid of science, return to natural childbirth, and let its pangs scare off the weaklings and the cowards. In this matter the charges of our critics are conflicting. They condemn us for having sought the aid of science to mitigate our suffering, and in the same breath tell us that a return to natural child-bed will bring back a primitive exhilaration and freedom from pain lost for thousands of years. I do not believe that for any comparatively civilized race, any race really worthy the name of man, childbirth has ever been painless. The upright position, held by eighteenth-century divines to be a source of pride in man, was the first injustice to women. Nor do I believe that the sufferings of modern women are any worse or their confinements any more difficult than those of women in the past. They are more closely observed and the difficulties known, and, where skill is available, the dangerous ones are less likely to be fatal. In the past the fragile woman died, or continued ailing, unobserved by a doctor and afraid to complain. People who live and breed in a state of nature are by no means so healthy and vigorous as our modern Rousseaus would have us believe: more children die than survive, and those who are left have physical defects and deformities which could have been remedied by knowledge and care. These and the ravages of smallpox and other diseases, and the deformities due to the natural accidents of life unmitigated by medical care, produce far more ugliness than the mark of an obstetric instrument on temple or forehead. Then, again, youth passes more quickly. The men and women we see in modern life, still reasonably young and fresh with rounded faces and teeth stopped or supplemented by art, would in a more primitive community be dead, or else crouching useless and despised, toothless and with sunken cheeks by the fireside of their sons and daughters.

Decay and pain belong to nature. To arrest the one and mitigate the other has been the task which the sciences that deal in physiology have set themselves. Remedial at first, they pass on to the stage of prevention. Already the principle of intelligent medicine is to strengthen what is weak in the body by nourishment and exercise rather than provide artificial substitutes. Paralysed limbs return to life; women retain their teeth white and strong through several pregnancies. This is not done by a return to nature, but by an increase in civilization and knowledge. In that way our very landscapes have been formed. We prune, we nourish the soil, we cross-breed our plants. The vegetables upon which the enthusiast for nature urges us to live are the product of science and artifice: thousands of years of cultivation, nitrates from Chile, skill of the experimenter, skill of the gardener's hand. The same is true of the animals we breed for meat, eggs, or milk supply. Agriculture and stock-breeding seem natural to us—they were not natural in the distant past. As regards the human body, to me at least it seems that we are now beginning to

35 [Author's note] The anti-feminists who see in emancipated women nothing but persecuting spinsters should take comfort from the fact that voluntary motherhood will ultimately destroy feminism, if they are right. The children of women passionately desirous of maternity will inherit strong parental and survival instincts, the occasional feminist "sport" not reproducing herself!

approach the right attitude. There was more dosing and doctoring of petty ailments among intelligent people in the last century. To-day we try to learn how best to live in order that such ailments may not occur, and substitute a well-balanced diet for aids to digestion and the normal functioning of our bodies. We do the same in rearing our children. And this attitude would become more general if those who rule us, Press, Church, rich men and politicians, would consider it really important that every man, woman, and child in the State should have health and happiness, and therefore supply broadcast the necessary rules of life and sufficient of healthy and staple foods for all, in place of advertisement of quack remedies and patent substitutes prepared by profiteers.

To return to the application of science and nature to maternity. A special sentimentality and superstition inherited from the completely savage periods of history cling about this, as about sex. The avoidance of suffering in childbirth is taboo in the Japanese moral code, as it was until recently in Christian morals. Religion has persisted in regarding the female body as unclean when engaged in its most important functions, and purifying it afterwards by special prayers to the Deity.[36] We find this savagery current in Judaism[37] as in Christianity, together with an exhortation to be fruitful and multiply, and therefore to pass through shame and uncleanness as often as we can. It was thought a horror and an outrage when chloroform was used to help us. It is a still greater horror when means are discovered of not having children at all. To this day most doctors and dentists refuse to give an anæsthetic and draw a rotten tooth which is wearing down a pregnant mother's strength by sleepless nights and days of agony. Yet this can be done with reasonable care and skill. Behind all this there is the mystic belief that somehow or other nature does the work best unaided and unhindered; and this mysticism is rooted in a savage taboo. Life is, indeed, so pertinacious that somehow some of us will survive whatever we do, but this does not seem to me an adequate attitude for the rational mother.

The truth is that it is not desired or expected that mothers should be rational. Maternal instinct is so wonderful, maternal devotion so sublime, cry our sentimental brutes. Whatever we may have known of life and the outside world, it is still expected in modern times that, once married, we shall descend into a morass of instinct and ignorance from which we shall never, if the male and the vindictively-minded spinster can prevent it, emerge again. We are privileged, so we are told, in that we may bear each year a child for the State, rock the cradle, wash, mend, and make, pass on the lore of housekeeping and infant-care to our daughters just as we received it from our mothers. It is such a beautiful picture: a pity it is entirely false. The old fashioned mother had no lore, and her instinct was inadequate. She succeeded by luck rather than by knowledge. She adored, or disciplined; she killed by kindness or by severity and neglect. She would coddle when she should have hardened, harden when she should have coddled; she would

36 [Author's note] See the service for the *Churching of Women* in the Prayer Book.

37 [Author's note] *Leviticus*, xii, 1–8.

over-feed and under-feed, or give the wrong kind of food. Since it has been the fashion for women to have minds, the books for mothers have become more scientific and our intelligent inquiries have been met by research and more adequate replies. Every mother with any intelligence who has reared one or more children through the first year of life and up to five years of age would admit nowadays that scientific knowledge was of more service to her than all the instinct and adoration at her command. Indeed, I believe the so-called maternal instinct in handling and understanding babies consists of habit almost imperceptibly learnt in tending the first, blossoming into a smooth instinctive unity with the coming of the second. The fashionable mother, said to be devoid of maternal instinct because she neglects her child, has simply not learnt it, because necessity does not compel her to practical duties. This is even true, though less so, of well-to-do mothers who feed their babies at the breast.

People will persist in imagining that uncivilized women were always able to feed their children in natural fashion. Very often they were obliged to seek the help of another mother, and, when that was not forthcoming, the baby died. It is quite true that our adaptation to modern conditions of life, nerve-stress, combined with overwork for women in towns and industrial districts, has caused breast-feeding to be less common than it was in the past. But here again the way of life is not back to nature, which is impossible because we cannot at a blow destroy industrialism and the towns—but onwards, to greater knowledge. Instead of bullying the mothers and telling them it is wicked not to feed their babies at the breast, let them know how, by pre-natal care of health and strength, by diet, by deliberate nerve-control, they can feed their babies with comfort and delight and without detriment to their health and the work which they must necessarily do—or even to their beauty. Here again, if choice is free and the child therefore ardently desired, there is more chance of success with breast-feeding. And knowledge of the chemical constituents of cow's milk and patent foods as compared with human milk is more likely to induce the modern mother to suckle her child than volumes of abuse or sentimental twaddle.

Then as to the hygiene of pregnancy. Could our mothers have taught us about the different food-values, about protein, nitrogenous foods, calcium from the green foods for teeth and bones, avoidance of too many albuminous[38] foods? Knowledge of what diet can do to help us in pregnancy and our children in early youth is in its infancy, but it is there, none the less. Shall we fling it aside and return to pure instinct? What massage and remedial exercises have taught can be applied to our bodies during pregnancy and after childbirth. It is probable that closer study of the functions of the muscles of the back and the abdomen would enable us to teach women to exercise and control them in a way that would make childbirth almost painless, and the recovery of poise and activity afterwards more rapid and more thorough. Under present conditions, muscles that are often too rigid or too feeble expand and never recover their tone; others—the back muscles, it may be—go out

38 Albumen is egg-white or the endosperm of some crop seeds such as maize or wheat.

of use temporarily and similarly do not recover. In the middle-class woman laziness is often the cause of difficult confinements and poor recovery of the figure; in the working mother a too speedy return to work which is too hard and does not exercise the body harmoniously; in both the ignorance which leads to wrong kinds of nourishment during pregnancy, and fear of doing harm to the child which leads to rigid and over-careful movement, are responsible for a good many troubles. Psychological effects may be serious. Most women develop during pregnancy sensitiveness and a timidity protective to the child. From this the very fertile mother has no opportunity to recover. Hence many of the silly old ladies who cannot cross roads unaided by a policeman. With birth-control, in two years a determined mother can completely restore her nerve, her joy in life, and her full muscular powers.

The author of *Lysistrata* suggests that by diet we may produce thin babies and therefore have easier confinements. This may be true, but it is a curious fact that the experience of some mothers and doctors goes to show that much protein (which Mr Ludovici suggests we should avoid) produces a thin baby and a corpulent mother; that, on the other hand, light and nitrogenous foods, while keeping the mother slim and supple, yield a plump 8 to 9 lbs. baby. I think the size of our babies is perhaps not so much under our control as many might wish to suggest. Heredity enters in. Children sometimes have large fathers. The sheep-breeder knows that he dare not mate certain larger types of rams with small-made ewes.

In all these problems, however, it is the frankness and intelligence which feminism has made possible for women which will bring solution and progress, rather than a return to the unguided instincts of our forefathers. The lore of motherhood is a science which is now beginning, but it is not following the lines which convention and the moralists expect. It defies sentiment, ridicules unnecessary and unintelligent sacrifice, is not content to suffer, but makes demands. It begins with birth-control, which to many seems the negation of motherhood, but which to the creative mother is the keystone of her work.

Suppose we have educated our young women sanely about physical matters, as suggested earlier in this chapter. As they reach the age of maturity and activity, what will they find? If they are middle- or upper-class, an existence that is not too intolerable. Feminism has won for them the right of entry to most professions and, provided they are fairly able, they can get work. None the less, it must be admitted that the years since the War have borne hardly upon wage-earning women of all classes. The lack of sexual freedom is a terrible burden, but the remedy ultimately lies in their own hands. Life in marriage still offers reasonable comfort and good food for man and wife and two or three children. But late marriages, from the lack of opportunity for men and the expense of living, cause girls' young bodies to be worn with longing unless they are bold enough to follow our modern Aspasias. This waiting to marry is a real danger to young women's health which conventional, unimaginative people refuse to face. It produces nervous disorders bordering at times on insanity.

As regards the care of her body in pregnancy and childbirth, and the feeding of her children, the middle-class mother is in a position to carry out what modern science has to teach. She cannot have a large family, it is true, and the cry goes up on all sides that it is very hard for the middle-classes to pay for the proper education of their children.[39] The best stocks are being penalized and extinguished, so we are told. This is part of a much bigger problem, and a problem that involves the class-war. All ambitious mothers, from miners' wives to the aristocracy, would like to breed the fine types who receive a thorough education and then enter one of the intellectual professions. Obviously this cannot be. And, given equal ability in two children of different classes of life, there is no just reason for driving the worker's child, who has less good food and conditions and is therefore less fitted to stand the strain, through the worry of the scholarship system, whilst the other child's path is made smooth to a ruling position. Man for man, woman for woman, the workers would be the equals of the middle-class in strength and ability, given the same nourishment, comfort and training. In actual fact, the middle-class is perpetually being replenished in one generation, or two at most, from below. Middle-class fathers and mothers have no right to claim the privilege of a large family unless their children, if they are strong but not clever, are prepared to work the railways or dig coal in the mines. Professional people, scientists, artists, research workers, pure mathematicians, as well as skilled engineers, are, indeed, the salt of the earth, and the community that fails to produce them and give them scope is doomed in this modern world. But they are supported by manual labour, and it cannot be denied that their number cannot be indefinitely extended except by an increase of productivity and wealth. A more equal system of society will diminish drudgery and make it possible for all to have a fine development of intelligence and understanding, whatever the work on which they are employed.

Feminism in the mother has led us far from maternity. That is what it is bound to do. The working mother to-day looks straight from her kitchen, if she is lucky enough to have one, on to one of the most complex situations in history. And the intelligent ones are not blind to the situation. That is why I suggested that, though middle-class feminism has conquered the professions, the feminism of working mothers might bring a new and powerful contribution to our work.

The life of the working woman who intends maternity is becoming well-nigh impossible, and she knows it. When she has found a husband the community denies them a decent house. Possibly they find one room or two at an exorbitant rent, with no water and a grate unsuited for cooking. There are no restaurants at which the pair can afford to feed. Therefore they exist on partially or casually-cooked food, innutritious bread, and food from tins. Things may not be so bad if the wife can go on with work at a mill and get food that is fairly good at the canteen, her wages helping the meals taken at home.

The coming of a baby too often means a search for another lodging. The Bishops and the Generals like babies, but landladies don't. Another room is found, perhaps.

39 [Author's note] An instance of the incredible snobbery surrounding this question is given by the decision of the conference of Headmasters of Secondary Schools, January, 1925, against free secondary education. While the middle-class parent groans against the cost of his children's education, he also refuses to take the obvious remedy of making education free, for fear the working class should get some of it. Class difficulties would not exist if health and education were adequately dealt with.

The mother works till the last moment, has a difficult confinement and inadequate attention, and gets up too soon. It is not easier for her than for a delicately-nurtured woman, and it is not less painful. Probably it is worse, because the working mother has from birth been underfed and has weaknesses and deformities—a contracted pelvis, perhaps—that a woman well-fed and cared for escapes. Then it goes on, baby after baby up to ten and eleven,[40] always in one room and no more money coming in. The mother works whenever she can to help keep the family. Frequently she is cursed or beaten by her husband for her fertility. Should the husband die, she must work continually and harder or send her children to the workhouse. In the opinion of the Bishops, she deserves the "stigma of the Poor Law,"[41] and, in the opinion of all right-thinking people, anything done for her by individuals or the State is in the nature of a charity.

If I but had the eloquence of Hecuba mourning her slaughtered sons! The crime of war is bad enough: this butchery of hope and promise and human lives is one so black that the heart and mind of every woman who has borne a child should revolt against it until it is tolerated no more. It is easy to escape into an aristocratic theory of society. It has been done before, and ends in the guillotine. These working mothers are the people who must be lied to and terrified by bogies for fear that they use their votes to help themselves. And it is they who, when they sit in conference, demand of the State the right to stem the tide of children, to endow mothers, to pension widows, to teach and tend maternity and ensure rest for pregnant and nursing women; to see that houses and schools are built, and to control and purify the food-supply. Here is the most serious problem for the mothers, and one which the middle-class politician does not touch, because for the middle-class pure and fresh food is almost always obtainable. It is for the working mother to tackle those tins. She cannot now destroy industrialism, which dragged her work and her after it to the mill; but she can claim her right to control it in the name of life and the destiny of her children. Control of the population is essential to solving the food-problem and improving national health. Women in small houses know it. They know, moreover, that contraceptives are better than infanticide and war. The survival of the fittest is a false doctrine in child-bearing as in fighting. Every child which starts with a reasonably good constitution can, by the right care up to one year and food up to five, grow up to be strong and well. And, if the weak and unhealthy are discouraged from breeding and healthy mothers given proper care, great improvements are possible. Poor food and over-crowding are the ladder down which we go to mental deficiency and ultimate complete feebleness of mind.[42] If we cared for life, the best

40 [Author's note] A woman of 45 years of age gave birth recently at Queen Charlotte's Hospital, Paddington, to her 23rd child. Ten children is not uncommon.

41 The British New Poor Law of 1834 was not abolished until 1948 in the expansion of the Welfare State after the Second World War. Dickens' *Oliver Twist* was an early critique of the old system.

42 [Author's note] Professor MacBride, dealing recently with the "Inheritance of Defect" (*Daily Telegraph*, January 8th, 1925) said: "The question of questions was whether the failure of the lowest strata of society was due to their surroundings or to their inborn characters. Such questions must be ultimately decided by experiment; and proper experimental work could only be done with animals; we were not entitled to make *corpora vilia* [bodies considered so worthless they can be

food would by law go to the pregnant and nursing mothers instead of, as at present, to clubs for fat old gentlemen and the frequenters of palatial hotels. It is probable that at present we do not produce enough milk, or produce and import enough butter and eggs to distribute adequately to all.[43] But, by stabilizing or decreasing our population, and by co-operation, intensive culture and control of marketing abroad and of marketing and purity at home, we could see to it that everybody had enough and that what they had was really good.

To feed an industrial population in a small island is a peculiar and special problem and one demanding expert care and advice. Food must come long distances and must "keep." Hence the preservatives and tins and the need to be watchful beyond measure against poisoning and the loss of what is vital to our well-being. With research, the problem would be easy; but we must make it clear that it is important. Science would easily enable us to produce more from the soil, and, as regards the food of mothers, since the assimilation of extra minerals, salts, etc., in their natural state is not always satisfactory or easy during pregnancy, we might find ways of growing food, through treatment of the soil, to provide for the special needs of their condition.

What then must feminist mothers demand? The right first of all to the recognition of their work—the most dangerous of all trades and the most neglected and despised. They should ask for endowment from the community. This is opposed by many on the ground that fathers delight to support their children, and it is they who should claim from the community an adequate family-wage. But, after all, it is the mother who bears and tends the child, and, although many women receive the whole of their husbands' wages, others must fight a humiliating battle against drink and tobacco for the wherewithal to build their children's bodies. This struggle is exemplified on a large scale in the spending of State revenue, most of which goes on armaments and the forces of destruction, and an infinitesimal portion to aid and support life. If Jason cannot give up his murderous playthings, let him have neither sons to destroy nor daughters to drag through misery. His children shall never be conceived. I have indicated that this is happening already, not as a deliberate revolt, but as a counsel of despair in a world which offers no hope, no joy, and no opportunity to the young.

The mother has a right to demand two years' rest between pregnancies; and the right to decide the number of her children. For some the call of motherhood is insistent and its charm grows with experience; they would be good mothers and might well have large families. They could help others by superintending

---

experimented upon] of our fellow human beings. For this reason he would direct attention to the common goldfish, whose weird monstrosities were all originally due to the starvation of the eggs with respect to light and air in the earliest stages of development. The result of this starvation was to weaken the developmental power and to produce a disharmonious arrest of growth of various organs. Similar arrests of growth occurred in human beings, and were the causes of mental and bodily defects. Their original cause, however, must be sought in the starvation and poisoning of the blood of the mother, but, once started, they were hereditary."

43 [Author's note] Working people live on tinned milk, margarine, and substitute eggs—all deficient in necessary vitamins.

nursery-schools in which children from one to five years might have their important meal of the day. But it is imperative that the woman who has children should not be shut out from public life. The ideal would be for a woman to continue her education at least till eighteen, have the first child at twenty-four, then perhaps three others at two-year intervals. This assumes that large numbers of women do not choose to breed. At thirty-five every mother of four children would, in a community of good schools, convenient houses, and well-run restaurants, be free again to take part in public life. It does not follow that she would be separated from her children: they would go to day-schools. But the mother would do the work for which she was best fitted in school,[44] kitchen, hospital, shop, mill, or Parliament. In this way her opinion would count, and her attitude to life help to permeate the community, which is otherwise left to be guided by the outlook of the single woman and the male. Problems of unemployment and competition due to married women's work are really questions of population pressure, muddled thinking, and bad organization. To discuss all this in close detail is hardly within the scope of this book.

In conclusion, it may be said that the community should never, except on the strongest grounds, deny parenthood to man or woman. Therefore marriages which after two years did not result in a child should be dissoluble at the wish of either party to the contract. This, apart from all other reasons for which the cancelling of marriage should be allowed. Partnership in marriage should in effect be regarded as a partnership for parenthood, and as such should not be entered upon lightly.

V. Jason and Admetus[45]

Men

BEFORE we pass on to an attempt at a summary and conclusion of the argument, it may be as well to re-state briefly what is the matter with men. Certainly they are not such fierce tyrants as when first we fought them; certainly they have some grounds to complain of the feminine arrogance which, not content with proving equality, wants to go on and prove women the superior sex. We might, on grounds of science perhaps, advance this claim, urging that, since a female being needs one more chromosome for its creation than a male, it must, therefore, be of higher importance. Should we do so, and seek to live alone on the planet, producing our

44 [Author's note] I am strongly of opinion that experience of maternity, even more than of marriage alone, would help the teacher. Some women, even teachers, are bored by children until they have one of their own, whereupon all children of all ages become interesting.

45 Admetus was one of the Argonauts and took part in the Calydonian Boar hunt. He was a favorite of Apollo, who often helped him. As told in Euripides' tragedy *Alcestis*, the greatest gift the god gave the mortal Admetus was to persuade the Fates (by getting them drunk) to reprieve Admetus of his fated death if he could find someone to take his place. Admetus thought one of his aged parents would volunteer, but neither did. Instead, his wife Alcestis agreed to die in his place. She is saved by the hero Heracles, who braves the underworld and wrestles with Thanatos (Death) to bring her back. In Russell's schema, Jason is the macho chauvinist, Admetus the timid ascetic intellectual.

children by parthenogenesis,[46] our pride would be doomed to a fall. Such children, there is reason to believe, would all be males. At least, that is what happens when the experiment is tried by sea-urchins. Men, on the other hand, like to pretend that our assumption of intelligence and independence is but a momentary spurt in a race which must end in masculine victory and feminine submission. They admit that the great development of our freedom in body and mind has given us a serious advantage, and the more discerning among them urge their fellows to press on and catch us up. Others trundle the golden apples beseechingly, but still Atalanta runs.

I believe it to be true that the education and outlook of men is more old-fashioned than that of women reared in the freedom of feminist traditions. Men have not yet realized how women's outlook is changing, nor attempted very seriously to adapt themselves to the change. They will do so, of a certainty; for, true as it may be that above all desires in woman is that to be pleasing to men, it is still truer that the desire of desires in man is to be pleasing to women. I believe that Puritanism or asceticism, of which they accuse us, is very strong in them. One of the compliments or insults that has been hurled during the sex-war is that the feminine mind is pervaded by the physical harmony of the feminine body. One may perhaps retort that the dualism of mind and matter is a very masculine philosophy; and one which, moreover, men have translated into their everyday lives by the sharp division they like to make between fighters and thinkers, games-playing idiots and thin intellectuals. Too often a woman of vitality and intelligence must choose between a soldier-gentleman and Chaucer's clerk.[47] Should she choose the former, she takes a plunge into the past. This man exults in murder, whether of animals or of his fellow-creatures; deep down within him he is still convinced that women are divided into good and bad—and both require the handling of a master. His wife must beware how she responds to his advances: she may be thought forward or impure. Decency must above all things be preserved. Though games and the classics may have taught the English gentleman the beauty of paganism and the joy of the naked body where man is concerned, he is still stuffy in his approach to sex. He rarely brings the freshness of the morning and the joy of the open skies to the love of mistress or of wife. Plush, gilt and silk stockings express the one; pipe, the armchair at the fireside, dinner, and a coldly furnished bedroom, the other. Conversation is a masculine monologue, punctuated by assent. He will be good to his children, provided they are not odd, and will protect his wife. He will never lift her to rapture. She fears, and will probably deceive him.

46 "Reproduction from a gamete without fertilization, occurring most commonly in invertebrates and lower plants" (*OED*).

47 [Author's note] A clerk ther was of Oxenford also
That unto logyk hadde longe y-go.
As leene was his hors as is a rake,
And he nas nat right fat, I undertake,
. . . .
And him was lever have at his beddes heed
Twenty bokes clad in black or reed.

The intellectual—perhaps by reason of the monastic tradition of learning, perhaps because he finds Jason so revolting—does all that he can to forget the needs of the body. Woman counts as one of them. She is a burden, a responsibility, a distraction, an incursion of the material into a world of contemplation. As for children and domestic life, they would make an end to all thought, to all art. An instinctive life—so he thinks—is possible only in spasms, if at all, for a man with serious mental work to accomplish. If woman persists in keeping him company, then she must shoulder the burdens, tend him and care for him, and leave him alone when he doesn't want her. It is this contempt for the natural play of instinct which eats the heart out of life for many intellectuals, men and women, of to-day. They dread the gift of themselves, the loss of independence which passion would bring, and therefore they never give freely. In part, they are cherishing the medieval tradition that to be worthy of spiritual or mental labour man and woman must go aside and renounce; in part, they are inspired by a tight conception of materialism, in which individuals are hurled like lumps of matter by dynamic forces through space, unable to do more than come near, but never mingle one with another. This view of life and the medieval are combining to destroy our world in lovelessness and despair.

The old-fashioned mind clings to spiritual duties and consolations and the framework of Church discipline as a bulwark against personal licence; the more modern mind is dominated by mechanism—which is, after all, no more than the rational control of matter—and seeks in an intelligent organization of the State, a framework within which each individual is to perform the duties for which he is best fitted. To neither conception is love between individuals, or sex-love between man and woman, important. In effect, personal relationships do not matter. The Christian doctrine of all-embracing love was once potent, but fails to-day because of the foundation of God, dogma, and Church on which it is built and which modern people cannot accept. "To love thy neighbour as thyself" is also inadequate without knowledge and understanding. But the rational materialists' attitude—such, for instance, is that of the Bolsheviks—to human relationships, in particular to women and sex, is as lacking in the sense of human dignity as the Christian. Monogamy and undiscriminating licence rest upon a common basis of contempt for love and personality, both asserting that the desire of a man is for a woman, of a woman for a man, but no matter whom. Dualism, as ever, is the culprit. Sex-love is to be no more than a physical need—no part of the serious business of life. Science has brought a more modern attitude to matter, which by its effect upon the imagination may change our conception of personality and sex. Force, struggle, solidity, contact, may yield to gentleness, non-resistance, intermingling and uniting—not by an ethical change, but by a change in scientific thought. We shall no longer think of mind and matter as wronging or thwarting one another, because they are not different forces; and we shall no longer be able to separate physical from mental virtue or depravity. We shall no longer value a love that suppresses or disregards the union of personality.

Taboos and superstitions, struggling dynamic individuals or States—how may we set up a new vision? Perhaps what I have written above seems far-fetched to the reader, but I do not think our life can be cut up into compartments. Philosophy and sex are more important in politics than General Elections. The revolt against the all-powerful Christian State began in the assertion of certain people that their love of good fruit and wine or their enjoyment of sex were not worthy of hell-fire. On personal conduct, on our standards of personal relationship, man to woman, parents to children, are built the customs and laws of States and ultimately their national and international policy. It is here, then, with man and woman, that we must begin. I have in mind, as I write, a piece of Chinese porcelain,[48] on which the sage or poet sits with his book and long pipe; a lovely and elegant lady peeps over his shoulder, and close at hand plays an impish child. I do not think that the Chinese who conceived it expected that poet to write bad verses, or, if a sage, to compose worthless philosophy. On the contrary, to love with devotion, to be learned, to have children, are ideas which have shaped the harmony of Chinese life. As compared with their generous acceptance of instinct our Christian dread of sex and horror of the body are obscene.

If we are to make peace between man and woman, and by their unity and partnership change the ideas that govern our politics and our outlook on the world, it is essential that men should make a more determined attempt to understand what feminists are seeking. It is useless to go on abusing, or pretending that this is a matter of minor importance. It is essential also that women should think clearly and continue in courage and honesty of word and action, neither abandoning all for the pursuit of pleasure nor glorying in opportunities for an oppressive morality belonging to past ages. First and foremost, man or woman, we are human beings. There is a great deal of the work in the community which we can each do with equal ability, given equal training and opportunity. There are other tasks which we must agree to delegate to one another, and neither despise the other for performing them. Life and harmony, generosity and peace are the ideals which the best thought of feminism has set before us. We believe that States and individuals can put them into practice. Will man not pause to understand before he continues on the path of destruction and strife, cupidity and war? Can we not persuade Jason from barbarity and Admetus to the abandonment of his fears? To live with vigour, body and mind and imagination, without fear or shame or dread of death; to drive these baser passions from the hold they have upon our morality and our politics—this is what we ask of modern men and women. They can come to it only in a reckless love of one another, a passion that gives again and again without fear of hurt or exhaustion. It is not an abandonment to nature and to instinct that we need. Pure and barbaric instinct is no more. Our bodies are too much impregnated by inherited habit and knowledge, too much surrounded in their growth by the findings of science. Men and women are not creatures of clay, nor disembodied spirits; but things of fire intertwining in understanding, torrents leaping to join in a cascade of mutual ecstasy. There is nothing in life to compare with this uniting of minds and bodies in men

48 See the frontispiece on p. 186 (Figure 9.1).

and women who have laid aside hostility and fear and seek in love the fullest understanding of themselves and of the universe. You cannot measure it in inches, nor turn it on and off like a tap. You cannot stay it now and indulge it another time. You cannot come to it by religion or by unaided reason, or by the brute courage of sheer physical vitality. Jealousy is death. Dualism is nonsense, compartments unavailing. You must have in you the thought that is creation; life's spring, and the daring of its unconquered waters—so may you transform the world and people it with gods who know no more the hates and littleness of men.

What I Believe[1]

Bertrand Russell (1925)

Headnote

Most of the volumes followed *Daedalus* and *Icarus* in taking their titles from Greek or Latin mythological figures. Some, like *Plato's American Republic*, played with the idea. Some took Latin phrases; some looked to Eastern mythology. A few had titles in English. The Latin for 'I believe' is *Credo*; but that would have sounded like an avowal of religious faith. Whereas the author of *Why I am not a Christian* (1927) two years later wants to set what he believes squarely against any religious or otherwise traditional beliefs.

******

Preface

In this little book, I have tried to say what I think of man's place in the universe, and of his possibilities in the way of achieving the good life. In *Icarus* I expressed my fears; in the following pages I have expressed my hopes. The inconsistency is only apparent. Except in astronomy, mankind have not achieved the art of predicting the future; in human affairs, we can see that there are forces making for happiness, and forces making for misery. We do not know which will prevail, but to act wisely we must be aware of both.

January 1*st*, 1925 B. R.

1 Russell published two different, shorter, essays with the same title, 'What I Believe.' The first, in the *Forum*, Vol. 82 (September 1929), 129–134, was written for the magazine's series called 'Credo.' The essay also appeared as 'How I Came by My Creed' in the British magazine which published several To-day and To-morrow writers (see 250, n. 18): the *Realist*, Vol. 1, No. 6 (September 1929), 14–21. The second was published in the New York *Nation*, Vol. 132 (29 April 1931), 469–471. The *Nation* reprinted it on 30 March 1940, 412–414, during a conservative legal challenge to his appointment to teach at the College of the City of New York. Since he had been accused of 'promulgating a moral code ranging from bolshevism to the advocacy of "eternal triangles",' the magazine wanted 'To refresh the public mind on Mr. Russell's real beliefs.'

DOI: 10.4324/9781032625522-16

Chapter I: Nature and Man

Man is a part of Nature, not something contrasted with Nature. His thoughts and his bodily movements follow the same laws that describe the motions of stars and atoms. The physical world is large compared with Man—larger than it was thought to be in Dante's time, but not so large as it seemed a hundred years ago. Both upward and downward, both in the large and in the small, science seems to be reaching limits. It is thought that the universe is of finite extent in space, and that light could travel round it in a few hundred millions of years. It is thought that matter consists of electrons and protons,[2] which are of finite size, and of which there are only a finite number in the world. Probably their changes are not continuous, as used to be thought, but proceed by jerks, which are never smaller than a certain minimum jerk.[3] The laws of these changes can apparently be summed up in a small number of very general principles, which determine the past and the future of the world when any small section of its history is known.

Physical science is thus approaching the stage when it will be complete, and therefore uninteresting.[4] Given the laws governing the motions of electrons and protons, the rest is merely geography—a collection of particular facts telling their distribution throughout some portion of the world's history. The total number of facts of geography required to determine the world's history is probably finite; theoretically, they could all be written down in a big book to be kept at Somerset House,[5] with a calculating machine attached, which, by turning a handle, would enable the inquirer to find out the facts at other times than those recorded. It is difficult to imagine anything less interesting, or more different from the passionate delights of incomplete discovery. It is like climbing a high mountain and finding nothing at the top except a restaurant where they sell ginger-beer, surrounded by fog but equipped with wireless. Perhaps in the time of Ahmes[6] the multiplication-table was exciting.

Of this physical world, uninteresting in itself, Man is a part. His body, like other matter, is composed of electrons and protons, which, so far as we know, obey the same laws as those not forming part of animals or plants. There are some who maintain that physiology can never be reduced to physics, but their arguments are not very convincing and it seems prudent to suppose that they are mistaken. What we call our "thoughts" seem to depend upon the organization of tracks in the brain in the same sort of way in which journeys depend upon roads and railways. The energy used in thinking seems to have a chemical origin; for instance, a deficiency of iodine will turn a clever man into an idiot. Mental phenomena seem to be bound up

2 The neutron was not discovered until 1932, by James Chadwick.

3 Niels Bohr had published his electron shell theory in 1922.

4 This judgment was premature. The standard model of particle physics currently recognizes 17 distinct elementary particles. Competing versions of string theory attempt to reconcile subatomic forces with gravity, and quantum mechanics with relativity.

5 On the Strand, London; it housed the General Register Office, recording births, deaths and marriage; and the Inland Revenue.

6 Ancient Egyptian scribe, living around 1550 BCE, and the first mathematician whose name is known.

with material structure. If this be so, we cannot suppose that a solitary electron or proton can "think"; we might as well expect a solitary individual to play a football match. We also cannot suppose that an individual's thinking survives bodily death, since that destroys the organization of the brain, and dissipates the energy which utilized the brain-tracks.

God and immortality, the central dogmas of the Christian religion, find no support in science. It cannot be said that either doctrine is essential to religion, since neither is found in Buddhism. (With regard to immortality, this statement in an unqualified form might be misleading, but it is correct in the last analysis). But we in the West have come to think of them as the irreducible minimum of theology. No doubt people will continue to entertain these beliefs, because they are pleasant, just as it is pleasant to think ourselves virtuous and our enemies wicked. But for my part I cannot see any ground for either. I do not pretend to be able to prove that there is no God. I equally cannot prove that Satan is a fiction. The Christian God may exist; so may the Gods of Olympus, or of ancient Egypt, or of Babylon. But no one of these hypotheses is more probable than any other: they lie outside the region of even probable knowledge, and therefore there is no reason to consider any of them. I shall not enlarge upon this question, as I have dealt with it elsewhere.[7]

The question of personal immortality stands on a somewhat different footing. Here evidence either way is possible. Persons are part of the everyday world with which science is concerned, and the conditions which determine their existence are discoverable. A drop of water is not immortal; it can be resolved into oxygen and hydrogen. If, therefore, a drop of water were to maintain that it had a quality of aqueousness which would survive its dissolution, we should be inclined to be sceptical. In like manner we know that the brain is not immortal, and that the organized energy of a living body becomes, as it were, demobilized at death, and therefore not available for collective action. All the evidence goes to show that what we regard as our mental life is bound up with brain-structure and organized bodily energy. Therefore it is rational to suppose that mental life ceases when bodily life ceases. The argument is only one of probability, but it is as strong as those upon which most scientific conclusions are based.

There are various grounds upon which this conclusion might be attacked. Psychical research professes to have actual scientific evidence of survival, and undoubtedly its procedure is, in principle, scientifically correct. Evidence of this sort might be so overwhelming that no one with a scientific temper could reject it. The weight to be attached to the evidence, however, must depend upon the antecedent probability of the hypothesis of survival. There are always different ways of accounting for any set of phenomena, and of these we should prefer the one which is antecedentally least improbable. Those who already think it likely that we survive death will be ready to view this theory as the best explanation of psychical phenomena. Those, who, on other grounds, regard this theory as unplausible will seek for other explanations. For my part, I consider the evidence so far adduced

7 [Author's note] See my *Philosophy of Leibniz*, Chapter XV.

by psychical research in favour of survival much weaker than the physiological evidence on the other side. But I fully admit that it might at any moment become stronger, and in that case it would be unscientific to disbelieve in survival.

Survival of bodily death is, however, a different matter from immortality: it may only mean a postponement of psychical death. It is immortality that men desire to believe in. Believers in immortality will object to physiological arguments, such as I have been using, on the ground that soul and body are totally disparate, and that the soul is something quite other than its empirical manifestations through our bodily organs. I believe this to be a metaphysical superstition. Mind and matter alike are for certain purposes convenient terms, but are not ultimate realities. Electrons and protons, like the soul, are logical fictions: each is really a history, a series of events, not a single persistent entity. In the case of the soul, this is obvious from the facts of growth. Whoever considers conception, gestation, and infancy cannot seriously believe that the soul is an indivisible something, perfect and complete throughout this process. It is evident that it grows like the body, and that it derives both from the spermatozoon and from the ovum, so that it cannot be indivisible. This is not materialism: it is merely the recognition that everything interesting is a matter of organization, not of primal substance.

Metaphysicians have advanced innumerable arguments to prove that the soul *must* be immortal. There is one simple test by which all these arguments can be demolished. They all prove equally that the soul must pervade all space. But as we are not so anxious to be fat as to live long, none of the metaphysicians in question have ever noticed this application of their reasonings. This is an instance of the amazing power of desire in blinding even very able men to fallacies which would otherwise be obvious at once. If we were not afraid of death, I do not believe that the idea of immortality would ever have arisen.

Fear is the basis of religious dogma, as of so much else in human life. Fear of human beings, individually or collectively, dominates much of our social life, but it is fear of nature that gives rise to religion. The antithesis of mind and matter is, as we have seen, more or less illusory; but there is another antithesis which is more important—that, namely, between things that can be affected by our desires and things that cannot be so affected. The line between the two is neither sharp nor immutable—as science advances, more and more things are brought under human control. Nevertheless there remain things definitely on the other side. Among these are all the *large* facts of our world, the sort of facts that are dealt with by astronomy. It is only facts on or near the surface of the earth that we can, to some extent, mould to suit our desires. And even on the surface of the earth, our powers are very limited. Above all, we cannot prevent death, although we can often delay it.

Religion is an attempt to overcome this antithesis. If the world is controlled by God, and God can be moved by prayer, we acquire a share in omnipotence. In former days, miracles happened in answer to prayer; they still do in the Catholic Church, but Protestants have lost this power. However, it is possible to dispense with miracles, since Providence has decreed that the operation of natural laws shall produce the best possible results. Thus belief in God still serves to humanize the world of nature, and to make men feel that physical forces are really their allies.

In like manner immortality removes the terror from death. People who believe that when they die they will inherit eternal bliss may be expected to view death without horror, though, fortunately for medical men, this does not invariably happen. It does, however, soothe men's fears somewhat, even when it cannot allay them wholly.

Religion, since it has its source in terror, has dignified certain kinds of fear, and made people think them not disgraceful. In this it has done mankind a great disservice: *all* fear is bad, and ought to be overcome not by fairy tales, but by courage and rational reflection. I believe that when I die I shall rot, and nothing of my ego will survive. I am not young, and I love life.[8] But I should scorn to shiver with terror at the thought of annihilation. Happiness is none the less true happiness because it must come to an end, nor do thought and love lose their value because they are not everlasting. Many a man has borne himself proudly on the scaffold: surely the same pride should teach us to think truly about man's place in the world. Even if the open windows of science at first make us shiver after the cosy indoor warmth of traditional humanizing myths, in the end the fresh air brings vigour, and the great spaces have a splendour of their own.

The philosophy of nature is one thing, the philosophy of value is quite another. Nothing but harm can come of confusing them. What we think good, what we should like, has no bearing whatever upon what is, which is the question for the philosophy of nature. On the other hand, we cannot be forbidden to value this or that on the ground that the non-human world does not value it, nor can we be compelled to admire anything because it is a "law of nature." Undoubtedly we are part of nature, which has produced our desires, our hopes and fears, in accordance with laws which the physicist is beginning to discover. In this sense we are part of nature; in the philosophy of nature, we are subordinated to nature, the outcome of natural laws, and their victims in the long run.

The philosophy of nature must not be unduly terrestrial; for it, the earth is merely one of the smaller planets of one of the smaller stars of the Milky Way. It would be ridiculous to warp the philosophy of nature in order to bring out results that are pleasing to the tiny parasites of this insignificant planet. Vitalism as a philosophy, and evolutionism, show, in this respect, a lack of sense of proportion and logical relevance. They regard the facts of life, which are personally interesting to us, as having a cosmic significance, not a significance confined to the earth's surface. Optimism and pessimism, as cosmic philosophies, show the same naive humanism: the great world, so far as we know it from the philosophy of nature, is neither good nor bad, and is not concerned to make us either happy or unhappy. All such philosophies spring from self-importance, and are best corrected by a little astronomy.

But in the philosophy of value the situation is reversed. Nature is only a part of what we can imagine; everything, real or imagined, can be appraised by us, and there is no outside standard to show that our valuation is wrong. We are ourselves the ultimate and irrefutable arbiters of value, and in the world of value Nature is only a part. Thus in this world we are greater than Nature. In the world of values,

8 Russell turned 53 in 1925. He lived to the age of 97.

Nature in itself is neutral, neither good nor bad, deserving of neither admiration nor censure. It is we who create value, and our desires which confer value. In this realm we are kings, and we debase our kingship if we bow down to Nature. It is for us to determine the good life, not for Nature—not even for Nature personified as God.

Chapter II: The Good Life

There have been at different times and among different people many varying conceptions of the good life. To some extent the differences were amenable to argument; this was when men differed as to the means to achieve a given end. Some think that prison is a good way of preventing crime; others hold that education would be better. A difference of this sort can be decided by sufficient evidence. But some differences cannot be tested in this way. Tolstoy condemned all war; others have held the life of a soldier doing battle for the right to be very noble. Here there was probably involved a real difference as to ends. Those who praise the soldier usually consider the punishment of sinners a good thing in itself; Tolstoy did not think so. On such a matter, no argument is possible. I cannot, therefore, prove that my view of the good life is right; I can only state my view, and hope that as many as possible will agree. My view is this:

The Good Life Is One Inspired by Love and Guided by Knowledge.

Knowledge and love are both indefinitely extensible; therefore, however good a life may be, a better life can be imagined. Neither love without knowledge, nor knowledge without love can produce a good life. In the Middle Ages, when pestilence appeared in a country, holy men advised the population to assemble in churches and pray for deliverance; the result was that the infection spread with extraordinary rapidity among the crowded masses of supplicants. This was an example of love without knowledge. The late War afforded an example of knowledge without love. In each case, the result was death on a large scale.

Although both love and knowledge are necessary, love is in a sense more fundamental, since it will lead intelligent people to seek knowledge, in order to find out how to benefit those whom they love. But if people are not intelligent, they will be content to believe what they have been told, and may do harm in spite of the most genuine benevolence. Medicine affords, perhaps, the best example of what I mean. An able physician is more useful to a patient than the most devoted friend, and progress in medical knowledge does more for the health of the community than ill-informed philanthropy. Nevertheless, an element of benevolence is essential even here, if any but the rich are to profit by scientific discoveries.

Love is a word which covers a variety of feelings; I have used it purposely, as I wish to include them all. Love as an emotion—which is what I am speaking about, for love "on principle" does not seem to me genuine—moves between two poles: on the one side, pure delight in contemplation; on the other, pure benevolence. Where inanimate objects are concerned, delight alone enters in: we cannot feel benevolence towards a landscape or a sonata. This type of enjoyment is presumably the source of art. It is stronger, as a rule, in very young children than in adults, who are apt to view objects in a utilitarian spirit. It plays a large part in our feelings towards human beings, some of whom have charm and some the reverse, when considered simply as objects of aesthetic contemplation.

The opposite pole of love is pure benevolence. Men have sacrificed their lives to helping lepers; in such a case, the love they felt cannot have had any element of aesthetic delight. Parental affection, as a rule, is accompanied by pleasure in the child's appearance, but remains strong when this element is wholly absent. It would seem odd to call a mother's interest in a sick child "benevolence," because we are in the habit of using this word to describe a pale emotion nine parts humbug. But it is difficult to find any other word to describe the desire for another person's welfare. It is a fact that a desire of this sort may reach any degree of strength in the case of parental feeling. In other cases it is far less intense; indeed it would seem likely that all altruistic emotion is a sort of overflow of parental feeling, or sometimes a sublimation of it. For want of a better word, I shall call this emotion "benevolence". But I want to make it clear that I am speaking of an emotion, not a principle, and that I do not include in it any feeling of superiority such as is sometimes associated with the word. The word "sympathy" expresses part of what I mean, but leaves out the element of activity that I wish to include.

Love at its fullest is an indissoluble combination of the two elements, delight and well-wishing. The pleasure of a parent in a beautiful and successful child combines both elements; so does sex-love at its best. But in sex-love benevolence will only exist where there is secure possession, since otherwise jealousy will destroy it, while perhaps actually increasing the delight in contemplation. Delight without well-wishing may be cruel; well-wishing without delight easily tends to become cold and a little superior. A person who wishes to be loved wishes to be the object of a love containing both elements, except in cases of extreme weakness, such as infancy and severe illness. In these cases benevolence may be all that is desired. Conversely, in cases of extreme strength admiration is more desired than benevolence: this is the state of mind of potentates and famous beauties. We only desire other people's good wishes in proportion as we feel ourselves in need of help or in danger of harm from them. At least, that would seem to be the biological logic of the situation, but it is not quite true to life. We desire affection in order to escape from the feeling of loneliness, in order to be, as we say, "understood". This is a matter of sympathy, not merely of benevolence; the person whose affection is satisfactory to us must not merely wish us well, but must know in what our happiness consists. But this belongs to the other element of the good life, namely knowledge.

In a perfect world, every sentient being would be to every other the object of the fullest love, compounded of delight, benevolence, and understanding inextricably blended. It does not follow that, in this actual world, we ought to attempt to have such feelings towards all the sentient beings whom we encounter. There are many in whom we cannot feel delight, because they are disgusting; if we were to do violence to our nature by trying to see beauties in them, we should merely blunt our susceptibilities to what we naturally find beautiful. Not to mention human beings, there are fleas and bugs and lice. We should have to be as hard pressed as the Ancient Mariner before we could feel delight in contemplating these creatures.

Some saints, it is true, have called them "pearls of God", but what these men delighted in was the opportunity of displaying their own sanctity.

Benevolence is easier to extend widely, but even benevolence has its limits. If a man wished to marry a lady, we should not think the better of him for withdrawing if he found that some one else also wished to marry her: we should regard this as a fair field for competition. Yet his feelings towards a rival cannot be *wholly* benevolent. I think that in all descriptions of the good life here on earth we must assume a certain basis of animal vitality and animal instinct; without this, life becomes tame and uninteresting. Civilization should be something added to this, not substituted for it; the ascetic saint and the detached sage fail in this respect to be complete human beings. A small number of them may enrich a community; but a world composed of them would die of boredom.

These considerations lead to a certain emphasis on the element of delight as an ingredient in the best love. Delight, in this actual world, is unavoidably selective, and prevents us from having the same feelings towards all mankind. When conflicts arise between delight and benevolence, they must, as a rule, be decided by a compromise, not by a complete surrender of either. Instinct has its rights, and if we do violence to it beyond a point it takes vengeance in subtle ways. Therefore in aiming at a good life the limits of human possibility must be borne in mind. Here again, however, we are brought back to the necessity of knowledge.

When I speak of knowledge as an ingredient of the good life, I am not thinking of ethical knowledge, but of scientific knowledge and knowledge of particular facts. I do not think there is, strictly speaking, such a thing as ethical knowledge. If we desire to achieve some end, knowledge may show us the means, and this knowledge may loosely pass as ethical. But I do not believe that we can decide what sort of conduct is right or wrong except by reference to its probable consequences. Given an end to be achieved, it is a question for science to discover how to achieve it. All moral rules must be tested by examining whether they tend to realize ends that we desire. I say ends that we desire, not ends that we *ought* to desire. What we "ought" to desire is merely what someone else wishes us to desire. Usually it is what the authorities wish us to desire—parents, schoolmasters, policemen, and judges. If you say to me "you ought to do so-and-so", the motive power of your remark lies in my desire for your approval—together, possibly, with rewards or punishments attached to your approval or disapproval. Since all behaviour springs from desire, it is clear that ethical notions can have no importance except as they influence desire. They do this through the desire for approval and the fear of disapproval. These are powerful social forces, and we shall naturally endeavour to win them to our side if we wish to realize any social purpose. When I say that the morality of conduct is to be judged by its probable consequences, I mean that I desire to see approval given to behaviour likely to realize social purposes which we desire, and disapproval to opposite behaviour. At present this is not done; there are certain traditional rules according to which approval and disapproval are meted out quite regardless of consequences. But this is a topic with which we shall deal in the next chapter.

The superfluity of theoretical ethics is obvious in simple cases. Suppose, for instance, that your child is ill. Love makes you wish to cure it, and science tells you how to do so. There is not an intermediate stage of ethical theory, where it is demonstrated that your child had better be cured. Your act springs directly from desire for an end, together with knowledge of means. This is equally true of all acts, whether good or bad. The ends differ, and the knowledge is more adequate in some cases than in others. But there is no conceivable way of making people do things they do not wish to do. What is possible is to alter their desires by a system of rewards and penalties, among which social approval and disapproval are not the least potent. The question for the legislative moralist is, therefore: How shall this system of rewards and punishments be arranged so as to secure the maximum of what is desired by the legislative authority? If I say that the legislative authority has bad desires, I mean merely that its desires conflict with those of some section of the community to which I belong. Outside human desires there is no moral standard.

Thus, what distinguishes ethics from science is not any special kind of knowledge, but merely desire. The knowledge required in ethics is exactly like the knowledge elsewhere; what is peculiar is that certain ends are desired, and that right conduct is what conduces to them. Of course, if the definition of right conduct is to make a wide appeal, the ends must be such as large sections of mankind desire. If I defined right conduct as that which increases my own income, readers would disagree. The whole effectiveness of any ethical argument lies in its scientific part, i.e. in the proof that one kind of conduct, rather than some other, is a means to an end which is widely desired. I distinguish, however, between ethical argument and ethical education. The latter consists in strengthening certain desires and weakening others. This is quite a different process, which will be separately discussed at a later stage.

We can now explain more exactly the purport of the definition of the good life with which this chapter began. When I said that the good life consists of love guided by knowledge, the desire which prompted me was the desire to live such a life as far as possible, and to see others living it; and the logical content of the statement is that, in a community where men live in this way, more desires will be satisfied than in one where there is less love or less knowledge. I do not mean that such a life is "virtuous" or that its opposite is "sinful", for these are conceptions which seem to me to have no scientific justification.

Chapter III: Moral Rules

The practical need of morals arises from the conflict of desires, whether of different people or of the same person at different times or even at one time. A man desires to drink, and also to be fit for his work next morning. We think him immoral if he adopts the course which gives him the smaller total satisfaction of desire. We think ill of people who are extravagant or reckless, even if they injure no one but themselves. Bentham supposed that the whole of morality could be derived from "enlightened self-interest", and that a person who always acted with a view to his own maximum satisfaction in the long run would always act rightly. I cannot accept this view. Tyrants have existed who derived exquisite pleasure from watching the infliction of torture; I cannot praise such men when prudence led them to spare their victims' lives with a view to further sufferings another day. Nevertheless, other things being equal, prudence is a part of the good life. Even Robinson Crusoe had occasion to practise industry, self-control, and foresight, which must be reckoned as moral qualities, since they increased his total satisfaction without counterbalancing injury to others. This part of morals plays a great part in the training of young children, who have little inclination to think of the future. If it were more practised in later life, the world would quickly become a paradise, since it would be quite sufficient to prevent wars, which are acts of passion, not of reason. Nevertheless, in spite of the importance of prudence, it is not the most interesting part of morals. Nor is it the part that raises intellectual problems, since it does not require an appeal to anything beyond self-interest.

The part of morality that is not included in prudence is, in essence, analogous to law, or the rules of a club. It is a method of enabling men to live together in a community in spite of the possibility that their desires may conflict. But here two very different methods are possible. There is the method of the criminal law, which aims at a merely external harmony by attaching disagreeable consequences to acts which thwart other men's desires in certain ways. This is also the method of social censure: to be thought ill of by one's own society is a form of punishment, to avoid which most people avoid being known to transgress the code of their set. But there is another method, more fundamental, and far more satisfactory when it succeeds. This is to alter men's characters and desires in such a way as to minimize occasions of conflict by making the success of one man's desires as far as possible consistent with that of another's. That is why love is better than hate, because it brings harmony instead of conflict into the desires of the persons concerned. Two people between whom there is love succeed or fail together, but when two people hate each other the success of either is the failure of the other.

If we were right in saying that the good life is inspired by love and guided by knowledge, it is clear that the moral code of any community is not ultimate and self-sufficient, but must be examined with a view to seeing whether it is such as wisdom and benevolence would have decreed. Moral codes have not always been faultless. The Aztecs considered it their painful duty to eat human flesh for fear the light of the sun should grow dim. They erred in their science; and perhaps they would have perceived the scientific error if they had had any love for the sacrificial victims. Some tribes immerse girls in the dark from the age of ten to the age of 17, for fear the sun's rays should render them pregnant. But surely our modern codes of morals contain nothing analogous to these savage practices? Surely we only forbid things which really are harmful, or at any rate so abominable that no decent person could defend them? I am not so sure.

Current morality is a curious blend of utilitarianism and superstition, but the superstitious part has the stronger hold, as is natural, since superstition is the origin of moral rules. Originally, certain acts were thought displeasing to the gods, and were forbidden by law because the divine wrath was apt to descend upon the community, not merely upon the guilty individuals. Hence arose the conception of sin, as that which is displeasing to God. No reason can be assigned as to why certain acts should be thus displeasing; it would be very difficult to say, for instance, why it was displeasing that the kid should be seethed in its mother's milk.[9] But it was known by Revelation that this was the case. Sometimes the Divine commands have been curiously interpreted. For example, we are told not to work on Saturdays, and Protestants take this to mean that we are not to play on Sundays. But the same sublime authority is attributed to the new prohibition as to the old.

It is evident that a man with a scientific outlook on life cannot let himself be intimidated by texts of Scripture or by the teaching of the Church. He will not be content to say "such-and-such an act is sinful, and that ends the matter". He will inquire whether it does any harm, or whether, on the contrary, the belief that it is sinful does harm. And he will find that, especially in what concerns sex, our current morality contains a very great deal of which the origin is purely superstitious. He will find also that this superstition, like that of the Aztecs, involves needless cruelty, and would be swept away if people were actuated by kindly feelings towards their neighbours. But the defenders of traditional morality are seldom people with warm hearts, as may be seen from the love of militarism displayed by Church dignitaries. One is tempted to think that they value morals as affording a legitimate outlet for their desire to inflict pain: the sinner is fair game, and therefore away with tolerance!

Let us follow an ordinary human life from conception to the grave, and note the points where superstitious morals inflict preventable suffering. I begin with conception, because here the influence of superstition is particularly noteworthy. If the parents are not married, the child has a stigma, as clearly undeserved as anything could be. If either of the parents has venereal disease, the child is likely to inherit it. If they already have too many children for the family income, there will be poverty, underfeeding, overcrowding, very likely incest. Yet the great majority of moralists agree that the parents had better not know how to prevent this misery by preventing conception. To please these moralists, a life of torture is inflicted upon millions of human beings who ought never to have existed, merely because it is supposed that sexual intercourse is wicked unless accompanied by desire for offspring, but not wicked when this desire is present, even though the offspring is humanly certain to be wretched. To be killed suddenly and then eaten, which was the fate of the Aztecs' victims, is a far less degree of suffering than is inflicted upon a child born in miserable surroundings and tainted with venereal disease. Yet it is the greater suffering which is deliberately inflicted by Bishops and politicians in the name of morality. If they had even the smallest spark of love or pity for children they could not adhere to a moral code involving this fiendish cruelty.

At birth, and in early infancy, the average child suffers more from economic causes than from superstition. When well-to-do women have children, they have the best doctors, the best nurses, the best diet, the best rest, and the best exercise.

9 *Exodus* 23:19. 'Seethe' meant to boil or stew.

Working-class women do not enjoy these advantages, and frequently their children die for lack of them. A little is done by the public authorities in the way of care of mothers, but very grudgingly. At a moment when the supply of milk to nursing mothers is being cut down to save expense, public authorities will spend vast sums on paving rich residential districts where there is little traffic. They must know that in taking this decision they are condemning a certain number of working-class children to death for the crime of poverty. Yet the ruling party are supported by the immense majority of ministers of religion, who, with the Pope at their head, have pledged the vast forces of superstition throughout the world to the support of social injustice.

In all stages of education the influence of superstition is disastrous. A certain percentage of children have the habit of thinking; one of the aims of education is to cure them of this habit. Inconvenient questions are met with "hush, hush" or with punishment. Collective emotion is used to instil certain kinds of belief, more particularly nationalistic kinds. Capitalists, militarists, and ecclesiastics co-operate in education, because all depend for their power upon the prevalence of emotionalism and the rarity of critical judgment. With the aid of human nature, education succeeds in increasing and intensifying these propensities of the average man.

Another way in which superstition damages education is through its influence on the choice of teachers. For economic reasons, a woman-teacher must not be married; for moral reasons, she must not have extra-marital sexual relations. And yet everybody who has taken the trouble to study morbid psychology knows that prolonged virginity is, as a rule, extraordinarily harmful to women, so harmful that, in a sane society, it would be severely discouraged in teachers. The restrictions imposed lead more and more to a refusal, on the part of energetic and enterprising women, to enter the teaching profession. This is all due to the lingering influence of superstitious asceticism.

At middle- and upper-class schools the matter is even worse. There are chapel services, and the care of morals is in the hands of clergymen. Clergymen, almost necessarily, fail in two ways as teachers of morals. They condemn acts which do no harm, and they condone acts which do great harm. They all condemn sexual relations between unmarried people who are fond of each other but not yet sure that they wish to live together all their lives. Most of them condemn birth-control. None of them condemn the brutality of a husband who causes his wife to die of too frequent pregnancies. I knew a fashionable clergyman whose wife had nine children in nine years. The doctors told him that if she had another she would die. Next year she had another and died. No one condemned him; he retained his benefice, and married again. So long as clergymen continue to condone cruelty and condemn innocent pleasure, they can only do harm as guardians of the morals of the young.

Another bad effect of superstition on education is the absence of instruction about the facts of sex. The main physiological facts ought to be taught, quite simply and naturally, before puberty, at a time when they are not exciting. At puberty, the elements of an unsuperstitious sexual morality ought to be taught. Boys and girls should be taught that nothing can justify sexual intercourse unless there is mutual inclination. This is contrary to the teaching of the Church, which holds that, provided the parties are married and the man desires another child, sexual intercourse is justified however great may be the reluctance of the wife. Boys and girls should be taught respect for each other's liberty; they should be made to feel that nothing gives one human being rights over another, and that jealousy and possessiveness kill love. They should be taught that to bring another human being into the world is a very serious matter, only to be undertaken when the child will have a reasonable prospect of health, good surroundings, and parental care. But they should also be taught methods of birth-control, so as to insure that children shall only come when they are wanted. Finally, they should be taught the dangers of venereal disease, and the methods of prevention and cure. The increase of human happiness to be expected from sex-education on these lines is immeasurable.

It should be recognized that, in the absence of children, sexual relations are a purely private matter, which does not concern either the State or the neighbours. Certain forms of sex which do not lead to children are at present punished by the criminal law: this is purely superstitious, since the matter is one which affects no one except the parties directly concerned. Where there are children, it is a mistake to suppose that it is necessarily to their interest to make divorce very difficult. Habitual drunkenness, cruelty, insanity are grounds upon which divorce is necessary for the children's sake quite as much as for the sake of the wife or husband. The peculiar importance attached, at present, to adultery is quite irrational. It is obvious that many forms of misconduct are more fatal to married happiness than an occasional infidelity. Masculine insistence on a child a year, which is not conventionally misconduct or cruelty, is the most fatal of all.

Moral rules ought not to be such as to make instinctive happiness impossible. Yet that is an effect of strict monogamy, in a community where the numbers of the two sexes are very unequal. Of course, under such circumstances, the moral rules are infringed. But when the rules are such that they can only be obeyed by greatly diminishing the happiness of the community, and when it is better they should be infringed than observed, surely it is time that the rules were changed. If this is not done, many people who are acting in a way not contrary to the public interest are faced with the undeserved alternative of hypocrisy or obloquy. The Church does not mind hypocrisy, which is a flattering tribute to its power; but elsewhere it has come to be recognized as an evil which we ought not lightly to inflict.

Even more harmful than theological superstition is the superstition of nationalism, of duty to one's own State and to no other. But I do not propose on this occasion to discuss this matter, beyond pointing out that limitation to one's compatriots is contrary to the principle of love which we recognized as constituting the good life. It is also, of course, contrary to enlightened self-interest, since an exclusive nationalism does not pay even the victorious nations.

One other respect in which our society suffers from the theological conception of "sin" is the treatment of criminals. The view that criminals are "wicked" and "deserve" punishment is not one which a rational morality can support. Undoubtedly certain people do things which society wishes to prevent, and does right in preventing as far as possible. We may take murder as the plainest case. Obviously, if a community is to hold together and we are to enjoy its pleasures and advantages, we cannot allow people to kill each other whenever they feel an impulse to do so. But this problem should be treated in a purely scientific spirit. We should ask simply: What is the best method of preventing murder? Of two methods which are equally effective in preventing murder, the one involving least harm to the murderer is to be preferred. The harm to the murderer is wholly regrettable, like the pain of a surgical operation. It may be equally necessary, but it is not a subject of rejoicing. The vindictive feeling called "moral indignation" is merely a form of cruelty. Suffering to the criminal can never be justified by the notion of vindictive punishment. If education combined with kindness is equally effective, it is to be preferred; still more is it to be preferred if it is more effective. Of course the prevention of crime and the punishment of crime are two different questions; the object of causing pain to the criminal is presumably deterrent. If prisons were so humanized that a prisoner got a good education for nothing, people might commit crimes in order to qualify for entrance. No doubt prison must be less pleasant than freedom; but the best way to secure this result is to make freedom more pleasant than it sometimes is at present. I do not wish, however, to embark upon the subject of Penal Reform. I merely wish to suggest that we should treat the criminal as we treat a man suffering from plague. Each is a public danger, each must have his liberty curtailed until he has ceased to be a danger. But the man suffering from plague is an object of sympathy and commiseration, whereas the criminal is an object of execration. This is quite irrational. And it is because of this difference of attitude that our prisons are so much less successful in curing criminal tendencies than our hospitals are in curing disease.

Chapter IV: Salvation: Individual and Social

One of the defects of traditional religion is its individualism, and this defect belongs also to the morality associated with it. Traditionally, the religious life was, as it were, a duologue between the soul and God. To obey the will of God was virtue; and this was possible for the individual quite regardless of the state of the community. Protestant sects developed the idea of "finding salvation", but it was always present in Christian teaching. This individualism of the separate soul had its value at certain stages of history, but in the modern world we need rather a social than an individual conception of welfare. I want to consider, in this chapter, how this affects our conception of the good life.

Christianity arose in the Roman Empire among populations wholly destitute of political power, whose national States had been destroyed and merged in a vast impersonal aggregate. During the first three centuries of the Christian Era, the individuals who adopted Christianity could not alter the social or political institutions under which they lived, although they were profoundly convinced of their badness. In these circumstances, it was natural that they should adopt the belief that an individual may be perfect in an imperfect world, and that the good life has nothing to do with this world. What I mean may become plain by comparison with Plato's Republic. When Plato wanted to describe the good life, he described a whole community, not an individual; he did so in order to define justice, which is an essentially social conception. He was accustomed to citizenship of a Republic, and political responsibility was something which he took for granted. With the loss of Greek freedom comes the rise of Stoicism, which is like Christianity, and unlike Plato, in having an individualistic conception of the good life.

We, who belong to great democracies, should find a more appropriate morality in free Athens than in despotic Imperial Rome. In India, where the political circumstances are very similar to those of Judea in the time of Christ, we find Gandhi preaching a very similar morality to Christ's, and being punished for it by the christianized successors of Pontius Pilate. But the more extreme Indian nationalists are not content with individual salvation: they want national salvation. In this they have taken on the outlook of the free democracies of the West. I want to suggest some respects in which this outlook, owing to Christian influences, is not yet sufficiently bold and self-conscious, but is still hampered by the belief in individual salvation.

The good life, as we conceive it, demands a multitude of social conditions, and cannot be realized without them. The good life, we said, is a life inspired by love and guided by knowledge. The knowledge required can only exist where governments or millionaires devote themselves to its discovery and diffusion. For example: the spread of cancer is alarming—what are we to do about it? At the moment, no one can answer the question for lack of knowledge; and the knowledge is not likely to emerge except through endowed research. Again: knowledge of science, history, literature, and art ought to be attainable by all who desire it; this requires elaborate arrangements on the part of public authorities, and is not to be achieved by means of religious conversion. Then there is foreign trade, without which half the inhabitants of Great Britain would starve; and if we were starving very few of us would live the good life. It is needless to multiply examples. The important point is that, in all that differentiates between a good life and a bad one, the world is a unity, and the man who pretends to live independently is a conscious or unconscious parasite.

The idea of individual salvation, with which the early Christians consoled themselves for their political subjection, becomes impossible as soon as we escape from a very narrow conception of the good life. In the orthodox Christian conception, the good life is the virtuous life, and virtue consists in obedience to the will of God, and the will of God is revealed to each individual through the voice of conscience. This whole conception is that of men subject to an alien despotism. The good life involves much besides virtue—intelligence, for instance. And conscience is a most fallacious guide, since it consists of vague reminiscences of precepts heard in early youth, so that it is never wiser than its possessor's nurse or mother. To live a good life in the fullest sense a man must have a good education, friends, love, children (if he desires them), a sufficient income to keep him from want and grave anxiety, good health, and work which is not uninteresting. All these things, in varying degrees, depend upon the community, and are helped or hindered by political events. The good life must be lived in a good society, and is not fully possible otherwise.

This is the fundamental defect of the aristocratic ideal. Certain good things, such as art and science and friendship, can flourish very well in an aristocratic society. They existed in Greece on a basis of slavery; they exist among ourselves on a basis of exploitation. But love, in the form of sympathy or benevolence, cannot exist freely in an aristocratic society. The aristocrat has to persuade himself that the slave or proletarian or coloured man is of inferior clay, and that his sufferings do not matter. At the present moment, polished English gentlemen flog Africans so severely that they die after hours of unspeakable anguish.[10] Even if these gentlemen are well-educated, artistic, and admirable conversationalists, I cannot admit that they are living the good life. Human nature imposes some limitation of sympathy, but not such a degree as that. In a democratically-minded society, only a maniac would behave in this way. The limitation of sympathy involved in the aristocratic ideal is its condemnation. Salvation is an aristocratic ideal, because it is individualistic. For this reason, also, the idea of personal salvation, however interpreted and expanded, cannot serve for the definition of the good life.

Another characteristic of salvation is that it results from a catastrophic change, like the conversion of Saint Paul. Shelley's poems afford an illustration of this conception applied to societies; the moment comes when everybody is converted, the "anarchs" fly, and "the world's great age begins anew".[11] It may be said that a poet is an unimportant person, whose ideas are of no consequence. But I am persuaded that a large proportion of revolutionary leaders have had ideas extremely like Shelley's. They have thought that misery and cruelty and degradation were due to tyrants or priests or capitalists or Germans, and that if these sources of evil were overthrown there would be a general change of heart and we should all live happy ever after. Holding these beliefs, they have been willing to wage a "war to end war". Comparatively fortunate were those who suffered defeat or death; those who had the misfortune to emerge victorious were reduced to cynicism and despair by the failure of all their glowing hopes. The ultimate source of these hopes was the Christian doctrine of catastrophic conversion as the road to salvation.

I do not wish to suggest that revolutions are never necessary, but I do wish to suggest that they are not short cuts to the millennium. There is no short cut to the good life, whether individual or social. To build up the good life, we must build up intelligence, self-control, and sympathy. This is a quantitative matter, a matter of gradual improvement, of early training, of educational experiment. Only impatience prompts the belief in the possibility of sudden improvement. The gradual improvement that is possible, and the methods by which it may be achieved, are a matter for future science. But something can be said now. Some part of what can be said I shall try to indicate in a final chapter.

10 Such incidents had been in the news from 1923, after the notorious case in which Jasper Abraham, the son of the Bishop of Norwich, flogged and kicked an African servant so violently that he died that night; his offence having been daring to ride his master's horse home after having to run along beside it for 17 miles on an errand.

11 The opening of the last Chorus in Shelley's verse drama *Hellas* (1822). The word 'anarchs' appear in several Shelley poems, including *Hellas*'s Chorus of Greek Captive Women: 'And all its banded anarchs fled,/Like vultures frighted [. . .].'

Chapter V: Science and Happiness

The purpose of the moralist is to improve men's behaviour. This is a laudable ambition, since their behaviour is for the most part deplorable. But I cannot praise the moralist either for the particular improvements he desires, or for the methods he adopts for achieving them. His ostensible method is moral exhortation; his real method (if he is orthodox) is a system of economic rewards and punishments. The former effects nothing permanent or important; the influence of revivalists, from Savonarola[12] downwards, has always been very transitory. The latter—the rewards and punishments—have a very considerable effect. They cause a man, for example, to prefer casual prostitutes to a quasi-permanent mistress, because it is necessary to adopt the method which is most easily concealed. They thus keep up the numbers of a very dangerous profession, and secure the prevalence of venereal disease. These are not the objects desired by the moralist, and he is too unscientific to notice that they are the objects which he actually achieves.

Is there anything better to be substituted for this unscientific mixture of preaching and bribery? I think there is.

Men's actions are harmful either from ignorance or from bad desires. "Bad" desires, when we are speaking from a social point of view, may be defined as those which tend to thwart the desires of others, or, more exactly, those which thwart more desires than they assist. It is not necessary to dwell upon the harmfulness that springs from ignorance; here, more knowledge is all that is wanted, so that the road to improvement lies in more research and more education. But the harmfulness that springs from bad desires is a more difficult matter.

In the ordinary man and woman there is a certain amount of active malevolence, both special ill-will directed to particular enemies and general impersonal pleasure in the misfortunes of others. It is customary to cover this over with fine phrases; about half of conventional morality is a cloak for it. But it must be faced if the moralists' aim of improving our actions is to be achieved. It is shown in a thousand ways, great and small: in the glee with which people repeat and believe scandal, in the unkind treatment of criminals in spite of clear proof that better treatment would have more effect in reforming them, in the unbelievable barbarity with which all white races treat negroes, and in the gusto with which old ladies and clergymen pointed out the duty of military service to young men during the War. Even children may be the objects of wanton cruelty: David Copperfield and Oliver Twist are by no means imaginary. This active malevolence is the worst feature of human nature, and the one which it is most necessary to change if the world is to grow happier. Probably this one cause has more to do with war than all the economic and political causes put together.

Given this problem of preventing malevolence, how shall we deal with it? First let us try to understand its causes. These are, I think, partly social, partly

12 The Dominican friar Girolamo Savonarola (1452–1498) was a religious reformer active in Florence. He denounced Papal corruption, and incited the 'bonfire of the vanities,' in which objects deemed sinful such as cosmetics, art, musical instruments and books were collected and publicly burned—Savonarola's own eventual fate after he was excommunicated by the Church.

physiological. The world, now as much as at any former time, is based upon life-and-death competition; the question at issue in the War was whether German or Allied children should die of want and starvation. (Apart from malevolence on both sides, there was not the slightest reason why both should not survive). Most people have in the background of their minds a haunting fear of ruin; this is especially true of people who have children. The rich fear that Bolsheviks will confiscate their investments; the poor fear that they will lose their job or their health. Every one is engaged in the frantic pursuit of "security", and imagines that this is to be achieved by keeping potential enemies in subjection. It is in moments of panic that cruelty becomes most wide-spread and most atrocious. Reactionaries everywhere appeal to fear: in England, to fear of Bolshevism; in France, to fear of Germany; in Germany, to fear of France. And the sole effect of their appeals is to increase the danger against which they wish to be protected.

It must, therefore, be one of the chief concerns of the scientific moralist to combat fear. This can be done in two ways: by increasing security, and by cultivating courage. I am speaking of fear as an irrational passion, not of the rational prevision of possible misfortune. When a theatre catches fire, the rational man foresees disaster just as clearly as the man stricken with panic, but he adopts methods likely to diminish the disaster, whereas the man stricken with panic increases it. Europe since 1914 has been like a panic-stricken audience in a theatre on fire; what is needed is calm, authoritative directions as to how to escape without trampling each other to pieces in the process. The Victorian age, for all its humbug, was a period of rapid progress, because men were dominated by hope rather than fear. If we are again to have progress, we must again be dominated by hope.

Everything that increases the general security is likely to diminish cruelty. This applies to prevention of war, whether through the instrumentality of the League of Nations, or otherwise; to prevention of destitution; to better health by improvement in medicine, hygiene, and sanitation; and to all other methods of lessening the terrors that lurk in the abysses of men's minds and emerge as nightmares when they sleep. But nothing is accomplished by an attempt to make a portion of mankind secure at the expense of another portion—Frenchmen at the expense of Germans, capitalists at the expense of wage-earners, white men at the expense of yellow men, and so on. Such methods only increase terror in the dominant group, lest just resentment should lead the oppressed to rebel. Only justice can give security; and by "justice" I mean the recognition of the equal claims of all human beings.

In addition to social changes designed to bring security, there is, however, another and more direct means of diminishing fear, namely by a regimen designed to increase courage. Owing to the importance of courage in battle, men early discovered means of increasing it by education and diet—eating human flesh, for example, was supposed to be useful. But military courage was to be the prerogative of the ruling caste: Spartans were to have more than helots, British officers than Indian privates, men than women, and so on. For centuries it was supposed to be the privilege of the aristocracy. Every increase of courage in the ruling caste was used to increase the burdens on the oppressed, and therefore to increase the grounds for fear in the oppressors, and therefore to leave the causes of cruelty undiminished. Courage must be democratized before it can make men humane.

To a great extent, courage has already been democratized by recent events. The suffragettes showed that they possessed as much courage as the bravest men; this demonstration was essential in winning them the vote. The common soldier in the War needed as much courage as a Captain or Lieutenant, and much more than a General: this had much to do with his lack of servility after demobilization. The Bolsheviks, who proclaim themselves the champions of the proletariat, are not lacking in courage, whatever else may be said of them; this is proved by their pre-revolutionary record. In Japan, where formerly the samurai had a monopoly of martial ardour, conscription brought the need of courage throughout the male population. Thus among all the Great Powers much has been done during the past half-century to make courage no longer an aristocratic monopoly: if this were not the case, the danger to democracy would be far greater than it is.

But courage in fighting is by no means the only form, nor perhaps even the most important. There is courage in facing poverty, courage in facing derision, courage in facing the hostility of one's own herd. In these, the bravest soldiers are often lamentably deficient. And above all there is the courage to think calmly and rationally in the face of danger, and to control the impulse of panic fear or panic rage. These are certainly things which education can help to give. And the teaching of every form of courage is rendered easier by good health, good physique, adequate nourishment, and free play for fundamental vital impulses. Perhaps the physiological sources of courage could be discovered by comparing the blood of a cat with that of a rabbit. In all likelihood there is no limit to what science could do in the way of increasing courage, by example, experience of danger, an athletic life, and a suitable diet. All these things our upper-class boys to a great extent enjoy, but as yet they are in the main the prerogative of wealth. The courage so far encouraged in the poorer sections of the community is courage under orders, not the kind that involves initiative and leadership. When the qualities that now confer leadership have become universal, there will no longer be leaders and followers, and democracy will have been realized at last.

But fear is not the only source of malevolence; envy and disappointment also have their share. The envy of cripples and hunchbacks is proverbial as a source of malignity, but other misfortunes than theirs produce similar results. A man or woman who has been thwarted sexually is apt to be full of envy; this generally takes the form of moral condemnation of the more fortunate. Much of the driving force of revolutionary movement is due to envy of the rich. Jealousy is, of course, a special form of envy: envy of love. The old often envy the young; when they do, they are apt to treat them cruelly.

There is, so far as I know, no way of dealing with envy except to make the lives of the envious happier and fuller, and to encourage in youth the idea of collective enterprises rather than competition. The worst forms of envy are in those who have not had a full life in the way of marriage, or children, or career. Such misfortunes could in most cases be avoided by better social institutions. Still, it must be admitted that a residuum of envy is likely to remain. There are many instances in history of Generals so jealous of each other that they preferred defeat to enhancement of the other's reputation. Two politicians of the same party, or two artists of the same school are almost sure to be jealous of one another. In such cases, there seems nothing to be done except to arrange, as far as possible, that each competitor shall be unable to injure the other, and shall only be able to win by superior merit. An artist's jealousy of a rival usually does little harm, because the only effective way of indulging it is to paint better pictures than his rival's, since it is not open to him to destroy his rival's pictures. Where envy is unavoidable, it must be used as a stimulus to one's own efforts, not to the thwarting of the efforts of rivals.

The possibilities of science in the way of increasing human happiness are not confined to diminishing those aspects of human nature which make for mutual defeat, and which we therefore call "bad". There is probably no limit to what science can do in the way of increasing positive excellence. Health has already been greatly improved; in spite of the lamentations of those who idealize the past, we live longer and have fewer illnesses than any class or nation in the eighteenth century. With a little more application of the knowledge we already possess, we might be much healthier than we are. And future discoveries are likely to accelerate this process enormously.

So far, it has been physical science that has had most effect upon our lives, but in the future physiology and psychology are likely to be far more potent. When we have discovered how character depends upon physiological conditions, we shall be able, if we choose, to produce far more of the type of human beings that we admire. Intelligence, artistic capacity, benevolence—all these things no doubt could be increased by science. There seems scarcely any limit to what *could* be done in the way of producing a good world, if only men would use science wisely. I have expressed elsewhere my fears that men may not make a wise use of the power they derive from science.[13] At present I am concerned with the good that men could do if they chose, not with the question whether they will choose rather to do harm.

There is a certain attitude about the application of science to human life with which I have some sympathy, though I do not, in the last analysis, agree with it. It is the attitude of those who dread what is "unnatural". Rousseau is, of course, the great protagonist of this view in Europe. In Asia, Lao-Tze has set it forth even more persuasively, and 2400 years sooner. I think there is a mixture of truth and falsehood in the admiration of "nature", which it is important to disentangle. To begin with, what is "natural"? Roughly speaking, anything to which the speaker was accustomed in childhood. Lao-Tze objects to roads and carriages and boats, all of which were probably unknown in the village where he was born. Rousseau has got used to these things, and does not regard them as against nature. But he would no doubt have thundered against railways if he had lived to see them. Clothes and cooking are too ancient to be denounced by most of the apostles of nature, though they all object to new fashions in either. Birth-control is thought wicked by people who tolerate celibacy, because the former is a new violation of nature and the latter an ancient one. In all these ways those who preach "nature" are inconsistent, and one is tempted to regard them as mere conservatives.

13 [Author's note] See *Icarus*

Nevertheless, there is something to be said in their favour. Take for instance vitamines, the discovery of which has produced a revulsion in favour of "natural" foods. It seems, however, that vitamines can be supplied by cod-liver oil and electric light, which are certainly not part of the "natural" diet of a human being. This case illustrates that, in the absence of knowledge, unexpected harm may be done by a new departure from nature; but when the harm has come to be understood it can usually be remedied by some new artificiality. As regards our physical environment and our physical means of gratifying our desires, I do not think the doctrine of "nature" justifies anything beyond a certain experimental caution in the adoption of new expedients. Clothes, for instance, are contrary to nature, and need to be supplemented by another unnatural practice, namely washing, if they are not to bring disease. But the two practices together make a man healthier than the savage who eschews both.

There is much more to be said for "nature" in the realm of human desires. To force upon man, woman, or child a life which thwarts their strongest impulses is both cruel and dangerous; in this sense, a life according to "nature" is to be commended with certain provisoes. Nothing could be more artificial than an underground electric railway, but no violence is done to a child's nature when it is taken to travel in one; on the contrary, almost all children find the experience delightful. Artificialities which gratify the desires of ordinary human beings are good, other things being equal. But there is nothing to be said for ways of life which are artificial in the sense of being imposed by authority or economic necessity. Such ways of life are, no doubt, to some extent necessary at present; ocean travel would become very difficult if there were no stokers on steamers. But necessities of this kind are regrettable, and we ought to look for ways of avoiding them. A certain amount of work is not a thing to complain of; indeed, in nine cases out of ten, it makes a man happier than complete idleness. But the amount and kind of work that most people have to do at present is a grave evil: especially bad is the life-long bondage to routine. Life should not be too closely regulated or too methodical; our impulses, when not positively destructive or injurious to others, ought if possible to have free play; there should be room for adventure. Human nature we should respect, because our impulses and desires are the stuff out of which our happiness is to be made. It is no use to give men something abstractly considered "good"; we must give them something desired or needed if we are to add to their happiness. Science may learn in time to mould our desires so that they shall not conflict with those of other people to the same extent as they do now; then we shall be able to satisfy a larger proportion of our desires than at present. In that sense, but in that sense only, our desires will then have become "better". A single desire is no better and no worse, considered in isolation, than any other; but a group of desires is better than another group if all of the first group can be satisfied simultaneously while in the second group some are inconsistent with others. That is why love is better than hatred.

To respect physical nature is foolish; physical nature should be studied with a view to making it serve human ends as far as possible, but it remains ethically neither good nor bad. And where physical nature and human nature interact, as in the population question, there is no need to fold our hands in passive adoration and accept war, pestilence, and famine as the only possible means of dealing with excessive fertility. The divines say: it is wicked, in this matter, to apply science to the physical side of the problem; we must (they say) apply morals to the human side, and practise abstinence. Apart from the fact that every one, including the divines, knows that their advice will not be taken, why should it be wicked to solve the population question by adopting physical means for preventing conception? No answer is forthcoming except one based upon antiquated dogmas. And clearly the violence to nature advocated by the divines is at least as great as that involved in birth-control. The divines prefer a violence to human nature which, when successfully practised, involves unhappiness, envy, a tendency to persecution, often madness. I prefer a "violence" to physical nature which is of the same sort as that involved in the steam engine or even in the use of fire. This instance shows how ambiguous and uncertain is the application of the principle that we should follow "nature".

Nature, even human nature, will cease more and more to be an absolute datum; more and more it will become what scientific manipulation has made it. Science can, if it chooses, enable our grandchildren to live the good life, by giving them knowledge, self-control, and characters productive of harmony rather than strife. At present it is teaching our children to kill each other, because many men of science are willing to sacrifice the future of mankind to their own momentary prosperity. But this phase will pass when men have acquired the same domination over their own passions that they already have over the physical forces of the external world. Then at last we shall have won our freedom.

Halcyon; or, the Future of Monogamy

Vera Brittain (1929)

Headnote

Vera Brittain (1893–1970) was an English Voluntary Aid Detachment (VAD) nurse during the First World War, and a writer, feminist, and socialist. Returning to Oxford to study history after the war, she met the novelist Winifred Holtby, and a lifelong friendship developed (Holtby also contributed an excellent volume to the series: *Eutychus; or, the Future of the Pulpit*; 1928).

Brittain's first published book, the novel, *The Dark Tide* (1923) caused a scandal for its portrayal of Oxford dons. *Halcyon* was followed by her best-selling memoir *Testament of Youth* (1933) which recounted her traumatic experiences during the First World War and her evolution toward pacifism. The British politician Shirley Williams was her daughter.

******

> Alcyone, or Halcyone, in ancient Greek legend, a daughter of Aeolus and Enarete or Aegiale. When her husband Ceyx was drowned, Alcyone cast herself into the sea. The gods pitied the devoted pair, and changed them into birds, the alcyons or halcyons (kingfishers); it was believed that during the breeding-time of these birds the sea was always calm.

Nelson's *Encyclopaedia, sub voc.*[1]

Introduction

Most of us have read, or at least have heard of, the now celebrated work of Mr J. W. Dunne, entitled *An Experiment with Time*. Those students who accept his theory of a many-dimensional world are already familiar with his belief that the fabric of dream is an interweaving of the past and the future. But few can have met with such confirmatory evidence of his thesis as that which was vouchsafed to me a short time ago.

1 Latin: under the specified word.

DOI: 10.4324/9781032625522-17

In preparation for a fair-sized work on Marriage which I hope soon to publish, I had been reading and taking notes from a number of volumes dealing with matrimonial relations and the organization of the family. I had worked late, and spent the first half of the night in the restless borderland between dream and waking, rearranging these notes and quotations in my mind.

Towards morning, however, I fell soundly asleep, and was at once plunged into a dream of extraordinary clearness. In this dream I spent what appeared to be about an hour and a half in reading part of a social and moral history destined to be published in the far-distant future. The contents of many of its pages, which as I completed them were turned before my eyes by an unseen hand, linked up in a quite remarkable fashion with the theories and prognostications of several of the authors whose works I had studied the evening before.

Judging from a biographical note which I saw in a flash on the cover of the book before my dream hurried me direct to its two hundred and thirty-seventh page, the author was a certain moral historian, by name Professor Minerva Huxterwin,[2] who lived and wrote about the middle of the twenty-first century. Her mother, if my memory of the publisher's note is correct, was an American economist from the University of Wisconsin, and her father a naturalised Scandinavian physiologist resident in London. This Professor Huxterwin seems likely to emerge as an important figure among the academic scholars of her future era, for her biography stated that at the time of writing she had already held for ten years the Chair of Moral History at Oxford, founded in 2001 through the generosity of the third Baron Morris of Cowley. The book itself appeared to be a monumental *History of English Moral Institutions in the Nineteenth, Twentieth, and Twenty-first Centuries*.

Unfortunately, since my dream took me right into the middle of this volume, I was able only to see what looked like the latter part of a large section, dealing, in four short chapters, with the fate of the so-called institution of monogamy in the period in question. The hand of my unknown guide was about to turn on to the next section when the dream, which cannot have begun long before the hour at which I am usually awakened, was abruptly ended by the sharp ringing of the telephone bell.

The degree of concentration demanded by the message to which I listened awoke me so effectually that, although I hastened back to my room in the hope of again falling asleep and continuing my study of Professor Huxterwin, I have never, either then or since, been able to return to the point at which I was interrupted, or to discover what the subject of the next chapter was to be.

I am only able, therefore, to publish Professor's Huxterwin's pages on the history of monogamy in the twentieth and twenty-first centuries, exactly, to the best of my recollection, as I read them in my dream.

V. B.

Four Chapters of a Section on the Institution of Monogamy, from a History of English Moral Institutions in the Nineteenth, Twentieth, and Twenty-first Centuries: by Minerva Huxterwin, Professor of Moral History in the University of Oxford

2 Minerva is the Roman goddess of wisdom and law or justice. The name 'Huxterwin' combines those of Huxley and Darwin with that of Edward Westermarck, described by the fictional author as her ancestor.

Chapter I: Morals in the Post-Victorian Era, 1900–1930

(1)

The Victorians and their immediate successors, as I have already indicated, were at least consistent in their endeavours to conceal from themselves the very existence of sex except as a remote medical indecency.

We are told by a magazine-writer[3] at the close of the epoch dealt with in this chapter that in 1889, when a book called *The Evolution of Sex* was published by two young Scottish scientists,[4] its mere title was considered extremely shocking. Any allusion to this book in Edinburgh society caused "the ladies" (a term, highly reminiscent of the obnoxious gentility of the period, which was then commonly used in speaking of women) to blush and look away, while its exhibition in his window by a leading Edinburgh bookseller was regarded as a bold and dangerous step.

Even my distinguished ancestor, Professor Edward Westermarck, author of *The History of Human Marriage*, who by venturing to state that "the concealment of truth is the only indecorum known to science" [5] proved himself to be one of the most daring thinkers of that time, was obliged to compromise with the pruderies of his age by reverting to Latin whenever he had to mention the sexual organs of the body.

One of the most interesting pieces of evidence that we possess of the false delicacy characteristic of those days is the collection recently made by Dr Hoover C. Pedersen, great-grandson of the New York specialist who protested nearly a hundred and fifty years ago against the pusillanimous recourse by "polite"' society to indirect terminology,[6] of expressions used in the first decade of the twentieth century to obscure the real significance of physiological and pathological facts.

This little volume, which was compiled by Dr Pedersen in order to help students to read the newspapers and novels of the period, shows that nothing having the remotest sexual connotation was known by its proper name. Menstruation, for instance, was described as "a certain time", and pregnancy as "a certain condition"; the menopause was known as "change of life", and coitus as "intimate association", while venereal disease, with astonishing boldness, combined with an even more astonishing ignorance of human predisposition to curiosity, was for some years referred to by the Press as "the hidden plague".

One of the most deplorable results of this inability to mention or even to contemplate sexual relationships except in terms of something quite different, was the ruin of many marriages by the misleading identification of wifehood with the

3 [Author's note] *The New Generation*, March, 1929.

4 The biologist, sociologist, geographer, and town planner Patrick Geddes (1854–1932) and the naturalist J. Arthur Thomson (1861–1933).

5 *The History of Human Marriage* appeared in 1891. The quotation is from the penultimate sentence of the introduction. Westermarck's *Ethical Relativity* (1932) was published in Ogden's International Library.

6 [Author's note] Victor Cox Pedersen, A.M., M.D., F.A.C.S., *The Woman a Man Marries* (New York and London, 1928): "It is a hopeful sign that many women are now no longer afraid to utter, in the midst of a dignified discussion, such words as pregnancy, prostitution and venereal disease".

activities involved in the management of a house. In those days "preparation" for marriage consisted, for a girl, in the teaching of cooking and domestic technique. Little attention was given even to mothercraft, since babies were considered an improper subject for discussion until they were not merely born, but had survived to the stage when their mothers were up and about again. This undue concentration upon the kitchen type of training provided society with a convenient excuse for completely omitting any instruction in the technique of sex intercourse, the knowledge of which, as we ourselves have long been aware, is something quite other than domestic proficiency, and is far more necessary to marital happiness than the most expert culinary skill.

We little realise, until we begin to examine the strange social delusions of our great-great-grandfathers, how fortunate we are to live in an age which regards marriage purely as a personal relationship based upon instructed sex union, of which the success in no way depends upon the economic occupation selected either by the husband or the wife.

(2)

The Great War of European Disintegration had naturally, for the time being, a catastrophic effect upon English moral complacency. Perpetually faced by the cruder realities of life and death, the youth of the War generation began, to the horror of their elders, to call the facts of sex by their names. Released from sheltered dependence by war-work, and from the trivial cowardice of shame by the constant threat of disaster, young women, as we learn from the conversations of the heroines in the fiction of the period, did not shrink from admitting both to themselves and to their lovers the mutual character of sex-desire. As the conflict gradually extended from military to civilian ranks, the enormously curtailed expectation of life, combined with the rapid subterranean spread of birth-control information, produced in the young of both sexes a tendency to seize immediately such physical and emotional satisfactions as offered themselves, without waiting for the conventional blessings of Church, State, or family.

A subtle change was even noticeable in the conservative opinion catered for by the cheaper sections of the Press. In the endeavour to offer their readers sentimental consolations for the alarming increase of illegitimacy, some of the more passionate newspapers of the period gave way to quite hysterical outbursts on the fruitful topic of war-babies. The mothers of these infants were said to have given "all" (whatever "all" may have meant; no doubt the meaning varied with the enthusiasm of the journalist) for the love of their country and of some noble soldier, though which particular noble soldier happened to be the one in question was not always easy to determine, even by the mothers themselves.

This characteristically British type of toleration proved to be quite ephemeral, for whenever, after the War ended, a woman indulged in a peace-baby with an exactly similar unorthodox origin, her neighbours again considered her conduct to be highly reprehensible.

(3)

As is indicated by their return to the vicarious pleasures of that emotion which used to be so strangely known as "righteous indignation", the moralists of the post-war period endeavoured to treat the war years as an excusable but quite temporary interlude of sexual laxity, and did their best to restore the conventional shibboleths of pre-war self-deception.

Particularly noticeable was the attempt to perpetuate, both in law and in custom, the Victorian illusion (to which the eighteenth century had barely even pretended) that the English-speaking peoples were monogamous in practice as well as in precept. Since the social ideal represented by monogamy had always been biologically far too grown up for all but a few quite exceptional individuals on both sides of the Atlantic, this endeavour compulsorily to restore, in spite of the facts, an appearance of national monogamous marriage, naturally led to a deep division of opinion upon the subject of the institution itself.

On the one hand, a small minority of writers and scientific investigators—mostly honest, courageous persons, belonging to, or at least in sympathy with, the war-generation—regarded monogamy as a synonym for incurable hypocrisy, and therefore as an arrangement which, whatever sexual reforms were to be made in the future, might well be abandoned as a workable proposition.

On the other hand, a large, loose-thinking majority gave lip-service to the non-existent monogamy assumed by English law and custom to be satisfactorily established, and clamorously endorsed all those restrictions which endeavoured to force it upon a people still too far from civilized adulthood to exercise much voluntary control over the unlicensed satisfaction of its instincts. At the same time this majority shrilly opposed, in the name of morality, all such changes and experiments as might in time conceivably lead to the institution of a true monogamy, and resolutely condemned that free discussion and honest education which alone enables monogamous marriage to be founded upon intelligence, instead of upon the improvident gamble of the old hit-or-miss method.

(4)

Counterparts of both majority and minority groups existed also in the United States, where the exuberant spokesman of the latter was a certain Juvenile Court Judge, Benjamin B. Lindsey of Denver, Colorado, who in the third decade of the century produced two volumes of hectic sanity incorporated in Middle Western journalese.[7] His opinions, however, with regard to the ultimate possibilities of monogamy differed somewhat from those of the English minority, which included

7 Lindsey (1869–1943) published (with the writer and journalist John Wainwright Evans; 1883–1957) *The Revolt of Modern Youth* (1925) and *The Companionate Marriage* (1927). The latter controversially advocated one-year trial cohabitation without children before a legally binding marriage.

such celebrated figures as the writer and thinker H. G. Wells and the philosopher Bertrand Russell, whose wife, Dora Russell, also held and expressed similar views.

"Monogamy and undiscriminating licence", wrote Mrs Russell in *Hypatia*, one of a famous series of twentieth-century political and sociological essays then called *To-day and To-morrow* but now popularly known as *The Fireworks Pamphlets*, "rest upon a common basis of contempt for love and personality, both asserting that the desire of a man is for a woman, of a woman for a man, but no matter whom".[8]

This writer apparently regarded a belief in the ideal of monogamy as tantamount to the view that in sex-intercourse personality does not count, and that, since any woman (or man) suffices for the purpose of marriage, the same woman (or man) will suffice always, a change of body being a mere work of supererogation.

Her theory, sincerely held and effectively stated as it was, took no account of the longing for permanence which exists in human nature side by side with its desire for variety. It disregarded the psychological fact that intellectual variation and experiment need a basis in emotional stability; it treated as non-existent man's innate craving for a foundation and a root, and as trivial the idea of loyal and lasting friendship—itself the product of highly individualized selection—embodied in marriage.

This view of monogamy, though professing itself positive, joy-loving, and constructive, was in reality the expression of a profound pessimism, which was fully justified by the majority opinions and actions of the time.

(5)

The decade immediately following the Great War was aptly described by a contemporary young political scientist as "An epoch of cocksure dictators from Moscow to Madrid and of cocksure censors from Westminster and Dublin to Tennessee—exponents of a censorship no more absurd when it is called religious than when it is called moral, and no more scientifically informed when it is called moral than when it is called religious".[9] The repressive spirit of this era, arising from the moral exhaustion produced by the War and from the consequent fear of facing the uncomfortable realities concealed beneath the serene surface of Western domestic life, resulted in a widespread obstructiveness to open sex-teaching throughout the English-speaking world.

The superstitious identification of virtue with ignorance obscured the fact, now familiar to all of us, that adequate pre-marital sex-education is able, with many young men and women, to take the place of cloying pre-marital experience. Few dared to admit, though nearly everyone secretly knew, that the vast majority of nineteenth and twentieth century divorces were caused by avoidable sex-maladjustments.

8 See p. 212. [Author's Note] To this opinion the future was to prove remarkably unkind. Recent events, however, have more than justified many of the other theories, beliefs, and prophecies set forth in Mrs Russell's suggestive little treatise.

9 [Author's note] G. E. G. Catlin, Review of Bertrand Russell's *Sceptical Essays* in the New York *Nation*, March 13th, 1929. [Brittain had married George Catlin in 1925.]

Though successful monogamy can obviously be founded only upon a thorough knowledge of sex-technique and therefore of the workings of the body and its entire range of peculiarities, insuperable obstacles were placed in the way of almost all attempts to give popular instruction in normal or abnormal physiology. Only one or two examples are needed to illustrate the psychology of this period, now so well known owing to the recent research work of the Historio-Telepathic Institute.[10]

It was about 1924 that the old London County Council, after consulting a number of representative teachers who shared their concern for the moral irresponsibility of contemporary youth, rejected a proposal for the teaching of sex-hygiene in elementary and secondary schools, thus depriving both themselves and the children of the only remedy for the laxity that they deplored.

Two or three years later, an outcry was raised in certain quarters against the already limited opportunities for co-education among medical students in London hospitals, on the ground that it was improper for young men and women, in the natural course of their daily work, to acquire in common such knowledge as only by being held in common enables marriage to succeed.

Perhaps, however, the most typical product of post-Victorian morality was the Irish Free State Censorship Bill of 1928.[11] As the celebrated Irish poet and Senator, W. B. Yeats, prophesied at the time, this endeavour to institute what the Bill's promoters apparently regarded as the Kingdom of God upon earth succeeded only in establishing a reign of moral cowardice in Ireland.

(6)

The attitude of the Free State Bill towards contraception was but a more lurid reflection of that maintained by English Ministers and Bishops. The country in general, and the so-called Ministry of Health in particular, appeared to live in constant terror of a bogy which they called Decline of Population, in spite of the fact that, notwithstanding the Great War, the population of Great Britain had increased by two and half millions between 1914 and 1927. Neither this rapid growth, nor the existence of a million and a half unemployed at the beginning of 1929, persuaded these tribally-minded Jeremiahs to see this bogy in its true guise as an angel of light.

In our own happier age we have long realized that nothing is more conducive to real monogamy than a timely knowledge of contraceptives, which not only permits earlier marriage but reduces in number the pregnancies that once drove the majority of husbands from the arms of their perpetually ailing spouses to seek consolation

10 Compare Brittain's conceit of using psychic research to study history, with A. M. Low's joke in *Wireless Possibilities*, 69, about hearing echoes of Henry VIII's conversation using 'a delicate microphone.'

11 [Author's note] The Free State Bill set up a Censorship Board of five persons, with power to prohibit the sale of any book or periodical which in the Board's opinion was deemed subversive of public morality. It also prohibited the issue of advertisements relating to venereal disease, as well as the publication of any book, periodical or advertisement advocating, or giving instruction in, the prevention of conception.

from an outcast class of women, who used for economic gain the knowledge forbidden to legitimate wives for the preservation of their health and dignity.

It is now, therefore, hardly credible that in the first quarter of the twentieth century the English Ministry of Health, although composed of professed monogamists, refused to allow birth-control information to be given at public welfare centres, even at the request of married women with large families whose health obviously required relief from the strain of excessive child-bearing. Though even in those days a few scientific methods of contraception had been introduced, the guardians of the national health and the protectors of the future generation preferred that women should injure themselves and their children by the use of clumsy, unhygienic devices, presumably with the idea that this was an effective way of keeping them ignorant of the existence of that which they already endeavoured to practise.

In the United States such confusion of thought prevailed that several legislatures made it legal for teachers to advocate family limitation but illegal to explain how this should be managed—a situation similar to that which would arise if medical men and women informed their patients of the medicines that would cure their diseases, and then refused to write the prescriptions.

To such lengths was this inconsistency carried, that in 1929 an elaborate police-raid was organized in New York upon a highly respectable Birth Control Clinical Research Bureau, founded some six years previously by Mrs Margaret Sanger, a pioneer teacher of contraceptive methods. This raid, now famous in the annals of preventive medicine as typical of the blind and superstitious hostility against which a handful of enlightened men and women had constantly to contend, resulted in the arrest and trial of two doctors and three nurses, and in the wholesale seizure of private case records.[12]

Various highly moral reasons were of course brought forward on both sides of the Atlantic to explain these anomalies. A favourite ground of opposition to contraception—the fear that it would be used by frivolous persons to escape the responsibilities of matrimony—was apparently advanced quite sincerely, under the strange impression that the forcing of unwanted children upon ignorant parents would somehow compensate for the unborn infants of the selfish and sophisticated.

(7)

The prohibition of sex-teaching was frequently associated with the condemnation and even the persecution of sex-teachers. The attack of the so-called British Comstockians[13] in 1898 upon Havelock Ellis, the great sexologist, for the volume on *Sexual Inversion* in his *Studies in the Psychology of Sex*, will not be surprising to readers with whom the gymnastic hypocrisies of the Victorian era itself have become proverbial. As I have already explained in a previous chapter, Ellis was one of the first who dared to instruct men and women in the detailed working of their reproductive functions, and it was not unnatural that such teaching should be

12 [Author's note] The case was ultimately dismissed, and the police director responsible for the raid removed from her post.

13 [Author's note] Followers of the militant nineteenth-century American Puritan, Anthony Comstock.

resented by a generation which regarded those functions as an immodest aberration on the part of the Creator.

It is, however, more astonishing to find the same type of rampageous obstructionism still in evidence 30 years later, and to learn that the works of the birth-control expert, Dr Marie Carmichael Stopes, who is now regarded as an eminent pioneer of the great social and moral revolution shortly to be described, were relegated in large numbers to disreputable little back-street stores.

The worst intolerances of the Ellis trial were amazingly repeated in 1928, in which year a certain female zealot named Radclyffe Hall published a somewhat sentimental but quite innocuous novel entitled *The Well of Loneliness*. This book set out, with a humourless sincerity worthy of a conscientious Sunday-school teacher, to describe homosexual tendencies among women, and to plead that those thus pathologically inconvenienced should at least be spared the additional bitterness of persecution and ostracism,

Though this did not seem to be too much to ask of a society which treated with genial complacency the seducers of girls and the corrupters of boys, and which included amongst its most respected citizens many men responsible for infecting both their wives and children with venereal disease, a terrific outburst of pious wrath resulted. For fear that the book might enable women with homosexual tendencies to recognize and understand them, and thus cease to threaten the normal sex-life of heterosexual persons, it was savagely suppressed by the joint efforts of a child-like Home Secretary, an ingenious Sunday-paper editor anxious to increase his circulation in the silly season, and a now forgotten magistrate belonging to the Victorian school of obscurantist morality.

Some months later this novel temporarily suffered a similar fate in the United States, where its ultimate triumph over its traducers led to the retaliatory suppression of the works of Rabelais and of Voltaire's *Candide*. In the same country also, in the following year, a certain Mrs Mary Ware Dennett was tried by the Brooklyn Federal Court for sending through the mail a pamphlet entitled *The Sex Side of Life*. The prosecution of this benevolent middle-aged woman was instigated by an association then familiarly known in America as 'D.A.R.' (Daughters of the American Revolution), whose existence so late as 1929 shows how remarkably favourable was the United States atmosphere even in those days to curious archaic survivals.

Although the pamphlet in question was originally written for the instruction of the author's young sons, and had been used for ten years by State Boards of Health and by the American schools of the Young Men's Christian Association, Mrs Dennett was found guilty of sending obscene matter through the mail, and was faced with a sentence of five years imprisonment or a fine of five thousand dollars, on the ground that her publication "was dangerous because it did not teach repression".[14]

A remark made by her during the course of the trial—"At present sex-knowledge is being conducted on a bootleg basis"—has since become famous, not merely for its reference to the extraordinary and now almost forgotten experiment known as

14 [Author's note] Assistant Federal Attorney James E. Wilkinson, as reported by the New York Times of April 26th, 1929

"Prohibition", but as describing in a phrase that blind suppression of fundamental facts which so seriously impeded the attainment of true monogamy by American youth.

(8)

The detailed work of the Historio-Telepathic Institute on the relation in the early twentieth century between sadism and the opposition to marital reform renders it unnecessary for me to dwell at length upon the hostility maintained by certain classes of society to any alteration in the laws relating to marriage and divorce.

In spite of the courageous attempts of Lord Buckmaster, one of the most liberal-minded peers of the age, to initiate changes permitting an honourable escape from marriages which provided neither physical intercourse nor mental companionship, a number of influential persons perpetuated the astounding belief that sham monogamy based upon a mistake was preferable to real monogamy attained after the acknowledgment of that mistake. Incredible as it now appears to us that compulsory association with a drunkard or a lunatic should actually have been thought more conducive to a noble life than the freedom to search afresh for an ideal companion, the Institute and other bodies dedicated to social and scientific investigation possess conclusive documentary evidence, not merely that these extraordinary views were held, but that they formed part of the majority opinions of the time.

Closely associated with this resistance to easier divorce was the refusal, more emphatic in England than in America, officially even to consider the advisability of legalizing such matrimonial experiments as became so common fifty years later, when the epoch of sexual reform was at its height.

Though these experiments, a number of which were made in the conscientious endeavour to discover how marriage could be rendered happier and more lasting, provided evidence directly favourable to the prospects of monogamy as an institution, they were summarily dismissed as immoral and socially subversive. Respectively repudiated as "free love" and "polygamy", the useful expedients of trial-marriage and second contract were alike taboo to the majority of post-War moralists.

Even the mild experiment of semi-detached marriage, as first described in detail by a contemporary writer, Vera Brittain, in her work *The Liberation of Marriage*,[15] provoked from certain quarters quite abusive criticism of the author.

(9)

This section may fitly conclude with a reminder to my readers that only at a comparatively recent date did the state of public opinion with regard to the position

15 The unfinished typescript of a work with this title is among the Brittain papers at McMaster University Library. Brittain attended the 1929 London Congress of the World League for Sexual Reform, along with several other To-day and To-morrow writers. The second of its ten-point plans was 'The liberation of marriage (and especially divorce) from the present Church and State.' See Ivan Crozier, '"All the World's a Stage": Dora Russell, Norman Haire, and the 1929 London World League for Sexual Reform Congress', *Journal of the History of Sexuality*, Vol. 12, No. 1 (January, 2003), 16–37.

of wives make possible the now quite usual selection of a Cabinet composed of a number of married women and headed by a married woman Prime Minister.

The best we can say of the determined enemies of matrimonial experiment less than a century and a half ago is that they displayed a certain stubborn consistency in their prolonged opposition to the feminist demands, then widely regarded as extravagant and unwarranted, for equal pay for men and women workers and for the freedom of women to continue their employment after marriage.

Even with beginner-students in moral history, it is now a commonplace that our present monogamous habits bec[a]me general only after the complete removal of all restrictions upon the paid work of wives, for this reform not only made possible earlier marriage for the male partner, but allowed the woman wisely to choose a husband whom she loved, instead of being driven by social and economic pressure to take the first comer who could provide her with subsistence.

In the early twentieth century, however, the various Governments and the majority of local authorities actually believed that by dismissing a woman teacher or doctor on her marriage, they were protecting the institution of monogamy and guarding the sanctity of the home. In reality, as recent research into the private lives of our ancestors has shown us, these misguided authorities, by establishing a number of celibate professions for women, made the sacrifice of normal human relationships the intolerable condition of professional success and economic security.

They thus provided many women who had no natural taste for life-long virginity with a direct and powerful incentive to the very irregular unions which were held in such abhorrence by contemporary society. Had local authorities in general not been devoid of commonsense, this result might easily have been foreseen, since the more intelligent members of the female sex had long ceased to be influenced by that sanctimonious hypocrisy which insisted that the place of every woman, irrespective of her tastes or talents, was the under-financed home to which she must not be permitted to contribute by her earnings.

It is scarcely surprising that the anomalies of this astonishing morality seemed to present almost insoluble problems to the organizers of a Conference on Sexual Reform which was held in London for the first time at the close of the post-Victorian era.

Chapter II: The Period of Sexual Reform, 1930–1975

(1)

A serious determination on the part of an enlightened minority to remove at least the outstanding legal and social irrationalities from the moral codes of the world followed the widespread revival of interest in sexual problems which brought the post-Victorian era to an end.

The London Conference mentioned at the close of the last chapter was only the first of many which took place during the succeeding years in various capital cities. At these Conferences such topics as the liberation of marriage, the control of conception, prostitution, eugenics, homosexuality, venereal disease, sex education, and the equal status of men and women were discussed with that complete freedom from prurience and sadism which was attainable only in the guaranteed absence of Purity Leaguers, Fundamentalists, Home Office officials, Federal Attorneys, members of the Police Force, Daughters of the American Revolution, and Ministers of any Established Church.

The majority of the Sex-Reformers, whose numbers increased more rapidly than had been thought possible even in the Scandinavian countries, proved to be uncompromising critics of monogamous marriage, though their private lives were seldom as dramatic as their theories. Their equal objection to both the name and the idea of monogamy appears to us not unnatural, since it was in the interests of the supposed protection of this non-existent institution that some of the worst oppressions had been committed.

For several years, that apathetic and conservative section of the population which in those days hung like an immense millstone round the necks of would-be liberal legislators in all countries remained sublimely unaware of the very existence of the Sex-Reformers. When, however, the real nature of their programme, with its rational attitude towards homosexuality and its advocacy of divorce by consent, was at last realized, the strident indignation of Sunday-paper journalists, suburban aldermen, provincial matrons and other self-appointed guardians of the public morality, was poured upon the men and women responsible for drawing it up.

So violent was the opposition from those very persons who most hindered the achievement of that monogamy which they acclaimed so shrilly, that the beginnings of moral reform might well have been postponed for yet another decade, had not the shocking Downing Street murder, which is rightly regarded as one of the most tragic, yet in its results one of the 'most important, crimes of history, caused a sudden revulsion of public feeling against the existing divorce laws.

(2)

One winter's evening in 1942, some ten years before the Metropolitan General Electrification Scheme had provided London with its present midday light throughout the hours of darkness, a woman Member of Parliament, who since her election at the time of the great Socialist victory in 1936 had become the object of exceptional popular respect and admiration, was brutally murdered by a drunken husband from whom she had for a long period vainly endeavoured to obtain release. The motive for the crime was proved to be a maniacal jealousy, since it was well known that this Member of Parliament had long been on terms of the deepest affection with a Cabinet Minister of the highest rank, at whose house she had been dining on the evening of her death.

Since the Minister in question was a widower, their marriage was prevented only by the artificial bond which tied this noble and gifted woman to a husband with whom no matrimonial intercourse had been possible for years, and who, in the intervals of release from the institutions where he had periodically to be confined, acted as a constant menace to the peace and safety of his wife and her young children. The tragedy seemed the more complete in that these highly-placed lovers, knowing only too well that any unorthodox union on their part would mean the sacrifice to popular prejudice of their unusually valuable services to their country, had for many years endured the strain and frustration of a purely platonic relationship.

Up to that time, as my last chapter will have shown, the majority attitude towards divorce reform had been such as to make only too sadly inevitable this surrender of personal happiness to impersonal obligations. The strange imperviousness of right-wing opinion to the obvious connection between private contentment and public efficiency was clearly demonstrated during the General Election of 1929, when a number of voters in the Borough of South Kensington—at that time still one of the most obstinate strongholds of self-righteous Victorianism—made an unsuccessful attempt to force the sitting member into retirement owing to his recent participation in divorce proceedings.[16]

The circumstances of the Downing Street crime, however, threw so lurid a light upon the worst injustices of the divorce law, and showed so unmistakeably that the worthiest citizens were liable to suffer most heavily from its penalties, that the pioneer efforts of the Sex Reformers began to be supported by a rising clamour for inquiry and improvement. Though scientific students of history well know that epochs of reformation can never be attributed to isolated occurrences, this tragic murder, like so many of the more dramatic historical events, undoubtedly hastened the extension of those minority opinions which first became articulate soon after the Great War. It was thus largely responsible for the remarkable social changes which began in the fourth decade of the twentieth century.

These reforms, as some of us may remember having heard in our youth from our grandparents, reached their height in the latter half of the long reign of Edward VIII. By the time that Elizabeth II ascended the throne, the greatest moral revolution for centuries had virtually been achieved.[17]

(3)

My readers, who live in an era which has grown accustomed to the instantaneous application of reason to moral problems, need not imagine that the agitation aroused by the Downing Street murder led immediately to the reform of those

16 [Author's note] This Puritanical endeavour to penalize a man's public career on account of his private misfortunes led to the curious result that, although the sitting member's record on feminist questions was notoriously unsatisfactory, quite a number of progressive women gave him their votes on this occasion, owing to their disgust with the intolerable marriage and divorce laws which were ultimately responsible for the unedifying squabble. [Editor's note] The reselection of the Conservative MP Sir William Davison was contested by members of his own party on the grounds of his divorce that year.

17 When *Halcyon* was published the monarch was George V, and the future Edward VIII was the Prince of Wales. As King he was to have the shortest reign (327 days) since that of Lady Jane Grey, abdicating in 1936 in order to marry the divorcee Wallis Simpson. Brittain's prediction that it would be Edward's niece, then the three-year-old Princess Elizabeth, who would become the future monarch, rather than a child of Edward's, was to prove the more accurate prediction. Elizabeth became Britain's longest-reigning monarch.

inhuman divorce laws to which the tragedy was due. Emphatic as the minority demand at once became for a new Matrimonial Causes Act embodying revised legislation throughout the entire field of marriage and divorce, it was obliged to give way to the universal insistence of the cautious majority upon "preliminary measures", which were of course intended to act as a brake upon the zeal of those who advocated dangerous alterations in the British Constitution.

In the earlier years of the twentieth century, the realization of an urgent need for reform in any department of public life—such as, for instance, the solicitation laws or the police regulations—was apt to give rise to a form of procedure which was almost identical on each occasion. Instead of immediate fresh legislation embodying those improvements upon the need for which everybody who dared to discuss the subject at all was agreed, a demand was made for a Government Commission to undertake a detailed inquiry into the whole situation.

After an interval of a few months, during which the promise of a Commission was deemed sufficient to satisfy even the most impatient, the names of the Commissioners—usually about ten men and two or three acquiescent women—were announced. The Commission then proceeded to sit at intervals for two or three years, hearing witnesses and discussing their evidence, after which it passed another year or so in writing its Report. This, when a further lengthy interval had elapsed, was published and criticised in detail by the Press, after which a sense of righteous achievement pervaded the country, and everything continued exactly as before.

Those who are familiar with the circumambulatory methods of that epoch will, therefore, be but little surprised to learn that the first result of the popular feeling aroused by the crime of 1942 was not an alteration in the Divorce Laws, but the institution in the following year of Domestic Advisory Courts for the hearing, investigation and settlement of matrimonial disputes among both rich and poor.[18] This innovation was followed, after the astonishingly short interval of two months, by the appointment of a Government Commission to study for three years the cases discussed by the Courts and to hear the evidence of the magistrates and other officials appointed to conduct them.

18 [Author's note] The demand for these courts, which already existed in the United States, was not altogether new in England, having been mildly advanced for some years by individuals of informed judgment. In 1929, for instance, so typical a representative of current views as a Metropolitan magistrate, by name J. A. R. Cairns, wrote an article in a new publication called *The Realist*, which shortly became the most influential organ of scientific opinion, advocating the establishment of such Courts for poor persons. "If the domestic differences and matrimonial strains of Society", he remarked, "were ventilated side by side with those of the poor, in Police Courts, the need for Courts of Domestic Relations would be regarded as imperative." [Editor's note: the progressive monthly *The Realist* had been launched in April 1929. George Catlin, Arnold Bennett, H. G. Wells, Rebecca West, Herbert Read, Aldous and Julian Huxley and Naomi Mitchison (who was J. B. S. Haldane's sister) were on the editorial board; and its contributors included many To-day and To-morrow writers, such as Haldane, Bertrand Russell, Norman Haire, Winifred Holtby, and Brittain herself. Her essay 'Monogamy and Censorship' appeared in December 1929: II:3, 361–69. The magazine ceased publication after Jan. 1930, a victim of the Depression. Cairns, 'Domestic Squabbles in the Courts', *Realist*, 1:1 (April 1929), 48–53 (p. 52).]

(4)

The Domestic Advisory Courts (Establishment) Act of 1943 was the first of that long series which ended only with the State Guardianship of Children Act in 1974. When the Report of the 1943 Royal Commission on the Present State of Matrimony, set up "to study and investigate the problems dealt with by the Domestic Advisory Courts in the first three years of their jurisdiction and to report upon the causes thereof," was published in 1947, it contained overwhelming evidence that, with the exception of a few unusually fortunate partnerships, true and happy monogamous marriage was virtually non-existent throughout the country.

This failure in the great majority of cases was attributed by the Commissioners to physiological and psychological maladjustments due to the complete or partial ignorance of sex-technique, combined in the poorer classes of the community with the absence or inadequacy of birth-control knowledge. They gave as a secondary but nevertheless important cause of disaster, since it affected the more intelligent and influential section of the population, the restrictions and prohibitions still placed both by the Government and by local authorities upon the work of married women in professions.

The illogical notion, it was pointed out, that family life could be protected by forcing women who had been expensively trained for other careers to spend the rest of their days against their will upon the detail of household organization, with permanent spinsterhood as their only alternative, had been responsible for a most distressing number of both irregular and unhappy unions. In other cases, owing to the straitened circumstances which followed the reduction by at least one half of incomes that would have been largely spent upon the rearing and education of a family, the very children in whose supposed interests these arbitrary restrictions upon the private lives of wives and mothers were maintained had never been born at all.

So obvious were the necessary moral reforms which now lay in the hands of the House of Commons, that remarkably little opposition was raised to the two new Bills in which these reforms were embodied. The unexpectedly complete acquiescence of the public, however, was attributed by political reporters of that era, not only to the powerful influence of the Royal Commission's Report, but to the high mortality during the last great influenza epidemic in the winter of 1947–1948 amongst the older inhabitants of Hampstead, Chelsea, Kensington, Putney, Streatham, and the majority of the large Cathedral cities.

Whatever the full explanation may have been, these two Bills received the support of the majority of the men and of practically all the women M.P.s. Progress was no doubt assisted by the fact that the latter were now comparatively numerous and effective, since after the famous Women Candidates' Deputation to the Prime Minister in 1941, the practice had markedly diminished of sending Conservative women to fight industrial towns and mining districts, and Labour women to contest health resorts, Cathedral cities and agricultural areas.

Both these Acts, which were passed by large majorities of the House, came to be identified, greatly to the satisfaction of twentieth-century feminists, with the four women Members of Parliament who had made themselves responsible for piloting the measures through the House.

The first to be passed was the Sexual Instruction (Schools and Welfare Centres) Act of 1948. This piece of sane and wise legislation, associated as it was with two women representatives of London constituencies who throughout their public lives had been pioneers of the birth-control movement,[19] not only removed the prohibition upon instruction at State Welfare Centres in the use of contraceptives, but made graduated classes in sex-hygiene and birth-control a necessary part of the curriculum in both elementary and secondary schools.

The second measure, which followed in 1949 and was known as the Married Women's Independence Act, removed all outstanding restrictions upon the employment of women after marriage. It was ably carried through by the two well-known feminists who, after unsuccessfully fighting several elections, had by that time sat for many years as Labour Members for the once Conservative strongholds of Chiswick and St Albans.[20]

These legislative reforms were followed almost immediately by a marked drop in the slum birth-rate and an equally conspicuous increase of efficiency in the public services, while the Registrar-in-Chief of the Domestic Advisory Courts reported that in 1949–1950 the number of recalcitrant cases brought before them had decreased by 20%. In spite, however, of this initial evidence that these liberal changes were likely to be conducive to the ultimate attainment of monogamy, further reform was postponed by the unexpected fall of the Government in 1953 and by the five years of retrogression which followed. Fortunately, however, the working of the Sexual Instruction and Married Women's Independence Acts had already proved so successful that no attempt was ever made to repeal these particular measures. Thus no obstacle stood in the way of an immediate return to moral reform when the progressive parties came back to power with an overwhelming majority in 1958.

(5)

The Matrimonial Causes Act of 1959 was passed at the crest of that wave of reformist enthusiasm which swept through every department of English life from domestic relations to international affairs. We have only to recall the censorious pettiness a mere thirty years earlier of the South Kensington "Divorce Election"[21]—of which the central figure became the subject of further episcopal animadversions on the occasion of his second marriage a few weeks afterwards—in order to realize how startling such a measure must have appeared to our ancestors. Not only did the Act

19 Only 21 women had been elected as MPs to 1929, and of these the only two to represent London constituencies were the doctor and suffragist Ethel Bentham and the Fabian Arabella Susan Lawrence.

20 Both seats were held by the Conservatives till 1945 (St Albans had a Liberal by-election interlude from 1904 to 1906).

21 See note 16 above.

give drunkenness, insanity, permanent invalidism, desertion, and penal servitude as legitimate causes for the dissolution of marriage, but it brought about the greatest change of all by legalizing divorce by consent. The necessary procedure for such mutual terminations of matrimonial contracts involved only the production of evidence of incompatibility before a Magistrate of the Domestic Advisory Courts, followed by a written statement to the effect that they wished to separate signed by both parties to the marriage.

The customary objections that this Act was revolutionary, subversive of family life, and highly dangerous to national institutions inevitably arose from the usual non-progressive groups. Its passage through the House was followed by the ludicrous incident known to history as the Bishops' Strike, when every Bishop throughout the kingdom resigned the charge of his diocese, in the apparent belief that the enforced association of those who hated each other was likely to further the attainment of that monogamy which had always been regarded by the Anglican Church as an essential factor in the British Kingdom of Heaven.

This occurrence lasted, however, only for a fortnight, for on the discovery that the national religious institutions were continuing to function quite perfectly in their absence, all the Bishops hastily returned to work.

(6)

It is perhaps surprising that a generation which had accepted the principle of divorce by consent should have required a further decade in which to reconcile itself to the idea of experimental union. The explanation probably lies in the fact that in this case the hostility of the clergy was reinforced by that of teachers and other educationists, who feared, particularly in the Universities, that trial marriages might exercise a catastrophic influence upon the results of Final Examinations. Ultimately persuaded, however, by investigation and experience that frustrated love affairs and clandestine associations were equally effective in reducing First Class Honours to mere Thirds and Fourths, they withdrew their opposition after five years of adverse propaganda.

There can be no doubt that the bold experiment in companionate marriage carried through in 1968 with full publicity by one of the youngest adult members of the Royal Family gave material assistance to the passage in 1969 of the Trial Marriages Act, which legalized companionate and other forms of experimental matrimony. That the country as a whole, however, was now ready, thanks to the continued work of the Sex-Reformers, for social and moral innovations which in the first half of the century would have been thought perilous in the extreme, was proved by the fate of the second Contracts Act in 1973.

This measure permitted, by licence from the Domestic Advisory Courts under certain specified conditions, such as the prolonged and unavoidable separation of a well-disposed married couple, permanent or temporary marriage to a second partner without divorce from the first. Not only did the Act pass its Third Reading in the House of Commons without the cry of "Polygamy!" being raised even by the Members for Oxford City and South Kensington, but on the motion of the leading Peeresses, one or two of whom had belonged to the original group first admitted to the Upper House by the Macdonald Government, it was immediately adopted, in the intentional absence of the majority of Bishops, by the reformed House of Lords.

Owing, as I shall show in my next chapter, to the phenomenal development of the transport system in the last quarter of the twentieth century and the consequent diminution of the worst disadvantages of undesired separation, the applicability of the Second Contracts Act was limited to a comparatively few years. It lasted, however, long enough to give special point in 1974 to the State Guardianship of Children Act, by which the increasing tendency of the State during the preceding fifty years to adopt the doctrine of *parens patriæ*[22] was embodied in English law, and the State became a third, or super-Guardian, in addition to the parents, of any child born either in or out of wedlock.

This Act, by acknowledging the direct responsibility of the State for every child, while leaving undisturbed the now almost entirely private character of all marriages without or before the birth of children, completed the differentiation between marriage and parenthood which had been first popularly emphasized by Benjamin B. Lindsey so long ago as 1928.

(7)

To us this distinction between helpless infants who depend for their very existence upon benevolent support, and their potential parents who as responsible adults are entitled within the wide limits of State regulations to regard the sexual relationship into which they may choose to enter as their own private concern, appears so self-evident that we find almost incredible the degree to which public interference with married couples, and particularly with married women, was once thought to be justified by the mere circumstance that such a relationship had been established. It is necessary, therefore, to emphasize the fact that the legislation described above was accompanied by that gradual modification of matrimonial custom which has changed English marriage, as distinct from parenthood, from a status warranting every kind of impertinent intrusion, into a private contract which has become at least as safe from self-righteous prying as the banking accounts of those who make it.

It is indeed amazing to recall how complacently the typical twentieth-century matron continued to accept the surviving badges of her property-status, thus perpetuating in a so-called modern world the Old Testament attitude towards women, which classed a man's wife with his house, his servant, his ox and his ass, and did not even place her first among these. We ourselves may perhaps be forgiven for identifying her with the last of the four when we remember that she actually regarded the wearing of a wedding ring and the adoption of her husband's name and

22 Latin: 'parent of the nation'; legal policy allowing the state to assume responsibility for children abused by their parents, guardians or carers.

nationality as testimony both to her skill in man-catching and her impeccable virtues as a wife. Nor shall we be surprised to learn that the average married woman of that era suffered conspicuously from the mental and moral confusion invariably characteristic of a servile group or class, so that the more persistently the man whom she had caught was unfaithful to her in his body and disloyal to her in his mind, the more ostentatiously did she flaunt the symbols of her appendage-relationship.

This customary attitude of militant matron-hood began to moderate soon after the establishment of the Domestic Advisory Courts, since the discussion of matrimonial problems in an atmosphere of dignified privacy removed the opportunity of rampageous triumph over couples who acknowledged their mistakes on the part of those who merely concealed them. Shortly afterwards, when the Married Women's Independence Act had deprived employers of the right of inquisition into the private affairs of their women employees, the habit of wearing a wedding-ring and the derogatory custom of taking a husband's name after marriage rapidly declined. The prejudices of both local authorities and individual employers could not, it was recognized, be immediately removed by Act of Parliament, and the mere knowledge that a woman employee was married too often aroused in them an irksome suspicion of non-existent shortcomings.

Thus the principle of private contract so long preached by the Sex-Reformers was reinforced by considerations of wise expediency, and marriages were solemnized with no more publicity than was involved in the double signature of the Marriage Registration Form at the nearest Domestic Advisory Court. Ecclesiastical services, with full panoply of veil, ring and vegetable decorations, were still available for those who desired them, but their use soon came to be thought almost as archaic as the utterance of the word "obey" had been considered by intelligent women in the nineteen-twenties. Marriage-vows, it was recognized, were only the expression of a couple's honourable intentions, and the simpler they were, and the less complicated by ulterior obligations to the Deity, the greater was the chance that the monogamous ideal that they represented would come within the range of possibility.

(8)

The increasing recognition of marriage as a private contract-relationship automatically eliminated the prejudiced hostility which had long precluded a wider use of the convenient arrangement now known as semi-detached marriage. The actual practice of this expedient began in the early twentieth century in both England and America with the realization of a few enlightened couples—such as the United States novelist, Fanny Hurst, and her husband[23]—that marriage had too long been the name for one domestic ménage, instead of representing the union of two careers and two sets of ideals.

23 The novels and short stories of Fannie Hurst (1889–1968) were popular between the wars. She was secretly married to the Russian émigré pianist Jacques S. Danielson; they agreed to renew the contract every five years if both agreed.

Once the rigid tradition was shaken which regarded matrimony as a day-by-day, hour-by-hour, unbroken and unbreakable association, marriage contracts were entered into in which the partners, for economic and professional reasons, each continued to live under his and her own roof in different towns and even different countries, and to meet periodically at week-ends or during other holidays. The original outcry against such intermittent unions as "unnatural" and "immoral" of course lost much of its substance when the privacy of the Domestic Advisory Courts removed marriage, just as it had removed divorce at an earlier date, from the range of public curiosity and criticism.

(9)

My summary of the legislation which characterized this remarkable period of sexual reform would be incomplete without a final word regarding the effect of the Married Women's Independence Act upon the general attitude towards maternity in this country. It also had some little influence in America, where a few outstanding anomalies remained in spite of the fact that since the beginning of the century the work of married women, particularly in the business world, had encountered comparatively little opposition.[24]

From the point of view of our own generation, which looks upon the provision of healthy and intelligent children as one of the most important services that any man or woman can render to the community, the penalization of parenthood in the first half of the twentieth century seems equally to lack explanation and excuse. There can also be no doubt that it was one of the circumstances most prejudicial to monogamy, since for many young couples, who could not afford even to contemplate matrimony unless both continued to earn their respective small salaries, it rendered inevitable the postponement of marriage, often for several years.[25]

Even, too, when the financial risk had at last been taken, the difficulties of child-bearing and rearing were so great, and encountered so few attempts at amelioration on the part of society, that after the appearance of the first baby a man's home was usually transformed into the place that he avoided rather than sought. Far from becoming the unifying influence that it now is in all our normal families, parenthood in the twentieth century appears to have been one of the chief sources of marital misery and disruption.

The explanation of course lies in the lunatic values, already mentioned on several occasions, which were common throughout the post-Victorian era. Instead of efficient maternity being treated by our ancestors as a service which put society under an obligation to those who performed it, the very capacity for motherhood

24 [Author's note] The U.S. Census of 1920 showed that two million women workers—i.e., one quarter of all those "gainfully employed"—were married.

25 [Author's note] Cf. an incident related by Abraham Flexner, one of the greatest authorities on the social evils of that era, in his work *Prostitution* (Grant Richards, 1919), p. 39: "Of ninety physicians interrogated by Professor Neisser respecting their sexual history, only one denied all intercourse prior to marriage, and he attributed his exemption to an early engagement alone."

was regarded as a kind of hereditary disability from which women were unfortunate enough to suffer.

As if, moreover, this obstructive attitude on the part of a race which ostensibly desired both to continue and to improve were not illogical enough in itself, it seems to have been directed with especial violence against precisely those women who, by reason of sound physique, trained intelligence and natural idealism, were undeniably the best fitted for maternity.

When the slum mother indulged her disability to the extent of ten or twelve unwanted children, she was not only pitied, but even negatively encouraged by the withholding from her of such information as might have enabled her to limit her family. At the same time various social alleviations, such as pensions, allowances, maintenance grants, and welfare supervision, lifted a considerable part of the burden of responsibility from her shoulders.

But when, at the other end of the intellectual scale, the woman doctor or teacher or journalist showed any disposition to produce, at her own expense, her small, carefully planned family, society treated her as if she ought to have known better, and threatened her with the removal of her life-work and the loss of her income as penalties for the folly of giving birth to a child whose standard of living and behaviour were certain to be amongst the highest in the country. So low, indeed, was the estimation of first-rate human life in comparison with the value set on property, that the most virulent opponents of married women's employment were the ratepayers in rich boroughs, who feared that a policy of encouraging the best women to have children might extract a few more paltry pence from the depths of their well-lined pockets.

(10)

The gradual reversal of this attitude after the passage of the Married Women's Independence Act was directly attributable, not only to the increase of married women in hitherto "celibate" occupations, but to the election of mothers to responsible positions, which were formerly attainable only by married men, bachelors, and spinsters, in professional, commercial, and industrial organizations. It was, of course, the influence of these mothers which was originally responsible for the now self-evident but then quite startling notion that, instead of adapting the biological needs of the race to professional regulations, it might be quite possible to adapt professional regulations to biological needs.

I need not here describe in detail the methods by which, with a view to making the motherhood of the best women as easy as possible, and to eliminating by the removal of outstanding disadvantages the old idea that maternity was both a misfortune for the mother and a tiresome handicap to commercial production, the various industries and professions were gradually adapted to meet the needs of temporarily part-time workers. This process is described *in extenso* in the 1960–1980 records, which are accessible to all my readers, of the Reorganized Institute of Industrial Psychology. They will also find there an account of the adoption of the custom, now regarded as a commonplace, but then thought of as an innovation liable to wreck the industrial prosperity of the British Empire, of an optional "child-birth year" for all women who, without wishing to abandon their economic occupations, desired or required—perhaps on more than one occasion—to give special care to their children in early infancy.

It is hardly necessary to point out that this changed estimate of parenthood, combined with the greater efficiency in birth-control and sex-technique which the Sexual Instruction Act had made possible for everyone, led not only to the more enthusiastic study of mothercraft by mothers, but provided fathers with an incentive to practise that art of fathercraft which came to be regarded as one of the most essential qualifications for marriage.

With the effect of earlier marriage, better-adjusted sex-relationships, efficiently managed children and mutually co-operative parents upon the institution of monogamy, I must refrain from dealing until I have shown how closely the moral revolution of the later twentieth century was related to contemporary progress in scientific development.

Chapter III: Scientific Progress, 1950–2000, and Its Relation to the Moral Revolution

(1)

We are most of us familiar with the inscription from his own writings on the base of the statue of the philosopher Bertrand Russell, which stands facing the river in the lovely Chelsea garden of the Institute for the Coordination of Moral Problems and Biological Interests (popularly known as the M.P.B.I. Institute). The two simple lines run thus: "As science advances, more and more things are brought under human control".[26]

Nothing could more effectively suggest the mastery over his own nature which scientific progress enabled man to acquire during the seventy years that followed the close of the post-Victorian era. Most striking of all, perhaps, was the assistance given by new developments and inventions to those who wished to reverse the nineteenth century practice of clinging to the name of monogamy while repudiating the substance.

Dealing, for the sake of convenience, with twentieth century innovations in chronological sequence, we shall find radiology to be the first department of science in which sufficient data exist to enable us to estimate its effect upon contemporary morals. The rapid perfecting of broadcasting and television between the years 1925 and 1940 is of course a matter of scientific history, but only those who have access to the detailed research tables of the M.P.B.I. Institute are aware how close was the relationship a century ago between the popular use of wireless and the success of monogamous marriage.

In 1937, for instance, it was discovered from the replies to an immense questionnaire sent out by the Sociology Department of the London School of Economics

26 [Author's note] *What I Believe* (To-day and To-morrow Series), p. 11 [See p. 218 in the present volume].

to the inhabitants of Greater London, that in the previous decade approximately 850,000 marriages which had been heading for disaster through sheer monotony had been saved from ruin and restored to happiness by the installation of loud speakers and televisors in the home.

Had it occurred to the authorities at any institution in the preceding ten years (1917–1927) to organize a similar inquiry into the effects of the cinematograph, the result would no doubt have been equally remarkable. The 1937 questionnaire forms, of course, only one of the many pieces of evidence possessed by the M.P.B.I. Institute which indicate how frequently broadcasting, television, and still later the so-called "home talkies" apparatus put an end to that wide-spread domestic boredom which so often in the past could find alleviation only in the excitement of a clandestine love-affair.[27]

(2)

It is now a well-established belief that a number of the matrimonial breakdowns which occurred between 1850 and 1950 were due to the stage reached during those hundred years in the development of the system of transport and communications.

Though the collection of the data showing the close relationship between marital happiness and the conquest of distance was begun by the Historio-Telepathic Institute only a short time ago, every intelligent student will realize that the partial annihilation of time and space constitutes a threat to human relationships which is absent when man is either completely the servant or completely the master of these elements.

At the close, for instance, of the eighteenth century, when New York was a six-weeks voyage from Southampton, and the stage-coach journey from London to Edinburgh occupied almost as much time as the sea-route from London to Calcutta in 1900, and was fraught with far greater dangers, the opportunities for travel were limited to a very small proportion of the whole population. When the long and difficult journeys to foreign lands were undertaken, it was unusual, except on military occasions, for husbands and wives to travel separately, since these lengthy and complicated voyages were generally the prelude, not to a short holiday or a business trip, but to a fresh start and permanent settlement in a new country.

So complete was the immovability of the average family, and so deeply sunk its roots, that whatever excuses might be alleged for moral license, the separation of the married partners could seldom be included amongst them. By the time, indeed, that the airless morality of the mid-Victorians dominated English social life,

27 [Author's note] The M.P.B.I. Institute is fortunate enough to contain amongst its records a copy of an ancient article by two American writers who recognized as long ago as March, 1929, the relation between scientific invention and matrimonial contentment. Writing in the old New York *Survey*, the authors, Ira S. Wile, M.D., and Mary Day Winn, declared that in those days married life, especially for the "home woman", was "often cursed with a terrible monotony, from which either she or her husband may seek escape in the thrill of an affair. The degree to which this goes on in small, dull little towns is almost proverbial [. . . .] No one will ever know how many marriages have been saved from dissolution by the second-hand thrill of the movies" ['Romance Outside the Pale', *Survey*, 61 (1 March 1929), 716–717, 845 (p. 717).]

conjugal virtue had come to be identified with "never alone waking, never alone sleeping". Many of our nineteenth-century ancestors were actually wont to boast that they had been married for thirty or forty years without spending so much as one night under different roofs.

Dreadful as this example of the degree to which human beings can enslave both themselves and one another by their own traditions must appear to us in these freer and better days, it is nevertheless instructive to remember that in 1930—little more than a century ago—there were husbands and wives still surviving who regarded even brief holidays apart from one another as almost immoral.

Long before the twentieth century, however, the obligatory monogamy—or at least the substantial appearance of it—forced upon many married couples by isolation and immobility, was ceasing to be inevitable. The reduction by 1900 of the journey between London and Edinburgh to ten hours, and of that between New York and Southampton to six days, meant that the science of transport had reached the stage, so perilous to conjugal fidelity, when lengthy but not permanent separations between husbands and wives for the purposes of business or pleasure had become feasible.

(3)

With the introduction of aerial transport, and its rapid establishment on a commercial basis, the dangerous intermediate period of man's conquest of space reach its crucial stage. While it was not yet possible, as it is to-day, for married couples to rejoin one another at a few hours' notice from opposite ends of the earth, the reduced time of railway journeys, sea voyages, and aerial flights, as well as the increased rapidity of communication by telephone and cable, removed the necessity of travelling either together or not at all which had formerly kept husbands and wives in such constant association throughout their lives.

From nearly every point of view, the greater elasticity of domestic life which resulted from the increased possibility of temporary separations was an unmitigated good. It held, however, one threat to the institution of monogamy which about 1927 was summed up in a few sentences by a well-known contemporary specialist from Harley Street, a district that remained the headquarters of London medical men and women until 1970, when the development of aerial transport made possible the establishment of our present Doctors' Colony and Medical Research Foundation on the South Coast.

Contemplating, not the future of material conditions but only the restricted means of transport and communication with which he was familiar, this specialist saw in the prolonged marital separations of that time sufficient reason for the permanent survival of the prostitute. "She will always find", he wrote, "a certain demand, say, from men who are travelling . . . as well as from men who find themselves temporarily separated from their wives or mistresses."[28]

28 [Author's note] Norman Haire, *Hymen, or the Future of Marriage* (To-day and To-morrow Series), p. 55.

It is one of the frequent ironies of moral history that the Second Contracts Act, which so wisely endeavoured to substitute legally recognized second partners for the gradually diminishing but still numerous outcast class of professional prostitutes, was permitted by public opinion to be passed only when the real necessity for it had almost disappeared.

(4)

For all ordinary social purposes—such as the practice of monogamous marriage by those who preferred this type of sexual relationship—the virtual annihilation of space had been accomplished by 2000. It had, however, been foreseen by one or two imaginative writers, such as H. G. Wells, as early as the later eighteen-hundreds, and by the third decade of the twentieth century quite a number of scientists with a taste for prophecy had begun to foretell those conditions of transport which over fifty years ago had already eliminated compulsory separation.[29]

The attainment by 1965 of the 500 mile-per-hour average for aircraft did not of course mean the immediate undertaking of such rapid voyages by commercial aeroplanes. By 1985, however, this speed had been successfully combined with safety by the British-American Oceanic Airways Company, which guaranteed a journey from London to New York in under seven hours.

The physical discomforts suffered by pioneer air-voyagers had of course been remedied long ago by the invention of speed-neutralizing helmets for the protection of the eyes and ears, as well as by the Croydon Silencers which were made legally compulsory for all aeroplanes soon after the abandonment of the antiquated Channel sea routes in 1943. For travellers, however, with whom considerations of speed were not urgent and who maintained their prejudice in favour of ocean-voyages for long-distance journeys, the fleet of two-day Atlantic Tourist Cunarders, completed in 1980, combined the maximum of comfort with the minimum of expense, while the so-called Lightning P. & O.s offered similar conveniences to those desiring to reach India, South Africa, or Australia.[30]

Although by the end of the twentieth century the habit of spending week-ends in New York or Sydney had not become as general as it is in our own era, when week-end air-tickets to all parts of our tiny earth are obtainable for a few shillings, the inaccessibility of the partner as an excuse for promiscuity had ceased to exist

29 [Author's note] Cf. J. B. S. Haldane, *Daedalus* (To-day and To-morrow Series, 1924), p. 19: "We are working towards a condition when any two persons on earth will be able to be completely present to one another in not more than 1–24th of a second. We shall never reach it, but that is the limit which we shall approach indefinitely." [See pp. 25–6 of the present volume].

Also Professor H. Levy, article on "Implications of Aerial Development", *The Realist*, April, 1929 [pp. 86–95]: "In aircraft we are already within reach of a 300 miles-per-hour average. The record of last year at the Schneider Cup Race will most certainly become the commonplace seven or ten years hence... Yesterday England was a week wide, to-day a few hours can span it, to-morrow we shall stride it in half-an-hour."

30 The Cunard line was the best-known British transatlantic passenger steamship company of the period. P. & O. (Peninsular & Oriental Steam Navigation Company) specialized in travel to the Iberian Peninsula, the Mediterranean, India and Australia.

at least amongst the well-to-do classes. To-day the rapidly approaching equalization of incomes, combined with the cheapness and efficiency of our World Aircraft Services, has rendered the failure of marriage through the unavoidable absence of husband or wife a social phenomenon of the uncivilized past.

(5)

Accustomed as we now are to the carefully controlled numerical equality of the sexes, it is difficult to re-live in imagination those ages in which the accidental excess of women was often put forward as evidence that monogamy was not only unattainable, but also undesirable.

In the nineteenth and early twentieth centuries, when the proportion of men and women between twenty and fifty had reached about 104 women to every 100 men,[31] the problem of obtaining normal sexual relationships for all women gradually emerged into the open out of the stifling Victorian reticence in which it had been assumed that no respectable woman could possibly be conscious of physical desire. After the great European War, with its further drastic reduction of the numbers of marriageable males, the Press itself gave considerable attention to the position of those women who, though often amongst the most valuable members of society, were contemptuously described as "superfluous" merely because they happened to remain unmated.

Undoubtedly a close relationship did exist between the camouflaged promiscuity of post-Victorian males and the statistical preponderance of females, although the effect of the latter, which was more marked in the declining years of life, seems sometimes to have been exaggerated.[32] As my worthy ancestor, Professor Westermarck, pointed out at the end of the Victorian era, the form of marriage depends to a great extent upon the proportion between the male and female population, and polygamy has usually prevailed whenever women constitute the majority in countries unaffected by European civilization.[33] It was not, therefore, unnatural that in Europe itself monogamy should have been at that time a form rather than a fact.

Students of social history will of course be aware that this excess of females—so typical a phenomenon of semi-civilized humanity still largely at the mercy of natural forces—arose not so much from the disproportion of the sexes at birth as from the higher mortality amongst men owing to their greater exposure to the accidents of war and dangerous occupations, as well as to the inferior vitality of the male during infancy. Until the elimination a century or so ago of these avoidable causes of female preponderance, monogamy of course

31 [Author's note] Cf. Westermarck, *History of Human Marriage*, 2nd edition [London: Macmillan, 1894], p. 147.

32 [Author's note] In Germany, however, owing to war casualties and post-war impoverishment, the number of unmarried men between the ages of 20 and 40 had by 1929, according to the Reich Health Bureau records of that date, been reduced to 4,526,000, whereas the unmarried women between the ages of 15 and 35 numbered 7,142,000. This fact is said to have accounted for the large percentage of female suicides which occurred in Germany about that time.

33 [Author's note] *History of Human Marriage*, 2nd edition, p. 482.

lacked that rational foundation in numerical sex-equality which it possesses to-day.

(6)

As long ago as the nineteenth century, scientists had begun to investigate the determining causes of sex. In the light of our present knowledge, it is interesting to recall one or two of their more imaginative theories.

Göhlert, for instance, believed that sex was influenced by the absolute age of the parents, while Platter, after examining 30 million births in 1875, concluded that the less the difference between the ages of the parents, the greater was the possibility of boys. Düsing, on the other hand, maintained that women predominated where nourishment was abundant and men where it was scarce.[34] Yet further theories gave consanguinity as a cause of males and illicit unions of females, while by others the temporarily superior parent was thought to produce infants of the opposite sex.[35]

In the post-war decade of the twentieth century, when the average young married couple believed, with their customary illogicality, that an excessive production of boys for the next generation would somehow replenish the shortage of men in their own, interest in sex-determination had reached such a pitch that self-constituted specialists without scientific credentials could acquire extensive reputations by merely writing ingenious treatises on the subject.[36]

Happily, however, for the past three or four generations knowledge has replaced speculation, for the principles of sex-determination began to be taught at all Public Health Centres immediately after the discovery of the main sex-determining factors by the young Polish-American biologist, Eugenia Walinowski, in her Baltimore laboratory in 1955. Had this discovery been made a century earlier, during the Victorian apotheosis of the male, it would doubtless have resulted in an overwhelming preponderance of men, with the attendant evils of homosexual practices and male prostitution. Fortunately the twentieth-century triumph of feminism, which culminated in the same decade as Walinowski's researches, had rendered the social and economic value of women at least equal to that of men, and had thus eliminated the ancient preference of conventional parents for male infants.

In 1957 an alarmed attempt was even made by the small group of Masculists in the House of Commons to push through a Bill to Compel the Equal Distribution of the Sexes Within Families, for fear the already numerous feminist mothers might seek to reinforce their hard-won triumphs by producing an over-large percentage of carefully educated daughters. Drawn up by the veteran opponents of feminism, Meyrick Booth and Anthony Ludovici,[37] this Bill was defeated at

34 Vincenz Goehlert, Julius Platter, and Carl Düsing are all authorities cited in Westermarck's book.

35 [Author's note] Westermarck, *History of Human Marriage*, 2nd edition, pp. 469–472.

36 [Author's note] Note, for instance, the large sales at that time of three popular books, *Sex at Choice*, *Nature's Law of Birth-Control*, and *Why be Childless?* by Mrs Monteith Erskine.

37 Educational psychologist Meyrick Booth published 'The Myth of the Modern Woman', in the *English Review*, Vol. 45 (December 1927), 705–709; and the book *Woman and Society* in 1929. Ludovici wrote the anti-feminist alarmist volume for To-day and To-morrow, *Lysistrata; or, Woman's*

its Second Reading on the ground that such interference with the liberties of individual parents would never be tolerated by public opinion.

Nevertheless, a feeling quickly arose that parents who produced an excessive number of children of one sex were guilty of an anti-social action which might well complicate the future relations of men and women. To-day, as all my readers are well aware, the numerical equality of the sexes in families has become a leading social principle in all civilized countries.

(7)

This numerical equality could not, of course, have been maintained by sex-determination alone. Other twentieth century changes of the first importance, such as the remarkable development of infant welfare, the virtual abolition of national war between civilized countries, and the further safeguarding of dangerous trades and industries, were equally responsible for the attainment of this result.

The now historic Infant Welfare Movement, which taught the poorer classes of twentieth century society to use their intelligence in the care of their children, had its first rudimentary beginnings a hundred and fifty years ago. It was not, however, until a quarter of a century later that a similar application of intelligence, initiated by co-operation amongst such small groups as the mothers who founded that celebrated pioneer experiment, the Chelsea Babies Club,[38] began amongst middle-class women.

Owing to the gradual multiplication of such groups, as well as to continual improvements in the system by which the health of working-class mothers and children was kept under constant supervision, motherhood had ceased by the middle of the twentieth century to be identified with the four S's—sentiment, suffering, sacrifice, and stupidity—which had so long been considered the appropriate adjuncts of maternity.

In consequence of the persistent demands made by mothers—now no longer imprisoned within four walls, but sharing, thanks to the Married Women's Independence Act, in ever-growing numbers in the life of the community—for better gynaecologists, more infant specialists, healthier food, labour-saving homes, and maternity benefits under specified conditions, motherhood had successfully established by 1975 its claim to that capital expenditure which child-bearing and rearing had lacked for centuries.

The result of the final incorporation of all children within a constantly modernized Infant Welfare Organization was most clearly shown in boys, for males, being constitutionally the frailer sex, had always been more liable than females to suffer from the ignorance and indifference displayed, particularly amongst the

Future and Future Woman (1924), which was countered by Dora Russell's *Hypatia* (included here, 185–214).

38 Brittain had written about the Club in 'Middle-Class Welfare: A Chelsea Experiment', *Manchester Guardian* (27 April 1928).

middle-classes, by accoucheurs,[39] nurses, and mothers. By the end of the century, the science of Infant Welfare had reached a point where the survival of his birth by a boy could be as confidently predicted as the survival of hers by a girl.

(8)

The abolition of "civilized" warfare by the growing use of the League of Nations and other instruments of arbitration—a movement which first took a permanently positive turn during the long period of Labour Government that began in 1929[40]—has already been dealt with at considerable length in a previous chapter of this volume, while the vast extension of safeguarding to industry between 1930 and 1960 will be described in a subsequent section.

Here I need only point out that both these important departments of moral progress contributed in a large degree to the attainment of numerical equality between the sexes, and hence, indirectly, to the growth of monogamy.

It is, of course, a matter of common knowledge that war in the past was a primary cause of furtive promiscuity, and even, in extreme cases, of openly recognized polygamy. Students of ancient history will remember that among the early Mohammedans polygamy was deliberately adopted as a protection for women against promiscuity, owing to the reduction of men by war to a proportion of one to five, while bigamy was legalized in several German states after the Thirty Years' War.

The death of soldiers in battle did not, of course, constitute the only threat to successful monogamy which war provided. Equally hostile to early and happy unions was the peace-time drafting, in some countries, of young men into the Army under conscription, which debarred them from marriage at the age of the greatest sexual urge, while rendering easily accessible other forms of physical relationship, often of the least desirable type.

Compared with the wholesale slaughterings of war, the accidents due to dangerous trades appeared at any given time to be comparatively slight, but, as the figures which I have subsequently quoted for the mining industries will show, their accumulation over a period of years proved equally unfavourable to the maintenance of numerical equality between the sexes. It was not, however, until the Consolidated Factories Act of 1959 that such protective legislation as men's and women's Unions had not already obtained for themselves began to be based exclusively upon the nature of their work and extended throughout industry to both sexes alike.

39 French: male midwives.

40 The General Election on 30 May 1929 was known as the 'Flapper Election'—it was the first time women aged between 21 and 29 were allowed to vote. It produced a hung parliament but Labour, with the largest number of M.P.s, formed a government, led by Ramsay Macdonald. *Halcyon* was published in September, so the Labour government was a fact, not a prediction. It lasted two and a half years—less than the 'long period' Brittain hoped for.

The results in the striking reduction of male accidents, as well as in the simultaneous improvement of physical health in both sexes and of women's economic status, fully justified, as I have elsewhere shown, the thirty-years struggle of a small group of long-misunderstood feminists against differential restrictive legislation in industry and the professions.

(9)

Until the second half of the twentieth century, one of the chief sources of matrimonial tragedy throughout history had been the fact that women, though they usually lived longer, aged biologically much earlier than men. In these days, when men and women alike, by submission to standardized treatment, preserve their sexual potency to the age of seventy and even beyond, it is difficult to realize how exceptional amongst our ancestors was the motherhood of women over forty-five. The fact nevertheless remains that, up to a century ago, a man's period of physical attractiveness was anything from ten to twenty years longer than a woman's—a circumstance which was apt to cause marital chaos whenever the husband was one of those semi-civilized males who set more value upon naiveté and freshness than upon the beauty which arises from intelligence and experience.

It must be added that biological differences were not solely responsible for this unfortunate contrast between men and women. In the strangely masochistic Victorian era, when an appearance of sober middle-age was regarded as a moral obligation for all mothers, however youthful, and for all spinsters over the age of thirty, the average woman began to look elderly for many years before her sexual activity ended.

One of the first results of the emancipation of women through the feminist movement appeared quite early in the twentieth century in a better adjustment of their appearance and activities to their physical vitality. This adjustment was accelerated by the Great European War, in which the services of women who could not only think but move quickly came to be urgently required by all the combatant nations. The liberation of women's legs and waists was, indeed, one of the very few beneficial changes wrought by this conflict. Ten years after the euphemistically-named Peace of Versailles, the youth of large numbers of women had already been noticeably prolonged by exercise, athletics, occupation, fresh air, hygienic clothing, and the limitation of families, and many of them even achieved the postponement of age more successfully than their husbands.

It was not, however, until later in the century that the possibility of controlling biological changes by science became generally known to the public, although experiments in so-called "rejuvenation" had been familiar to limited groups of scientists for many years.

Professor Eugen Steinach, of Vienna, whose classic work *Verjüngung* was published in 1920, began his investigations into the sexual characteristics of animals as early as 1894, while many experiments in testicle-grafting, ovary-transplantation and the tying of the sperm-duct were successfully carried out upon animals, and even in a few cases upon human beings, by Lichtenstern and Voronoff during the first quarter of the twentieth century.[41] At the same time the eminent biologist, J. B. S. Haldane—whose adventurous youth was so largely dedicated to sensational experiments, particularly upon himself—had foreseen the possibility of isolating and synthesizing the definite chemical substance produced by the ovary upon which he believed a woman's sexual potency to depend.[42]

Rejuvenating operations, though still regarded with fear and suspicion by the unenlightened, had been undergone by intelligent men and women for some years, when in 1973 Felix Rosenfeldt, working in co-operation with his master Haldane, succeeded in isolating the substance which the latter had indicated half a century earlier. This great discovery was fraught with tremendous social consequences, for it rendered immediately possible the composition of Rosenfeldt's Rejuvenating Serum, of which a simple injection at the first sign of the menopause restored the sexual activity of 'women for a further period of fifteen to twenty years.

In a few cases it was even found possible to revive the capacity for child-bearing in women whose periodic functions had long ceased. One of the most celebrated of these experiments was that performed in 1977 upon the first Queen of the re-established Kingdom of Hungary, who by means of three injections given by Rosenfeldt himself was enabled to produce an heir to the throne three days after her sixty-second birthday.

(10)

These revolutionary biological discoveries ensured for all time the prolongation of the youth of men, and to an even more astonishing degree of the youth of women, and provided for that simultaneous ageing of married partners which has done so much to eliminate the once frequent tragedies due to the premature sexual failure of women.

Needless to say, this happy result was not achieved without strong opposition on the part of English clerics and American Neo-Fundamentalists, the latter of whom suspected some subtle and sinister connection between rejuvenation and the doctrine of evolution.

The objections of the clergy, however, were based less on scientific than on moral grounds, for many of them still took for gospel truth the melancholic dictum

41 [Author's note] Cf. Two articles by Dr Norman Haire on "Rejuvenation" in *The Realist*, for April and May, 1929. [Editor's note] Steinach (1861–1944) believed that vasectomy could reduce or reverse aging. Robert Lichtenstern (1874–1952), working with him in Vienna, also implanted testicles to replace those lost in war or illness. Serge Voronoff (1866–1951) popularised the procedure in Paris, implanting slices of chimpanzee testes into humans. See Jonathan Charles Goddard, 'Monkey Glands', *Urology News*, 24:6 (September/October 2020): https://www.urologynews.uk.com/features/history-of-urology/post/monkey-glands

42 [Author's note] J. B. S, Haldane, *Daedalus*, p. 74; "Old age is perhaps harder on women than on men. They live longer, but their life is too often marred by the sudden change which generally overtakes them between forty and fifty, and sometimes leaves them a prey to disease though it may improve their health. This change seems to be due to a sudden failure of a definite chemical substance produced by the ovary. When we can isolate and synthesize this body, we shall be able to prolong a woman's youth and allow her to age as gradually as the average man".

of their model, Charles Kingsley, that "Men must work and women must weep".[43] The discipline of suffering, they maintained, had always been regarded as especially necessary for the female character, and since a great part of woman's unhappiness had been occasioned by the early cessation of her sexual activities, any attempt to remove this cause of sorrow must be regarded as profane interference with the Deity's determination to afflict females for their own good.

It need hardly be said that the men and women of a society still directly influenced by the veteran feminists who had been familiar in their early youth with remarkably similar arguments, countered these anachronistic doctrines with a healthy determination to sacrifice no available increase of joy to such moral shibboleths as had so often destroyed the potential happiness of past generations.

Closer attention, however, was paid to the more rational opposition of the Talkie Stars' Union, which pointed out that instead of prolonging the biological youth of women it would be better to curtail the physical potency of men, on the ground that few spectacles were more revolting than that of an elderly man suffering from senile sensuality.

It was, indeed, the supporting contention of a group of University Professors that an earlier release from their carnal impulses would render the good life more accessible to many men and women who were hindered from attaining it by the disproportionate part played in the average existence by sexual complications, which was responsible for the establishment at the Beach Head Rejuvenation Clinic of a special department for the treatment by glandular operation of those afflicted with excessive physical desire.

(11)

Though many of the biological inventions and discoveries so characteristic of the later twentieth century have necessarily been omitted from my summary because unrelated to the subject of this section, a concluding word must be said upon the attempts which began about 1950 to separate the sexual love of men and women from their generative processes. Since the preoccupation of the female with her tedious reproductive cycles is known to have accounted for a percentage of matrimonial failures, even the merely partial success of ectogenetic experiments has not been without its influence upon the practice of monogamy.

As long ago as 1923, in the remarkable little volume of predictions called *Daedalus*, which I have so often cited, J. B. S. Haldane prophesied the existence of 60,000 ectogenetic children by 1968, and went on to foretell the ultimate production of a seventy-per-cent. laboratory-grown population derived from the reproductive cells of a limited number of highly selected parents.[44]

The older generation may remember having heard from their grandparents of the tremendous sensation which occurred in 1971, when, only twenty years later

43 The refrain of Kingsley's 1851 poem 'The Three Fishers.'

44 [Author's note] *Daedalus*, pp. 63–69.

than the date predicted for this event by Haldane, an ectogenetic girl was successfully reared through the embryonic stages and brought to "birth" in Monet's laboratory. Heedless of the world-wide ecclesiastical *furore* let loose by this triumph, the leading ectogeneticians of England, France, Germany, Russia, and the United States at once went rapidly ahead with the rearing of the embryonic children supplied by a small but slowly increasing number of co-operative parents.

Although, however, their experiments showed the universal production of ectogenetic children to be quite feasible, the complete divorce of sexual relationships from their consequences was never found desirable, owing to a risk that it might in time lead to the demolition of the human race.

Students of the pioneer movement in Infant Welfare may remember that some of the first child-specialists were accustomed to point out in their lectures and writings that the intelligent "mothering" of small children by their own parents had a definite scientific value. The truth of this view, believed in 1970 to be an out-of-date relic of sentimental Georgianism, was demonstrated very clearly by the first laboratory-grown children, who suffered as much psychologically from lack of individual parental affection as they gained physiologically through being selected from the best stock. The majority of them, indeed, though most carefully exercised, dieted and exposed to sunlight, dwindled away and died about the fifth year.

These unsatisfactory consequences of ectogenesis, combined with the fact that the early adoption by women of Kettmann's system of muscular and digestive control rendered childbirth painless and pregnancy definitely pleasurable, led nearly all twenty-first century parents to return to natural methods of reproduction. In the case, however, of the few marriages where normal pregnancy was exceptionally inconvenient to the wife, or would involve a long separation from her husband, recourse was had to the expedient, subsequently familiar under the name of Dickensian Pre-Birth, which Sir Frederick Benedickens invented in 1984.

By this operation, now comparatively common, the fertilized embryo is removed from the mother a few weeks after conception, without damage to itself or deterrent effect upon her ability to conceive afresh, and is grown for eight months in Kolinovski's Gestative Solution. The majority of such children, however, are returned to the mother immediately after their first complete exposure to the air for the purpose of artificially induced lactation, and in no case are they allowed to remain in the laboratory for more than a year after "birth".

I must now conclude my survey of mankind's moral and physical liberation with a brief description of the type of marriage produced by these changes.

Chapter IV: The Triumph of Voluntary Monogamy, 2000–2030

(1)

The majority of my readers will already have realized that the rapid moral development of the English-speaking peoples between 1930 and 2030 had no historical counterpart for swiftness except in the scientific progress of the last two centuries. Those who so justifiably complained after the Great War of 1914 to 1918 that man had not civilized himself in proportion to the speed with which he had perfected his scientific instruments, little realized how quickly the turning point was to come.

The ultimately stabilizing effects of the moral revolution of 1930–1975 were not, it is true, immediately evident, for the first result of the Matrimonial Causes Act of 1959 was the practice in this country for one generation—though only, it is true, by a limited group—of the same type of successive polygamy as had been a recognized feature of American life for some years after the introduction of easy divorce in the early twentieth century.

Trial marriages and even second contracts were also entered into in considerable numbers after the Acts of 1969 and 1973, owing to the naïve belief, common to the less rational section of every community, that relationships which have been prohibited are bound to be more satisfactory than those which have been permitted.

The alarmed excitement which this outburst of social experimentation provoked in certain mid-Georgian circles was forcibly expressed in the famous Denunciatory Sermon of 1978. This celebrated address was broadcast, and relayed to New York, Chicago, and San Francisco, from St Paul's Cathedral on the First Sunday in Lent by the Archbishop of Canterbury, who took his text from Genesis xix. verse 24: "The Lord rained upon Sodom and upon Gomorrah brimstone and fire".

By the end of the twentieth century, however, a noticeable change from license to self-imposed restraint had already begun. It remained for the generation of 2000–2030 to complete this new tendency by the general adoption of that voluntary monogamy which is now so conspicuous a feature of our modern social life.

(2)

An outbreak of license among the less stable minority of a long-restrained population immediately the prohibitions are lifted can always be foreseen, and is invariably exaggerated when it comes. A subsequent reaction in favour of moderation, usually permanent, can also be predicted with similar certainty, and is equally attributable to the removal of restriction.

Though its effect upon the slum birthrate was obvious long before its disciplinary influence, there can be no doubt that the Sexual Instruction Act of 1948 was an initial cause of our present genuine monogamy. While the same type of person who had once concealed his or her sexual instability beneath the whited sepulchre of apparent marital fidelity began at least openly to experiment in trial marriage and frequent divorce, a quiet majority was adopting that dignified attitude towards moral problems which comes only from knowledge and frankness.

As soon as the teaching of sex hygiene caused future sex-relationships to be visualized in education, instead of being ignored with the embarrassed perturbation characteristic of youth's instructors in the nineteenth and early twentieth centuries, courses of instruction in the technique of marriage and parenthood in both their physiological and psychological aspects became an integral part of school and college curricula. In order, also, that these subjects should be approached by even the youngest children in a spirit of unprejudiced honesty, the Domestic

Advisory Courts were instructed severely to reprimand any mother who was found to have deceived her inquiring infants with the old type of ingenious and cowardly falsehoods.

As each external restraint and repression was abolished by law, girls and boys were taught by their sex educators the need for the additional exercisc of voluntary control by themselves. We have only to read a chapter or two of Dr Diana Trevelyan's impressive work, *Twentieth Century Progress in Educational Methods*, to see how rapidly the elimination of fear and ignorance in the second half of the century was followed by the disappearance of shame, furtiveness and that "curiosity concerning matters on which they could not obtain information otherwise" which Lindsey had alleged in 1928 to be the primary cause of promiscuous habits amongst young men.[45]

Enlightened education, combined with the removal of prohibitions and taboos, left no cause for that youthful contra-suggestiveness which immediately leads to a strong desire to do the thing forbidden. As soon as the old sentimental outlook on sex and parenthood was banished, it was found possible to inculcate in the majority of boys and girls a capacity for robust judgment between right and wrong in relation to their own experience.

(3)

A cause of past matrimonial disagreement only second to inadequate education was the difference in age which frequently existed between young wives and their husbands, who had postponed marriage until middle-age either for economic reasons, or because, being as yet unable to control biological changes, they preferred a wife whose sexual potency would outlast their own to one whose mental maturity offered some hope of satisfactory companionship. I am proud to think that my long-ago kinsman, Edward Westermarck, was one of the first to realize the folly of such a choice, pointing out in his still justly famous History that great difference of age is fatal to close sympathy, since similar objects and interests are associated with similar periods of life, and these with similar intensity of sexual desire.[46]

Fortunately for ourselves, the universal knowledge of birth-control, combined with the economic self-sufficiency which the Married Women's Independence Act first made accessible to all wives and which the growing privacy of marriage rendered still easier, has for almost a century enabled men and women to marry early and at similar ages. Young wives, wedded to young husbands whom they first meet at school or college, and in common with whom they often finish their education after marriage, have no longer any need to display that calculated ignorance of an older husband's pre-matrimonial record which so often in the past exposed women to venereal disease.

45 [Author's note] *The Revolt of Modern Youth*, p. 68.
46 [Author's note] *History of Human Marriage*, 2nd edition, p. 362.

With earlier marriage, indeed, the danger of such disease soon became negligible, while it is now only in Victorian and post-Victorian novels that we come across the Jehovah-like spouse who eclipsed and absorbed his young wife's personality, himself dictating her tastes, her occupations and even the number of her children. It would be almost equally difficult to-day to encounter the once quite customary selfish and cruel exploitation of an intelligent young woman by an exacting husband who, careless of the risk of her future resentment, persuaded her that the interests of the community were best served by the dedication of her talents to his career, instead of her own.

Those marriages have now proved to be the happiest in which husband and wife together face the hardships of youth, modify together, and together meet the old age which even in this century of prolonged life comes at last. The risk of tragedy due to over-young marriages has been reduced to a minimum since the Matrimonial Causes Act of 1959 legalized divorce by consent, but the successive improvements which have taken place in pre-marital education since this first became obligatory have rendered very small indeed the percentage of mistakes.

(4)

The economic independence of women which made early marriage practicable has also been responsible for removing other obstacles to successful life-long matrimony. No longer are women condemned to suffer from that sense of bitter frustration which once caused so much discontent and misery to wives for whom marriage had meant intellectual atrophy.

Those partners who, in spite of the conquest of space by science, are still obliged by their circumstances to practise semi-detached marriage, do so because they realize that the advantages of living under one roof do not outweigh the disadvantages of one partner being deprived of her chosen occupation. Waste of training and the thwarting of legitimate ambition have long been recognized as serious threats to the success of even the most promising union.

Modern husbands, again, are no longer goaded into seeking consolation outside their homes by the jealous possessiveness of a spouse whose imprisonment within her four little walls has deprived her of that toleration and experience which comes from contact with the wide impersonal world.

Man's need for diversion was indeed comprehensible when his wife resembled at best a well-groomed cow devoid of mental qualities. By the middle, however, of the twentieth century, the majority of wives had ceased to be either slaves, toys, exotic green-house plants, or carefully-reared animals with whom sex-intercourse was an humiliating concession grudgingly bestowed in return for an economic *quid pro quo*. They had already become reasonable, self-supporting human beings, in whose full and busy lives domestic relationships played only a proportionate part, and whose rich minds provided their husbands with that variety of experience which married men had once desperately sought from the physical novelties of mistress or prostitute.

Present-day husbands sincerely pity their illogical male ancestors, who by insisting upon both the financial and psychological dependence of their wives deprived themselves of that very free and equal co-operation which they had professed to seek in courtship, and who were so often accepted by cynical young women, not for what they were but for what they could provide. The men of the past three generations have been well aware that the best wives and mothers come from those women who have fully developed all sides of their humanity.

(5)

In contrast with the increasing number of happy monogamous marriages, the confused and constantly uprooted lives of those who took undue advantage of the divorce laws appeared sad and unenviable. As, moreover, the general standard of intelligence rose rapidly after the sterilization of the unfit became law in 1981,[47] men and women who devoted so large a proportion of their time to unprofitable sex-experiences soon found that their careers lagged behind those of their better-disciplined comrades.

A similar objection gradually arose to the practice of trial-marriage, which was also admitted by many who had practised it to deprive a later permanent marriage of much of its glamour. The idea, these experimenters discovered, that a partnership might not last was liable to engender a cynical expectation of failure which prevented it from doing so, and which also established sceptical habits of mind that were deleterious to the endurance of any relationship. Since the notion of intended permanence was found to be of the very essence of marriage, the majority of couples came deliberately to postpone sexual experience until they were reasonably sure that they had found the partner with whom they desired to share it for life.

Thus it came about that the scientific changes which rendered married partners both more accessible and more attractive to each other were accompanied by the development of a public opinion unfavourable to every type of promiscuity, which by 2030 was regarded as the fifth-rate pastime of lethargic individuals who provided themselves with insufficient occupation for body or mind.

Polygamous practices have never, of course, been made illegal, and much less punished, since the disappearance of those unenlightened generations which regarded excessive sexual appetite as a moral offence instead of a physiological disadvantage. The men and women who possess these inconvenient and unnecessarily developed instincts, which remain unsatisfied by the controlled delights of normal sexuality, are to-day merely encouraged in early youth to submit themselves—and indeed often submit of their own accord—to the already-mentioned minute glandular operations which modify the character of desire.

47 This troubling moment appearing even in a work like Brittain's devoted mainly to the progressive liberalization of institutions, is indicative of the pervasive reach of Eugenics in the interwar period.

Human marriage at last appears to have completed its long development away from the total indifference manifested between husbands and wives among primitive peoples, to the point at which life orients itself round one individual, and sexual intercourse is voluntarily reserved as the supreme and final expression of love for that person. While we regard modern matrimony as a relationship of which the beauty is not necessarily spoiled by an occasional and frankly-acknowledged extra-marital experience, permissible for some cogent reason such as that of honour to an old obligation, we recognize that this type of adventure should be indulged in only as the outcome of mutual agreement and full confidence between husband and wife if it is not to destroy that loyal attitude of two individuals towards one another which constitutes true monogamy.

(6)

The contrast between our present happily-ordered sexual relationships, and the dark confusion which prevailed little more than a hundred years ago, must without doubt be mainly attributed to the abolition by the later twentieth century reformers of those legal restrictions and penalties which, though aimed at the preservation of monogamous marriage, defeated the very ends that they set out to serve. The rigidity of Victorian and post-Victorian law and custom, which thrust all but one type of sex union outside themselves, so narrowed what was permitted as to create perpetually renewed groups of moral pariahs, who had hitherto done no injury to society, but who from the moment of becoming outcasts hampered by their actual position every effort made to introduce reason and toleration into contemporary standards of sexual behaviour.

The legalizing of experimental unions reduced the number of these social offenders to a minimum, while the experiments in their turn afforded men and women an opportunity of studying marriage with the same conscious intelligence that they had applied to other forms of control by civilization over the forces of nature. Liberation from external prohibitions, and confidence in humanity's ability to impose voluntary restraints upon its own conduct, have been justified in this twenty-first century with a completeness for which even the happy co-operation of science does not fully account. As several writers, such as the vociferous Benjamin B. Lindsey, had foreseen as far back as the nineteen-twenties, the monogamy that was never achieved when forced upon society became the voluntary choice of the majority as soon as they were free to form such relationships as they wished.[48]

It is not, indeed, altogether surprising that the social arrangement most conducive to happy marriage should have proved after all to be the monogamous relation which had been idealized so much and practised so little, for no type of union has yet been discovered in which the maximum of intellectual variety may be so effectively combined with the minimum of emotional disturbance. No other relationship appears so well fitted to leave the mind free from the fret of sexual urge

48 [Author's note] Cf. *The Revolt of Modern Youth*, p. 128–9.

by adequately satisfying the demands of the body without giving to them that undue degree of attention which renders sex an incubus rather than an inspiration to mankind.

(7)

I cannot conclude this long section more appropriately than by quoting two significant paragraphs from the recently-published volume, *Civilized Sex-Life in the Twenty-First Century*, by my esteemed colleague, Professor Pond-Hill:

"The scientific achievements of the later twentieth century", writes this distinguished sexologist, "liberated humanity for all time from its age-old enslavement to animal instinct, physical limitation, and biological change. With the assistance of contemporary modifications in moral values, these inventions and discoveries placed man's nature under his own control, and demonstrated beyond the possibility of contention that if, for instance, he desired true equality between men and women and hence genuine monogamy in marriage the attainment of both lay in his own hands and in them alone. Monogamous marriage has triumphed amongst us solely because mankind has realized that there are better reasons for its practice than the old fear of social consequences if its appearance were not maintained.

"Sex is now a recognized vital power, which is as much taken for granted as the meteorological forces that regulate climatic conditions. It would be a return to the unenlightened days of the earlier twentieth century to allow this power ever again to acquire the tyranny that it appears to have exercised when its repression by law and custom caused its manifestations to dominate the subconscious mind of humanity".

Excerpts

From: *Quo Vadimus? Some Glimpses of the Future*

E. E. Fournier d'Albe (1925)

Headnote

Edmund Edward Fournier d'Albe (1868–1933) was an Irish physicist, astrophysicist and chemist. He taught mathematics at University College, Dublin; physics at the University of Birmingham from 1910; and in 1914 he went to teach at the University of the Punjab in Lahore. He worked on electromagnetism and astrophysics, and experimented with improvements to radio and television. He invented the optophone (as described in *Automaton*; see 156) and worked as an assistant to the physicist Oliver Lodge. Fournier d'Albe also worked for three years as the secretary at the Dublin section of the Society for Psychical Research. He wrote another excellent volume for the series, also in 1925: *Hephaestus; or the Soul of the Machine.*

Most of the To-day and To-morrow books focused on a single subject each. In *Quo Vadimus* (the Latin tag asks 'Where are we going?') Fournier d'Albe follows Haldane's lead in *Daedalus*, and speculates about developments in a number of areas: Transport and Communications; Privacy; Clothing; Housing (in which 'All the "work" of the house will be done by machinery requiring but the turning of a switch and the aiming of implements resembling magic wands,' 47); Children; Education; Labour; and Government. Most of the writers were unwilling to risk venturing further ahead than a generation, a lifespan, a century. But Fournier d'Albe—again like Haldane, or like Bernal—ends by considering the much more distant future.

******

The Farther Outlook.—So far, we have looked but little ahead, a century at most. The prophet's task becomes more arduous as the time is extended. Historical guidance fails us. Familiar landmarks get blurred and disappear. We are in danger of getting lost in a bog of unreal speculation. Yet the task has often been essayed, and it is necessary and desirable that it be essayed now and again. Let me make my own humble attempt, in the light of what knowledge I have acquired and what great thoughts I have encountered in many lands and languages, and in discussions with many thinkers.

DOI: 10.4324/9781032625522-19

We must extend our time scale from centuries to millennia, and from millennia to geological eras. Above all, we must take into account not only the rapid advance of science and invention, but the constant *acceleration* of that advance.

The consequence of that constant acceleration is that new developments and achievements succeed each other with bewildering rapidity. Hardly have we got accustomed to the idea of telegraphy without wires when radiotelephony becomes an accomplished fact, and within a few years there is a rich crop of listeners with their wireless receiving sets counting by the million. An entirely new form of publicity comes into being, and a speaker on Savoy Hill[1] is able to speak to an audience of millions and sway them by his voice more effectively than he can do by cold print in the newspapers.

And this is only a beginning. Communication will become closer and more general. Already the earth is a network of lines and cables, linking continent to continent. Soon a speaker will have the earth for his sounding board and his hall of audience, and the privilege of addressing the human race will be prized above a coronation. Human sight and hearing will extend its range enormously, not only in space, but in time also. For the cinema film and improvements in the recording of sound will make it possible to make minute and comprehensive records of past sights and sounds for future reproduction, so that nothing of any value may be lost.

Other progress will go hand in hand with the rapid development of "signalling" communications, such as telegraphs and the like. The transport of goods and passengers will rapidly gain in speed, comfort, and safety, until the whole earth becomes accessible to all. It will not only become accessible, but also habitable. The tropics, the original cradle of the human race, will once more be reclaimed from our most formidable enemies of the insect world and the ever-present bacterium. The higher organism will assert its much-contested supremacy over those minute organisms which owe their influence to mere numerical superiority. Our descendants will pay an afternoon's visit to Timbuctoo or Mount Ararat, much as we should visit the British Museum or the Lake District. Everybody will be a globe-trotter, but the "globe" will not be confined to the ordinary tourist resorts. It will include every part of the world, even the Poles. And wherever they go they will find friendly voices, long familiar in the home through the service of radio-telephony. There will, of course, be an international auxiliary language, understood everywhere, a language artificial in its structure – every literary language is largely artificial – but utilizing those roots which have already become part and parcel of all cultured languages. This will not mean the displacement or loss of native languages which have proved their title to survival by their literature.

War will not cease for perhaps a century or more. But it will finally cease when the truth has sunk in that war is a loss to every belligerent and to the whole world. Human rivalry and competition will take other forms. There are many ways of killing men, women and children, besides suffocating them with chlorine. If a tribe is

1 Savoy Hill, off the Strand in central London, housed the BBC from 1923 until it moved to Broadcasting House in 1932.

to be exterminated, nobody will be killed, but all its members will be painlessly sterilized by X-rays or some such modern means, so that the next generation will know them no more. It will be more humane than the Biblical expedient of "dashing their children against a stone."[2]

The mass of interconnections between human nations and individuals will be like a closely-woven fabric. Even now, the digging up of a city road reveals a tangled network of water pipes, electric mains, gas pipes, and drain pipes suggestive of the dissection of an animal body. It is but a faint foreshadowing of what is to come. The substratum of life will become more and more complete as conscious life becomes simpler.

Nobody is conscious of the appalling complexity of his anatomical organization when using his body as a well-poised instrument of thought and intercourse. "The simple life" is not the old-fashioned country life of England or the primitive life of savage humanity. Real simplicity is constituted by the life in which most things are done by pressing a button, and a man can travel across a continent in such comfort that on arrival at his destination all memories of his journey are dimmed or lost, and he can hardly recall having travelled at all.

We may, therefore, expect that, as facilities for intercourse become more detailed and widespread, the effect will be, not to increase the tax on our nerve force until it becomes unbearable, but to increase our area of selection. There will thus be more consistency in our actual interests and activities and more real harmony and leisure.

The unification of the planet which is being accomplished before our eyes will have some astounding consequences. Mankind will assume a definite mastery of his home in the solar system. Attila could boast that when he plunged his spear into the ground, the whole earth trembled. The earth trembles even now to the electric signals of our powerful wireless stations. What will it be in a hundred or a thousand years? In a hundred years the unification of the human race will be complete. The earth and the fulness thereof will be under the full mastery of man. All animal, vegetable and bacterial life will be kept within strict bounds in the interests of humanity. The earth will be under one government, and one language will be written and understood, or even spoken, all over the globe. There will still be different races and perhaps allied nations, but travel and commerce will be free and unfettered, and calamities will be alleviated and dangers met by the united forces of all mankind.

And all the world will be *young*. The advances of medicine and surgery will have been such that most of the ailments and limitations of old age will have been eliminated. Life will be prolonged at its maximum of efficiency until death comes like sunset, and is met without pain and without reluctance. There will be no death from disease, and almost any sort of injury will be curable.

And in a thousand years? What will become of our globe and its dominant race, if no great catastrophe occurs to stop its exponential curve of progress. But for

2 Psalm 137, verse 9: 'Happy shall he be, that taketh and dasheth thy little ones against the stones.'

that exponential curve and its tendency towards constant acceleration, a thousand years would be no great period to foretell. Life has become world-wide in the last thousand years. The intellectual outlook has increased with the area of travel and communication. Dogmas and shibboleths have lost their force. Art and science have been emancipated from their ecclesiastical fetters. But the immense leap made since coal came into its own as a world-force belongs to our own age. The exhaustion of the coalfields might slow down progress for a time, but so long as mankind keeps its continuity, its past achievements and its rate of achievement will act as a stimulus and encouragement to further efforts, and new sources of energy will be discovered and utilized.

And so we may feel justified in expecting continual progress for at least a thousand years. Can we imagine the result? A globe laid out like a huge garden, with a climate under perfect control; the internal heat of the earth brought to the surface and utilized as a source of never-failing energy. Portions of the interior of the earth reclaimed and made habitable; all machinery and sources of power wisely distributed and made instantly available for all legitimate purposes. The earth's surface and the rippling ether in which it swims made into a vast playground of human thought and emotion, and all mankind throbbing in unison to every great thought.

The Earth will have become a sentient being.—It will be as closely unified and organized as the human individual himself. Mankind will be the "grey matter" of its brain. It may not resemble a sentient being high up in the scale of life, but it will be at least on the level of *protococcus*[3] or some other such humble plant-cell, which also consists of a minute proportion of material truly "alive" together with a greater bulk of stored foodstuff and waste products.

Man will be conscious of his closer attachment to the earth. He will feel towards it a sort of personal patriotism, or the sort of loyalty that a veteran feels towards the Old Regiment. Specially exalted or sensitive people may even indulge in a kind of Geolatry[4] animated by an old-world religious fervour.

Can we focus our mental telescope into yet farther depths of time? A million years or so?

It seems rather risky to extrapolate our curves so far. But a million years are but a span in the life of the earth. Its records speak to us of many millions. The chalk cliffs of Dover took several million years to deposit on a former sea-bottom, and many more to rise to their present eminence.

If there is a still farther advance in the life of the earth, what sort of direction can it take? Will a new race have arisen, as much above humanity as man is above the arboreal ape? Or will the further differentiation of man have come to a definite end, and progress be confined to an ever-increasing richness of intellectual, artistic, and emotional life? If there is any progress at all, it must be by *effort*, and the question could be answered with fair probability, if we could find an incentive to effort after the earth is entirely subdued. Such an incentive towards effort will lie in

3 Unicellular green algae.

4 Worship of the Earth.

the ever-present danger of a celestial catastrophe, such as a collision with one of the smaller wandering planets or other denizens of outer space. It may be that the earth will by that time be alive to its own peril, and will take precautions! Its "grey matter"—our own descendants—will be confabulating and organizing in some great scheme of defence. The earth will have to adopt a Foreign Policy, if it is to be the master of its fate for all time.

Here our dreams are checked by the realization that among the older planets of our solar system we can trace no activity attributable to a "foreign policy" of their own. But we must remember that ours is a small planet, which has very little influence upon its neighbours in space, and is certainly not a danger to them. If Mars, millions of years older than ourselves, has arrived at such a stage of advancement that it can think of transcending its own boundaries, it may make some attempt at communication, but the attempt might take the form which to us would be quite unrecognizable. Some observers thought that the persistent thunderstorms and magnetic disturbances experienced during the last opposition over wide areas were signs of such an attempt, but the coincidence may have been quite accidental.

In any case, we cannot find any sign of the Martians having succeeded in exercising any powers beyond the surface of their own planet, though, if the "canals" really exist, their engineering feats must be truly stupendous. It may be that the earth, owing to its position between torrid Mercury and ice-bound Neptune, enjoys conditions specially favourable to mental and physical advancement. And so it may happen that it will for immense ages of time be the only planet to burst into consciousness. Thus it may at some epoch find itself the undisputed master of the solar system, and may be able to influence the other planets and make them subservient in some way to its own needs.

Let nobody think I am unduly optimistic about the future of this earth of ours. The difficulty lies rather in visualizing what recent and current progress, and *accelerated* progress, must inevitably bring about when continued for a long time. The only doubtful element appears to be whether the magnificent élan vital[5] of our race, which has enabled it to conquer the world, will last through the vast ages to come.

But, fortunately, there is no sign of its exhaustion. Love is still the ruling passion and inspiration of humanity, which enables men and maidens to brave all the trials and dangers of life in unconscious devotion to a future as yet unimagined and unborn. [75–92]

5 See p. 102, n. 33.

From: *Kalki; or, the Future of Civilization*

Sarvepalli Radhakrishnan (1929)

Headnote

Sarvepalli Radhakrishnan, OM (1888–1975) was an Indian politician, philosopher, and statesman who became the first vice president of India from 1952 to 1962, and then its second president, from 1962 to 1967. He had a distinguished academic career first, as an influential scholar of comparative religion and philosophy, and interpreter of Hinduism to the west. He was King George V Professor of Mental and Moral Science at the University of Calcutta from 1921 to 1932 and held the Spalding Chair of Eastern Religion and Ethics at the University of Oxford from 1936 to 1952—during which time he also managed to serve as vice-chancellor of Banaras Hindu University.

******

I. Introduction

CIVILIZATION to-day seems to be passing through one of its periodic crises. The world is casting off its old garments. Standards, aims, and institutions which were generally accepted even a generation ago are now challenged and changing. Old motives are weakening and new forces are springing up. Anyone who has an insight into the mind of the age is vividly conscious of its restlessness and uncertainty, its dissatisfaction with the existing economic and social conditions and its yearning for the new order which is not yet realized. All this confusion of thought and unstable enthusiasms for ill-defined ideals show that humanity is about to take a new step forward.

One of the chief factors in this unsettlement is modern science. Though science is not peculiar to our civilization, its pace of progress has become latterly too fast and its range too wide and deep for our quick adaptation. If we take an animal out of its normal environment and thrust it into another, it is bound to feel restless and uneasy until it adjusts itself to the new situation. When the Bishop of Ripon suggested a scientific holiday for a short period of time, he meant to warn us that while science is progressing pretty fast furnishing us with new inventions, man the user of them is not refining himself at an equal rate.

The world is becoming outwardly uniform. Europe and America as well as Asia and Africa are moving in the same direction, only the former faster than the latter.

DOI: 10.4324/9781032625522-20

The motor-car, the aeroplane, and the pictures, which are the potent symbols of modernization, are visible even in the most backward countries. From China to Mexico there is increasing faith in the progress depending on the continued expansion of man's command over the resources and control of the powers of nature.

India and China are being sucked into the maelstrom. The unrest in the East is due to the new consciousness that, if the Oriental nations are not to decay and die, they must strive to stand as equals alongside of other nations whose spirit of enterprise and power of organization, have spread their sway to the uttermost parts of the earth. The East and the West are not so sharply divided as the alarmists would make us believe. The products of spirit and intelligence, the positive sciences, the engineering techniques, the governmental forms, the legal regulations, the administrative arrangements, and the economic institutions are binding together peoples of varied cultures and bringing them into closer reciprocal contact. The world to-day is tending to function as one organism.

The outer uniformity has not, however, resulted in an inner unity of mind and spirit. The new nearness into which we are drawn has not meant increasing happiness and diminishing friction, since we are not mentally and spiritually prepared for the meeting. Maxim Gorky relates how, after addressing a peasant audience on the subject of science and the marvels of technical inventions, he was criticized by a peasant spokesman in the following words: "Yes, we are taught to fly in the air like birds, and to swim in the water like the fishes, but how to live on the earth we do not know." Among the races, religions, and nations which live side by side on the small globe, there is not that sense of fellowship necessary for good life. They rather feel themselves to be antagonistic forces. Though humanity has assumed a uniform outer body, it is still without a single animating spirit. The world is not of one mind.

Spengler's thesis in his remarkable work on *The Decline of the West* that different nations have different cultures expressive of distinctive ideals is rather damaging to any hope of the development of a common culture or civilization. His fascinating hypothesis that races and cultures are units which undergo a rhythmically ordered sequence of birth, growth, decline, and decay does not seem to be quite adequate for the facts. In the past, perhaps, regional civilizations succeeded one another or passed through the phases of infancy, youth, maturity, and old age, and, when they decayed, they left their inheritance to the younger ones which sprang up after them. The possibilities in this direction are practically exhausted. The period of provincial civilizations confined to limited areas is drawing towards its close. Nor can we with any certainty look upon the history of humanity as a single continuous movement which later divided itself into separate streams on account of differences in milieu and native endowment. Ascertained facts suggest that the different cultures developed along their own lines for a very long time, inclined latterly towards each other, and are now tending to unite themselves in one mighty whole. When Spengler tells us that Western culture is passing through its old age in obedience to an inexorable destiny against which it would be futile to revolt, the truth that his view contains is a much larger one—namely, that all provincial civilizations are tending to pass away and we are about to make a new experiment in the art of life on a world-wide scale. No single historical culture or civilization can lay claim to an absolute universality, for each of them is the expression of the living forces of a group of individuals. In these matters there is no other logic than that of history, and history does not know an abstract universal man, and so there can be no universal civilization. The future civilization will have to rise to a universal vision of man and human life. The provincial cultures of the past and the present have not always been loyal to the true interests of the human race. They stood for racial, religious, and political monopolies, for the supremacy of men over women and of the rich over the poor. Before we can build a stable civilization worthy of humanity as a whole, it is necessary that each historical civilization should become conscious of its limitations and its unworthiness to become the ideal civilization of the world.

While the triumph of mechanical inventions provides a common basis for the civilization of the future, the breakdown of traditional systems of thought, belief, and practice is the necessary preparation for the building of a spiritual unity. The leaven is at work among all the peoples, especially among the youth who are unwilling to be mere clay in the hands of others, be they ever so old or wise. There is a quickened consciousness, a sense of something inadequate and unsatisfactory in the ideas and conceptions we have held and a groping after new values. Dissolution is in the air. The old forms of faith are tottering. Among the thoughtful men of every creed and country there is a note of spiritual wistfulness and expectancy.

If we leave aside the fanatics with whom no argument is possible, the leaders of every historical civilization to-day are convinced that mankind in all its extent and history is a single organism, worshipful in its growing majesty and capable of a progress to which none dare set any bounds. Dante proclaimed: "There is not one goal for this civilization and one for that, but for the civilization of all mankind there is a single goal."[1] If there is a single goal for all civilization, it does not mean that all shall speak a common tongue or profess a common creed, or that all shall live under a single government, or all shall follow an unchanging pattern in customs and manners. The unity of civilization is not to be sought in uniformity but in harmony. Every great culture is due to the blending of peoples of different ideals and temperaments. Egypt and Babylon, India and China, Greece and Rome, testify to this truth. To-day the circle of those who participate in the cultural synthesis has become wider and includes practically the whole world. The faith of the future is in co-operation and not identification, in accommodation to fellowmen and not imitation of them, in toleration and not absolutism. [7–13]

[. . . .]

III. The Problem

Philosophy in the larger sense of the term is the unseen foundation on which the structure of a civilization rests. It is the soul which slowly builds for itself a body. The organized habits and institutions of a community help its members to realize

1 Dante Alighieri, *De Monarchia* (sometimes translated as 'On World-Government'), Book 1, the conclusion of Chapter 2; possibly Radhakrishnan's translation from the Latin.

its chosen values, its reading of the meaning of life, and its purpose. When we praise or condemn a civilization we have in view its scale of values.

Ancient Hindu wisdom and Greek thought agree in regarding the human individual as a replica of the universe. He has a body which has weight and measure like the metals, organization like the plants, sensibility and movement like the animals, and in addition reason and spiritual aspiration. Broadly speaking the human self is a trinity of body, mind, and spirit. Our physical being, which is not much different from that of the ape, testifies to our animal and vegetable ancestry. Professor Elliot Smith asserts that the human brain shows no formation of any sort other than that of the brain of the chimpanzee. Some of our mental qualities also point to our kinship with the animal world, such as our inherent indolence, our tendency to vegetate, our clinging to the soil, and our subjection to passions like rage and fear. Our craving for the unseen, our spiritual aspiration and adventure, our striving to elevate ourselves are also a genuine part of our being, and are responsible for our mythologies and philosophies, religions, and arts. Throughout the course of human evolution, our spiritual longings have persisted and expressed themselves in varying forms from the crude stages of superstition, animism, and myth to the refined and complicated philosophical systems and ethical cultures of to-day.

Though there is much in us which may be regarded as our heritage from animal ancestry, man as man is different from the animal. Our vices and virtues are peculiarly human. When we make the pleasures of sense our aim in life, we are said to be more animal than human, but no animal can frame an ideal of a life of pleasure and pursue it in the way man can do. Again, there are cases where animals are more decent than men. There are certain things which are natural to the animals but which men have to acquire through deliberate effort and discipline. Animals rigidly subordinate sex to the function of reproduction. The savage and the primitive closely resemble the animal in this respect. The power of reflection and choice with which we are endowed opens great possibilities to us to rise on the animal heritage to the heights of heroism or fall into the depths of degradation. So when we talk about sinking to the level of the beast, we are using a figure of speech signifying our abuse of freedom in the interests of the needs which are common to men and animals.

The animal in us is ever striving to fulfil itself. When all impulses are perfectly satisfied, we have the full development of the animal being, the perfection of our animal nature. If we identify the self of man with the body and life-purpose with physical development, we are said to be barbarian, worshipping brute strength and power and idealizing the satisfaction of the passions. Such an exclusive culture of the body would coarsen the spirit and deprive it of its rights. The supremacy of physical prowess and development is the characteristic mark of barbarism. In such a society, men belittle and exploit women, for the latter are physically weaker, and women in their turn respect and pander to brute strength and prefer those known for their bravery and deeds of arms.

A community which gives greater importance to mind than to life and body is higher in scale. But unless the "mind" is taken in a larger sense so as to include aesthetic development and ethical perfection, i.e. unless the mind is interpreted as one with spirit, we have not reached the ideal of civilization. Our knowledge may be increased, but it is used not for the sake of the higher spiritual ends but for the vital and physical satisfactions. Our life becomes dominated by the desire for the satisfaction of wants which steadily increase and the accumulation of possessions. The sort of mental life which prevails is at a low level. Emotional thrills and intellectual sensations, aesthetic occupations and mental excitements attract us and not deep appreciation of great literature and noble art. Mechanical plots, detective stories, crossword-puzzles allure and amuse us. The individual belonging to a community in this second level does not think for himself but acts in a conventional unintelligent way. His moral nature is crude and undeveloped, being a mass of likes and dislikes, prejudices, and prepossessions. Shaping his life by a mere compliance with conventional codes, he has no other standards than comfort and seemliness. Education is valued for its utility in fitting a man for success in the competitive economic struggle, science is respected for its useful knowledge, the comforts and conveniences which it makes possible, its power of organization and its capacity to devise machinery for increased production. The outer acquisition does not, however, bring about inner refinement. We have still wars—only they are no more contests of human arms but contests of machines. We are not so much brothers as beasts of prey, and, so long as our selfishness is unsubdued, we are more dangerous because our power for mischief has increased a thousandfold. Being the slaves of conventions, we are servile in soul. When the group believes in the absoluteness of its dogma or the superiority of its culture, the individual is ready to fight for them. Faith in force is the first principle, as the history of religious persecutions proves. If in such a society a few rise above its average and think that the ultimate goal of humanity is the creation of a world-family expressive of a universal God of love and the winning of men freely to want the good of the whole and propose to rule out force, they are treated as rebels and heretics, and society has a short way with them. The timid are frightened into submission and the dissentients are destroyed. This stage of society is one of economic or intellectual barbarism, since it confuses civilization with comfort, morality with convention, religion with routine, and politics with trade, exploitation, and opening of markets.

A community which is almost entirely preoccupied with life and body, physical and economic existence, scientific and technical efficiency to the exclusion of the higher humanistic ideals of mind and spirit is not truly civilized. Body, mind, and spirit form distinguishable aspects of an inseparable unity. Human nature is all of a piece, and unification of the three is the true aim of civilization. The conflict and the friction of the parts are not to be deplored but demand conquest and adjustment. Excellence of the body, its soundness and health are necessary for a perfect manhood; a sufficient social and economic organization is necessary for good life, but these are not ends in themselves. The universe has spent so much pain and struggle to produce human individuals who adore the good, the lovely, and the true and who are not content with a finished animality. The difference between the half-civilized and the civilized is all the distance between a narrow self-regarding individualism controlled by the animal impulses of self-preservation, self-assertion, and self-satisfaction and the self-forgetful universalism devoted to the good of the whole.

It is the transformation of the individual into the universal outlook, the linking up of our daily life with the eternal purpose that makes us truly human. The process is costly, but when the redirection of our whole nature to this universal end takes place, the yoke is easy and the burden light. A new kind of life, a new order of consciousness would begin as different from that which men now have, even as human life and consciousness are different from animal life and consciousness.[2]

There has not been in the history of the human race any perfect embodiment of barbarism or of civilization. No community is strictly savage or completely civilized. No human organization failed to develop its group ethos, its religious rites, and social forms. It is difficult to find traces of a race which does not distinguish between right and wrong, which has not the rudiments of morality and art. Civilization seems to be as ancient as savagery. We regard the Eskimo, the Red Indian, the Basuto,[3] and the Fiji Islander as barbarous, simply because they had not risen to our conception of civil society with its schools and hospitals, law-courts and police-stations, but undoubtedly they exhibit an individuality in their modes of life, customs, and beliefs as much as the advanced Greeks and Romans or even the British and the Germans of to-day. We cannot call them savage or barbarian simply because their social organization was different, their knowledge of nature narrow, and their appliances crude. Even to-day we tend to regard politically backward nations as semi-civilized or semi-barbarian, on the assumption that the test of civilization is political success or economic prosperity or skill in the destruction of human beings. Japan came to be regarded as highly civilized when she fought and defeated Russia, though, on this assumption, the Tartars who overthrew the Sung dynasty and the barbarians who overran the Roman Empire are also to be held up as models of civilized humanity.

While the crude beginnings of civilization exist even in the most primitive communities, civilized societies have surviving in them a large mass of barbarism. We look upon the Hun, the Goth, the Vandal and the Turcoman as barbarous, but we cannot be sure that at some future date a more developed humanity will not look back upon much of our present-day civilization with amazement and disgust as the superstitions and atrocities of an imperfectly civilized society. Even as we condemn the gladiatorial shows of the Romans, our descendants will condemn our joy at the angry disputes of animals, our prize-fights, not to speak of our "refined" butchery called war.

Civilization is within ourselves, in our moral conceptions, religious ideas, and social outlook. We cannot call ourselves civilized simply because we use the steamship and the railway, the telephone and the typewriter. A monkey trained to ride a bicycle, drink a glass, and smoke a pipe is still a monkey. Technical efficiency has little to do with moral development. Though the achievements in exact

2 [Author's note] If we adopt the Hindu terminology, we may say that a society which idealizes physical strength is *tamasa*, that which gives the first place to satisfactions (vital, physical, and economic) is *rajasa*, and that which regards spiritual freedom and growth as the chief aim *sattvika*.

3 Basutoland was a British Crown colony from 1884 to 1966, in present-day Lesotho inside South Africa.

science and mechanical organization of Ancient India or Greece or Medieval Italy are immensely inferior to ours, it cannot be denied that they had a truer perception of spiritual values and the art of life. If civilization is not to be confused with a feverish thirst for new things or a mad race for wealth, there are many salutary lessons in the art of living which we can learn from India or China or Ancient Greece. Not that the latter had not their own defects. The leisure and enlightenment of the citizen-body in Greece was made possible by the exclusion from this privilege of the large body of artisans and slaves who followed the necessary but laborious pursuits of the community. Though the Hindu civilization by her wise toleration of local customs and beliefs gradually absorbed the indigenous races of the country into a free and organic synthesis, it neglected the education of the backward people. Hindu ideals, however admirable, did not penetrate the mass of the people. In later times there was a sad falling away from the high ideals consequent on the stunting of free manhood under autocratic rule.

Modern civilization is in the stage of economic barbarism. It is concerned more with the world and its power than with the soul and its perfection. It asks us to make the best of the business in hand, for first and final principles are beyond our ken. It is an endless, anxious striving to perfect the exterior forms of existence, to exploit the economic potentialities of the earth, to spread far and wide material well-being and master the forces of nature for the ends of man. We have the assertion of mind over life and matter and not yet of spirit over mind, life, and body. To control life and body we have understood their processes and possibilities. In the first triumphs of scientific progress it tended to cast aside philosophy, despise thought, and almost succeeded in slaying religion. Though we are more learned and scientific than our ancestors, we cannot say that we are less brutal and more humane. Our education has not freed us from intellectual bondage. It stimulates the mind without satisfying it. We read poetry, devour fiction, and attend the movies; and think we are cultured. Our rationality is a pretence. We use our reason to bolster up our instincts. We invent excuses for what we want to do and set forth arguments for what we want to believe. We have great faith in "going about doing good",[4] though it is more "going about" than "doing good". We live outwardly, talking glibly about the ideals of humanity and mouthing current phrases, though we are impervious to ideals and innocent of discipline. We are not very different from the old foolish, impulsive, credulous individuals capable of supreme heroism on occasions and incredible cruelties more often. Man, the combative animal, is still untamed. Economic success is our highest ideal, and almost all our wars are due to economic causes. Economics is our religion. Empire is big business. We wage wars to increase our trade, extend our territories and acquire colonies. For the sake of business and markets, we sacrifice our intellectual freedom, as it may produce doubt, our emotional sympathy for it may impair our efficiency in the exploitation of labouring-classes and government of backward races, and our

4 *Acts*, 10:38: 'How God anointed Jesus of Nazareth with the Holy Ghost and with power: who went about doing good.'

imagination as it may interfere with firmness. Our civilization is a conquering one based on the rivalries of individuals and races, the glory of combat and the joy of victory. It is compounded of speed and daring, adventure and excitement, eager activity and headstrong tumult. But its desire is unappeasable and it is fated never to be satisfied.

The mechanical virtues of speed, quantity, standardization, and absorption in things material have resulted in a spiritual hardening. We are suffering from an inner lack of unity and general mental anarchy. A spiritual ideal of a truly human life of freedom and beauty, love and virtue, is professed along with an unintelligent attachment to the life of the body, its vital needs and impulses, a low mental life of sense and emotions, and a grossly utilitarian code of practice. Even the cruder physical barbarism is not entirely absent. It betrays itself in our new feeling for the body and our appreciation of what are called "sacred lusts", the "divine fire", the "subterranean shrine", the "noble savage" who is near to nature, the "voice of the elemental world". Impulse is regarded as sacred and irrationality is cloaked in sanctimony. [37–49]

[. . . .]

From: IV. Reconstruction

[The section on 'Religion']

The supreme presents itself in a variety of aspects to the human mind. The bewildering variety of Hindu theism and pantheism, the Buddhist developments of an impersonal law and a personal saviour, the old classical paganism with its personal gods and goddesses, the Hebrew faith in an inflexible Lord of righteousness, the Catholic Christianity with its somewhat distant personal deity and more immediate minor divinities from the Madonna to the Patron Saints, the Protestant view of a personal God, and the Muslim creed of a one and only God are the different ways in which men have tried to orient their relations to the unseen reality conceived as something higher, better, and more sovereign than the individual self. If we admit the diversity of human nature, we can easily understand this variety of the appeal of God to it and the utter futility of reducing all dogmas to one. Underlying all the diversity of dogmas is the undefined and the indefinable conception of an ultimate reality.

Diversity of religion is often made the pretext for hostilities, and attempts to have a common religion for all mankind have contributed not a little to the unrest and sorrow of the world. The passion to impose one's own opinion on others is natural to the selfish temper. To believe that we have the exclusive possession of truth or that our reading of the meaning of the universe is accurate is an illusion of egotism. Every religion represents the soul of the people, the inner law of its being and aspiration. Each group has the divine within it into which it grows. When it comes into contact with others, it transforms their ideas and influences into something new. It is better to transform what we take from others than repeat what has been gained elsewhere. A single religion for all mankind will take away from the spiritual richness of the world. If we want to prevent the sterilization of the mind and the stagnation of the soul of humanity, we must not repudiate or refuse recognition to any one of the historical religions. "As many as are led by the spirit of God, these are the sons of God."[5]

In matters of spirit, each one must follow the lead of his own conscience. The acceptance of freedom means the risk of error, but the errors are of their own making, painful but not useless attempts to find the truth. They can be corrected only by the constant endeavour to think persistently on the deeper questions.

The religion of the future must be a comprehensive one embracing within its scope all those who are religious-minded in sentiment, allowing them full liberty so far as creeds and thought-pictures are concerned. For religion is not so much a theory of the supernatural as an attitude of spirit, a temper of mind.

We shall be judged not by technical points of belief and doctrine but by life and character. The religious, to whatever denominations they may belong, share the same outlook and spirit. They possess a serenity which is undisturbed by the strokes of fortune. The essence of all spirituality is the greatness of soul which is unconquered even in overwhelming catastrophe. The strong in spirit are superior to the world which they have conquered. They can speak the truth under a shower of bullets and refrain from reprisals even when they are literally crossed. They possess the universality of outlook which makes all attachment and self-interest stupid and superfluous. Sacrifice which does not count the cost, self-giving which does not ask for any return become their daily life.[6] Most of us dismiss their self-denying heroism and serenity as too much for human nature or at best regard them as admirable ideals fit to console the defeated peoples like the fishermen of Galilee or the feeble minds of India but impossible of realization. But every religion has insisted on strenuous life and self-denial. Religion is not mere good form. The consolations which are cheaply won are not truly religious. To look upon life as an uninterrupted pursuit of enjoyment is the mark of irreligion. Suffering is not an accidental accompaniment of life, but is central to it. In pain and travail all high achievement is wrought. The end of life is not pleasure (*preyas*) but the realization of the good (*sreyas*). A life of joy and the joy of life are not the same. If suffering leads us to the fulfilment of our ideal, it is as much happiness as a life of pleasure is. The most poignant pain can be joyously accepted if it is recognized as contributory to the realization of one's ideals. As Goethe says: "The heights attract us, not the steps thereto."[7] In minor matters we understand it. For most women physical discomfort is a lesser evil when compared with the defiance of fashion. They willingly face the ordeals of the beauty-parlour and the boot-shop. Man is self-willed and the exercise of self-will is the prime necessity of his life. But his self-will is to be reconciled

5 *Romans* 8:14.

6 Echoing the prayer of St Ignatius Loyola: 'Teach us, good Lord,/To serve thee as thou deservest;/To give and not to count the cost;/ To fight and not to heed the wounds;/To toil and not for seek for rest;/ To labour and not to ask for any reward/Save that of knowing that we do thy will.'

7 Goethe, *Wilhelm Meister's Apprenticeship*, Book VII Chapter IX.

with the demands of the universe. It means conflict, struggle, and suffering, and all these are in the day's work.[8]

Every individual soul is an undeveloped entity, which, though not wholly emancipated from the animal instincts, is yet capable of transmuting them. By a willing acceptance of the commanding claims of spirit and a discipline of our nature into conformity with its law, we achieve growth. A redirection of interest and reeducation of the unconscious mind alone can transform us from carnal-mindedness into spiritual-mindedness. It means restraint and discipline. There is real temptation, genuine struggle involved in every transcendence of the natural man. But this transcendence is the condition of the fulfilment of personality. Though the effort is costly, the reward is great. This transcendence is a phenomenon common to all progress. From the pursuit of its prey by the amoeba to the spiritual struggles of the striving soul, we have a continuously graded series of purposive efforts. We choose the goal everywhere and adapt the means. Only at the human level we can do so consciously and deliberately.

Man is neither the slave of circumstances nor the blind sport of the gods. The impulse to perfection working in the universe has become self-conscious in him. Progress *happened* in the subhuman world: it is *willed* in the human. Conscious purpose takes the place of unconscious variations. Man alone has the unrest consequent on the conflict between what he is and what he can be. He is distinguished from other creatures by his seeking after a rule of life, a principle of progress.

It is by transforming ourselves that we shall be able to transform the world. The soul of all improvement, it has been rightly said, is the improvement of the soul.[9] We need not leave the building of the new civilization to luck—it is a matter for cunning also. There is a good deal to be done yet. The world is unfinished in several respects. It lies with us to retard or hasten the progress. The lesson of evolution is clear. Life does not work according to any set programme, reaching satisfying climaxes in its progress. It wanders, hesitates, sometimes breaks off abruptly, and often returns upon itself. The processes of nature are by no means economical. The history of the past tells us that, if we choose wrongly, we shall sooner or later be eliminated. Each of us can participate in the process of creating a better world, by understanding the purpose of the universe and identifying ourselves with it. Every individual is a unique entity possessing specific features and capable of contributing a distinctive good to the world. Self-finding is the essence of all perfection. By seeing life steadily and whole, we find our place in it. Each human individual is an association of qualities and functions whose focus can be variously situated and according as this centre is higher or lower, the person's character is superficial or deep. Whatever be the empirical foundations, the choice of the centre redeems them. It is by means of reflection and meditation that we seek a centre which can

8 [Author's note] For a development of these views, see the writer's *Hindu View of Life*. George Allen and Unwin, London. 1927.

9 By the American Congregational minister and theologian Horace Bushnell (1802–1876): *An Address Before the Hartford Co. Agricultural Society* (Hartford: Brown and Parsons, 1847), 23.

unify our nature, realize our specific place in the universe, and acquire the strength to play our parts, however humble and however arduous they may be. Hence the necessity for quiet contemplation and solitude.

Moral life is one of significance and social value. Mere instinctive reaction or impulsive activity is not moral conduct which must embody an idea or be significant. Crude materials of human nature must be shaped to spiritual ends. It is the form given to the material that matters. All expressions of human life are significant and every one of them is judged by its meaning and purpose. We cannot have any one conventional type or pattern. Each man sees the world from a different angle, from that of his fellows. Whatever ideal we choose, we must recognize its nature and believe in its possibility, devise a plan of action, and with suffering and self-sacrificing ardour strive to realize it. Besides, the line of conduct to be moral must contribute to social preservation and harmony, which the process of evolution aims at. Any line of conduct which is destructive of the human race or promotes discord cannot be regarded as moral. Moral life requires us to respect the dignity of every individual. It is the enrichment of life that comes from the recognition of others and adaptation to them. If Moloch did not prevent the Phoenician mothers from feeding on their babies, some other god would have done so. All men are exceptional beings. Uniqueness is a quality which all of us share. The goal of the world process is a harmonious unity in which each individual finds his specific realization. [58–66]

[. . . .]

International Relations

The realization of the human commonwealth cannot be achieved by the utterance of conventional platitudes, pious hopes and beautiful phrases. Cultivation of brotherhood and fellowship among the nations is the indispensable prerequisite. Misunderstanding and pseudo-patriotism are now doing the deadly work. The nation we hate is the nation we do not know. When Newman as a child watched the French prisoners from the Napoleonic wars march through the streets of the city of London, he tells us how he noticed that men amongst the dense onlooking English crowds would quite seriously lift up the French soldiers' great coats from behind—to discover their tails. So simply and so seriously did the Englishmen believe the French to be apes. We see in our antagonists, if not monkeys' tails, at least their equivalents in special racial qualities. Our enemies are not human beings but fiends incarnate. We have had recently a sensational book about India indicting the whole people and justifying their perpetual subjection, if not removal from the earth's surface.[10] I am afraid that Carlyle's pungent metaphor, that nations, like dogs, seem to approach each other sometimes only to sniff at the shameful parts, is not out of date. The peace of the world depends not so much on signed documents and paper conventions, economic treaties and political combinations as on the drawing together of the minds and consciences

10 Probably the controversial *Mother India* (1927) by American journalist Katherine Mayo, criticizing Indian culture, religion, and society, especially its treatment of women and girls, and opposing the nationalist movement.

of cultured men and the growing commerce of knowledge and ideals. While we have differences with our neighbours, so long as we are devoted to material success and commercial prosperity, we feel the kindred throb of human nature when we turn to those treasures of the soul, those delights of mind which are not diminished by sharing. They alone will assuage the bitterness of economic strife and incline us to that understanding and sympathy which is the world's only security.

In addition to the fostering of oneness of thought and feeling, national psychology in its attitude to war requires to be altered. We need not believe the sceptic who says that the human race has always relied on wars and will always do so. So said the advocates of cannibalism. The human race has always eaten human flesh and always will. So said the supporters of slavery and duelling etc. etc. We can slowly change men's minds. Pacifism or internationalism is not a scientific device like the wireless or the telephone which the world can, all on a sudden, take to. It is a delicate plant which it takes long to rear. Patience and forbearance, mutual understanding and respect are necessary for its growth. From the nursery we must train people to believe in the oneness of humanity. We must develop in them the will to peace and declare that wars mean a house divided against itself. They set in artificial antagonism, what Providence and nature plainly intended to be one, "Since ye are brethren, why do ye wrong one another?"[11] The false glamour which human ingenuity casts on the primitive instinct of pugnacity, involving spoils to the few and suffering to the many, must be exposed. We must insist on a heightened sense of the sanctity of life. We must realize that wars, by accustoming us to violence, [rouse] in the hearts of the masses the bitter passions which civilization is struggling to subdue. Violence, physical or otherwise, arises from the lower, appetitive nature, is devoid of significance and is destructive of all social value. Patriotism which adopts the policy of our country, first, last, and all the time, is a soporific blinding us to the higher vision of humanity. *The Chicago Tribune* has for its motto "Our country. In her intercourse with foreign nations, may she always be in the right: but our country right or wrong." We are not disputing the rightful place of national pride and self-respect. It is national arrogance and intolerance that are dangerous. It is false patriotism that excuses everything and sets the state above the moral law. The impudent theory of state absolutism, which maintains that the state can do no wrong, that it is above criticism and all its wars are righteous is the substance of irreligion. And yet this is the religion of the modern world, which has asked so much of its adherents. No other religion ever demanded such gratuitous and wholesale human sacrifices. The Bolshevists aim at the redemption of the whole of humanity, irrespective of race or nationality. They look upon Russia not as a nation but as a mission for the propagation of their faith. So far as they loosen the strength of mere patriotism, they walk in the right direction. It is Lessing who said: "Patriotism is a heroic weakness which it is well to be without."[12] There is a higher law than that of might and

11 *Acts*, 7:26 (where 'Since' reads 'Sirs'; and 'one another' 'one to another,' in the King James Version).

12 This translation of the remark from the letters of Gotthold Ephraim Lessing (1729–1781) is from Romain Rolland's novel *John Christopher*, trans. Gilbert Cannan (London: John Lane, 1910), vol. II: *Storm and Stress*: 375.

a higher love than the conceit of patriotism. Loyalty to one's own country is not incompatible with the larger loyalty to mankind as a whole. Nations are the natural units of mankind and each of them is driven forward by an impulse derived from its own special past and historical tradition. Only an excessive particularist allegiance is dangerous. To praise one's own country must be regarded as no less vulgar than praising one's own self. It is no use praising our own institutions and opinions like inferior commercial agents. Nationalism is neither vain glory nor intolerance.

There are two ways in which the realization of the world-unity can be brought about, world-dominion or world-commonwealth. The former is impossible as nationalism is a standing barrier to its realization. This dream cannot be fulfilled without a fight to the death in which the world might perish. Why should the rest of the world be reduced to ruin, that a great race may prove its greatness? The methods of war are too costly, too drastic and no single empire has the means or the strength to overcome the rest of the world. After all, the author of the universe has not made mankind a homogeneous whole. He has made the nations of different races. But there is an easier and more reasonable way of merging national aims in a higher synthesis and international endeavour. Every people must play its part in this enterprise. Humanity must be made up of organized national wholes. The international order must be based on justice. To assume that the position of Europe will be menaced if Asia follows her example with success and resists economic exploitation and political invasion is a queer kind of logic. Our patriotism is the will of the gods; other peoples' patriotism is the work of the devil! If any country refuses to accept the existing forms of exploitation, we call her a peril. The security of the world is based not on the slavery of some nations but on the freedom of all. National freedom is the indispensable prerequisite of international co-operation. The genius of each people has a right to recognition and it must be the aim and endeavour of the whole world to grant to all nations freedom. It may be that some parts of the world are more backward than others. Their weakness should not be used as our opportunity. A nation which bullies a weak neighbour is just as guilty in the eyes of all honest men as the man who takes advantage of his helpless neighbour. Like individuals, nations need sympathy. The attitude of politically advanced nations to the peoples who are struggling to free themselves from age and service and not of patronage and control. The awakening of the Orient is not to be regarded as a menace to Europe. China is in a ferment. India's independence is only a question of time. Turkey, Persia, and Afghanistan are fast getting modernized, and all this is for the good of the world. No nation can live regardless of others. The interdependence of nations is increasing rapidly and faith in the destiny of humanity must fire us with the enthusiasm to fight against oppression and injustice wherever they are found.

Worthy ideals are not enough but a working technique is also necessary. The League of Nations and the Kellogg pact[13] are useful as means for educating public opinion. There is, however, a deep-rooted and widespread suspicion that the

13 The Kellogg-Briand Pact was an agreement to outlaw war signed by 15 nations on 27 August 1928 (a further 47 countries signed later). Frank B. Kellogg was US Secretary of State and Aristide Briand the French Minister of Foreign Affairs.

League is only an instrument of the *allied* nations devised for the purpose of maintaining and defending the *status quo*. It is not generally accepted that it is an institution pledged to work for providing better conditions of freedom and security for all nations. The value of the Kellogg pact is considerably diminished by the observation of its author that each nation "alone is competent to decide whether circumstances require recourse to war in self-defence." If we are to give up war, as an instrument of state policy, we should do so without any reservation, whatsoever. There can never be a just war. We must give up war as an instrument of state policy. We should not justify it even when undertaken for purposes of defence. Defence may be made to include prospective dangers and thus justify aggression. We cannot draw a line between light and darkness in a land where all is grey. Again, violence is bound to be followed by counter-violence and truth has little chance of prevailing. We are bound to have conflicts so long as nations move in orbits of their own. In their search for new and ever-expanding markets, the nations of the world are bound to meet one another competitively at every cross road on the earth's surface. But the conflicts should be settled by reason and not by force. We must develop a common code of law, establish a supreme court that would pronounce judgments that would command universal assent, and organize a police that would enforce its decrees. So long as the Great Powers are not prepared to relinquish a jot or tittle of their sovereignty which they are ready to uphold by force, whenever necessary, the League and the pact are all a mockery.

Religious idealism seems to be the most hopeful political instrument for peace which the world has ever seen. We cannot reconcile men's conflicting interests and hopes so long as we take our stand on duties and rights. Treaties and diplomatic understandings may restrain passion but they do not remove fear. The world must be imbued with a love of humanity. We want religious heroes who will not wait for the transformation of the whole world but assert with their lives, if necessary, the truth of the conviction "on earth one family", heroes who will accept the motto of the great Stadtholder: "I have no need to hope in order to undertake; nor to succeed in order to persevere."[14] [88–96]

14 William the Silent (1533–1584), the main leader of the Dutch revolt against Spain which initiated the Eighty Years' War. A Stadtholder was a nobleman appointed by a monarch as a governor.

From: *Chiron; or, the Education of a Citizen of the World*

M. Chaning Pearce (1931)

Headnote

Melville Chaning Pearce (1886–1969) was an educationalist, a writer on Kierkegaard and religion, and also a poet. He was founder and Headmaster of the experimental school Alpine College, Arveyes-sur-Bex, Switzerland. *Chiron* was his first book. In the preface he says: 'though I am and have been a schoolmaster for some 15 years, and for five years have directed my own experimental school, I have been many things besides this, a soldier, a political officer, a magistrate and a secretary to a High Commissioner in the East for ten years.'

A number of To-day and To-morrow books were devoted to education. Robert T. Lewis' *Romulus; or, the future of the child* (1929) discusses pre-school children and nursery schools; M. Alderton Pink's *Procrustes; or the Future of English Education* (1927) wanted to disengage school education from university requirements. Julian Hall's *Alma mater; or, the future of Oxford and Cambridge* (1928) and W. J. K. Diplock's *Isis, or The Future of Oxford* (1929) give the progressive and conservative views on the ancient universities. The theme of education plays across many of the other volumes, including Russell's *Hypatia* on women's education, as we have seen; or R. G. Gordon,' *Autolycus; or, the future for miscreant youth* (1928).

Chiron was written in response to *Eleutheros, or the Future of Public Schools* (1928), by J. F. Roxburgh, the Headmaster of Stowe School. Chaning Pearce says in his Preface:

> My only serious quarrel with that book, in which so much is so well and so wisely said, is that it seems to me to offer no adequate solution of the problem which its title poses. If the future of public schools is to be, in all radical respects, what their past has been, it is difficult to see how that future can be justified.
>
> (9)

******

DOI: 10.4324/9781032625522-21

To All Makers of Men

Chapter III: Citizenship in the Modern State

THE conclusion reached in the last chapter, that education should aim primarily at the training of good citizens, is not, we believe, one likely to bc disputed by the majority of modern educationists. It is chiefly the diversity of opinion as to the nature of good citizenship which divides their ranks.

What is good citizenship to-day, and—a question still more cogent—what is good citizenship likely to be to-morrow? Education is concerned, not with the good citizenship of yesterday, but with that of the world in which we move and in which our sons will move, a rapidly changing world.

Good citizenship is only good in so far as it is good for the 'city' or polity which it serves. Nor is good citizenship necessarily limited to one type; there are citizen leaders and citizen followers, citizen statesmen and citizen artisans. Leaders there must be in every state; the public schools have hitherto supplied the leader class, and it is therefore with the education of leader citizens that we are here concerned. Although it seems certain that the 'public-school classes' will not hold in the future that unique position in the State which they have held in the past, it seems equally certain that for some time to come, by virtue of their social and financial status, it is from these classes that the leaders of the new world may best be drawn. Our inquiry, therefore, narrows down to the consideration of the chief qualities necessary for leadership in the modern State. In order to perceive the qualities required for leadership in the new world we must first consider the salient features of that modern polity for which leaders are to be trained. The world of education turns upon the poles of politics and it is to the main trend of politics in the State for which citizens are to be trained that education must correspond. What are the chief characteristics of the modern world to which the qualities of its new leaders must be matched?

Such a debate is no new thing, and education cannot concern itself with the fluctuations of political opinion. But when political opinion, modifying or extending previous theories of citizenship, has become generally accepted political principle, it follows that education—if education be training in citizenship—must conform to such principles. Change of polity demands change in education. The method of education which will produce sound citizens for the city-State, for the nation-State, for the imperial State, for a 'Commonwealth of Nations,' must differ according to the difference of those polities.

It is not necessary or possible for our purpose to discuss here the validity or the reverse of modern political and social developments. It is necessary only to inquire whether any of these developments are so far facts of modern life as to demand a conformity of educational method and theory to them. The point at which political opinion, with which education cannot concern itself, passes into political principle, with which education must concern itself, is itself a matter of opinion. But it seems plain enough for our purpose that certain new political conditions and principles are now sufficiently established and commonly accepted to demand such a conformity.

The two most salient features of post-War society and politics are, firstly, a marked tendency towards the formation of new, and the increasing distinction of old, national, cultural and social groups, and, secondly, ideas of co-operation, internationalism and federalism which seek to combine those groups for purposes common to them all, but without sacrifice of the group-character which each possesses and cherishes—separatism and co-operation working side by side. Within the individual State these tendencies seem to be moving towards a conception of government by co-operation between parties and classes rather than government by any one 'governing class,' in inter-State politics towards co-operation between States rather than domination of one State by another. The conceptions of a 'governing class' and of a 'ruling power' are equally objectionable to the modern mind.

The key-word of the modern age is co-operation rather than domination, co-operation between employer and employed, between governors and governed, and even between teachers and taught. The clean division of mankind into ruling and subject races, masters and men, which, while it obtained, ensured so comfortably that every citizen 'knew his place,' is rapidly passing. The frontiers which formerly held fast between nation and nation, class and class, and even between sex and sex, are weakening under the corrosive influence of a score of forces. The old frontiers are going down before our eyes and new frontiers are forming. The social as well as the political map of the world was realigned in the post-War confusion. Such a realignment of frontiers is notable, too, in that world of knowledge with which education is chiefly concerned. There the former clear division between the 'cultured' and the 'uncultured' no longer holds good. Whereas, for our fathers, there was, in the main, only one type of culture generally recognized as such, a humanist and literary culture based on the classical languages, the modern man is confronted with many different types of culture which dispute the right to that honoured style.

The old culture still commands respect, and is still fostered and supported by the public schools, but alongside with it and growing yearly more competitive with it, are to be seen other forms of culture; a culture proceeding from science, a culture which can best be described as proletarian, based largely upon sociology, and, with the constant conquest of distance and consequent infiltration of conceptions of life hitherto sufficiently remote to be generally ignored, the cultures of other races, an eastern culture, a Slavonic culture, a negro culture, which have a totally different foundation and point of view from that western upper-class culture which once reigned supreme over the minds of western men.

Although these different cultures are often seen to overlap and to intermingle, they are distinct in origin and in tendency. They follow different gods. There is no longer one religion of culture bowing the knee in that pantheon of the classics where Homer and Virgil, Plato and Horace are revered; there are other 'high places' in the land with other prophets. One can no longer say that a man lacking in the older culture is necessarily without culture; his culture is probably different, different in origin and different in scope. The 'working-man' is often far more cultured than the public-school man; but his form of culture is his own and of another order. It is possible to take the somewhat pharisaical attitude that such heretical culture as this is not properly culture at all, but this is merely a quarrel of definition which can hardly be maintained if 'culture' is defined in its derived sense of the tilling of the mind. The tilling is there, but the seeds and the manner of tillage now vary where formerly they were invariable. There are groups, with an increasing group-consciousness, in the world of culture as in the social and political worlds, and these groups tend often to assume a partisan organization. That phenomenon is especially noticeable in the frequent *pour-parlers*[1] between the leaders of scientific and philosophical or religious thought which have so often to-day the character of a political round-table conference.

Everywhere there is to be seen the formation, the increasing separatism and consciousness of groups, and a general levelling-up of the rights and powers of groups, with a consequent tendency to impatience with any exterior control and to emphasis of group-codes and group-dialects. The day of the domination of one race by another, of one social class by another, of one form of culture by another is plainly passing. The fact may be deplored, but it can hardly be denied as a reality of modern life and politics. This being so, the only possibility of advance seems to lie along that line of co-operation which is its concomitant; any permanent return to methods of domination and absolutism, however efficient such methods may seem in comparison with a dilatory and still tentative régime of co-operation by a machinery which is still in the making, is a form of reaction for which, in the present state of affairs, it is difficult to see any issue but disaster.

The need for a mentality in our leaders capable of effective co-operation and free of insular and dominating instincts is therefore one of extraordinary urgency. But the theatre of post-War politics affords too ample proof that such a change of heart is not to be wrought by argument. The mould in which habit of mind and feeling is shaped must be re-formed. We all agree in some measure—the majority of the accepted leaders of modern political thought insist—that in some form of co-operation between classes and peoples rather than in domination the one by the other lies the future progress of our civilization and the prevention of such another catastrophe as the Great War which that civilization can scarcely survive. But while the prophets sing thus in tune, the people applaud with enthusiasm and straight-way plunge into a reactionary conservatism and nationalism.

The reason for this anomaly is not far to seek. We are not educated either for class-cooperation or for State-co-operation. We may agree in principle that we should think in terms not of class but of commonwealth, not of race or of empire alone, but of civilization. But, except for the rare mind which can free itself from the limitations of its education, in practice, with the best intentions in the world, the majority of us are as incapable of such mental pioneering as the baronial class of the Middle Ages was incapable of thinking of the interests of the serf, or the citizen of Renaissance Florence of thinking imperially. In the stress of practical politics we slip back into the well-worn grooves in which our minds were trained to run, so that it is possible to hear even now in post-War Europe, a glorification of that most malign fruit of nationalism, 'secret diplomacy,' and the League of Nations tends to

1 French: informal discussion, often prior to negotiation.

remain rather an academic experiment than a dynamic factor in world-politics, and the sentiments of Locarno[2] or the Kellogg Pact to perish stillborn.

The need for such a capacity for co-operation in the leaders of the new world and especially for co-operation between States, is, therefore, far more than ideally desirable; it is a necessity for practical politics for lack of which it seems that our civilization must inevitably be doomed. The first necessity for the leaders of the new world is that they should be men trained to think in terms, not of domination, but of co-operation, men who can think beyond the frontiers of class and race without betraying those frontiers, who have an understanding of the conditions and a living sympathy with the lives and manner of thought and feeling of other classes and peoples who compose the comity of our civilization—'citizens of the world' who can yet be loyal citizens of those closer and minor polities of which they are the members. For it is equally plain that a loose cosmopolitanism in this world of hardening groups is as useless for any effective co-operation as a rigid insularity. The '*déraciné*,'[3] the man without a State, is not a citizen at all, either of his own State or of the world. The need is not for men who represent no point of view at all or are racially colourless, but for leaders well set in their own traditions and group-consciousness who can recognize the right-to-be of other groups and traditions and seek with tolerance and accept with a sane and inclusive statesmanship formulas which can link groups and traditions in a wider and common purpose. The '*déraciné*' cosmopolitan no less than the insular and class-bound mind will equally fail that urgent demand. Do the public schools supply it? If they do not, can they adapt their system to meet it? [45–58]

Chapter V: The Public Schools Under Revision

It has been said earlier in this essay that if the defects of the existing public-school type were to be modified, its qualities would lose their efficiency. That is so for the purpose for which the type has hitherto been bred; insularity of mind was necessary for effective domination. Is that type therefore amenable to revision for the purposes of the new leadership which the times require? Would not this blade lose both temper and edge in the process of re-forging?

If the defects which now render it unfit for leadership under modern conditions were radical defects, or the qualities which have still their value for that purpose were superficial qualities, such an attempt must be in vain. But it is, I think, clear that, in reality, the reverse is the case. The defects, the tendency to domination, the exaggerated insularity, that ascetic philistinism which instinctively fears to confuse the capacity for clear decision and action by too much thought or feeling, are not necessarily inherent in the public-school process; they have been formed by the pressure of the conditions which the type has had to confront. The qualities of

2 The Locarno Treaties were seven agreements signed in 1925 to secure the territorial settlements of the Treaty of Versailles, and normalize relations with the German republic. Their success enabled Germany to be admitted to the League of Nations in 1926.

3 French: rootless or uprooted.

self-reliance, of reliability, of leadership, of endurance and determination, and of discipline, which the system has produced, are radical qualities, and there seems no good reason why those qualities should not continue to be bred, while those defects are eradicated. The problem is that of turning swords into ploughshares. The governing class which the public schools have bred in the past did indeed govern by the sword; the class of new leaders which our times require must put off the sword-belt if they are to lead. The public-school system forged a metal of splendid temper; conditions determined the final sword-shape which that metal assumed; but the steel which can be welded into swords can be wrought into sickles or plough-shares of no less a merit; the class which has ruled by domination in one age may lead by co-operation in the next. The change is not easy, but it is not impossible. When once the needs of the new situation are faced, the revision of educational method will follow. But the public-school world is, by its very nature, conservative and self-sufficient; recognition of the new realities will not readily or rapidly penetrate its fortifications, and the revision of the system from within seems likely to be too slow a process to be contemplated with patience while the new world burns. In the meantime it seems that the only forward move which can give prompt results must take place in some milieu detached from that world, the soil of which is calculated to modify in the desired direction the growth of the traditional seed.

The public-school system can be revised; the tradition can be retained and the tendency re-orientated; it can be done and it must be done if the needs of this new world are to be met. The disabling defect of the present public-school type has been seen to be that insularity from which the tendency to dominate rather than to co-operate ensues, and it has been seen that this insularity involves insularity of culture, insularity of class and insularity of political outlook. That mental insularity is inevitably bred by our present public-school system, and the tendency towards it is not lessened by our geographical position.

How is the requisite revision of the public-school system to be attained? We have arrived at a stage when concrete proposals will be claimed by every realist educationist, and the following tentative suggestions are, therefore, put forward as a basis of an educational scheme for the revision of prevalent public-school methods.

In order to counteract the cultural, social and political insularity which we have found to be the damning defect of the present public-school type, while retaining its traditional virtue, it is suggested, firstly, that the education of leaders of the new world should take place in some milieu outside our island walls, in schools organized on a federal rather than an international basis, where contact with other national groups and types will be constant and close with a view to the elimination of that 'foreigner' complex which now inhibits our statecraft, and cripples effective co-operation between national groups; secondly, that a far greater place should be given in the educational curriculum to the history of world civilization, to science and to sociology and civics than is at present the case, with a view to counteracting the existing culture-complex which inhibits co-operation between classes and cultures; and thirdly, since it is admitted that good education should be education for citizenship and should follow the track worn by the travelling race of man, that the scholastic organization should include some scheme of graduated self-government and the constant practice of civic responsibility in place of the autocracy of the traditional public-school administration.

In order that such destructive criticism of the public-school system as this essay contains may be complemented by definite constructive proposals embodying its conclusions, a note on the foundation of a centre for the education of leaders of the new world such as it contemplates is printed as an appendix to it. It will be obvious that this is no more than a '*ballon d'essai*'[4] inviting the winds of criticism. Let those who can criticize; let those who will promote or amend it.

But if the principle of federalism in education which is here proposed is to be carried to a logical conclusion, though the establishment of one such federal school may be the beginning, it cannot be the end of the matter. A federated system of such schools in the leading countries of the civilized world, working upon a common educational policy, with a free interchange of staff and pupils and an intimate inter-relation, is a necessary, if distant, corollary to it.

Such a conception will, no doubt, appear startling to most, impracticable to many, and to some, maybe, wrong in principle. With the latter (if they have persisted so far) we have no more to do than confidently to expect their condemnation. But those who, at first sight, criticize the conception, not as wrong in principle, but as too audacious, visionary or grandiose to be feasible in practice, we would ask to think well before condemning it on such grounds. Great dangers demand daring policies. Is the peril which plainly threatens our civilization less than great? Is our present system of insulated education that which is fitting the present generation or will fit the rising generation to meet it? Is there any radical remedy for our condition save in education at the roots? If such questions are squarely faced against the greatness of the issues, such an educational policy as that suggested may seem less grandiose than it would at first appear.

In the eye of history it may also seem less impracticable. Such a surmounting of frontiers as federalism in education implies, is, after all, less apparently impossible a consummation than when the ideas of monasticism or the Renaissance stormed frontiers more impregnable against odds more vast and an obscurantism more dark. What the mind of man deeply thinks, in the end it strongly wills, and what it wills with strength at the last comes to pass; that will does not know the impossible. But there is little sign that as yet it has thought deeply or willed strongly upon this urgent matter.

The problems of education and of the survival of our civilization are interlocked; the key of the future lies not at Geneva[5] but in the schools of the future. We must educate for peace and co-operation if we would avoid that final catastrophe which impends upon us. And that will not be done by a complacent acquiescence in things as they are or in a vague dissemination of mild international sentiments in schools. What Mr Madariaga has said excellently in his "Disarmament,"[6] rings

4 French: 'trial balloon'; an experimental proposal to gauge reactions; as in 'kite-flying.'

5 The seat of the League of Nations.

6 Spanish diplomat and author Salvador de Madariaga (1886–1978) was Director of the Disarmament Section of the League of Nations Secretariat: *Disarmament* (New York: Coward-McCann, 1929), 362–363.

home upon that problem of the education of the leaders of the new world which we have considered here, and with that astringent truth this essay may well conclude.

"The world is not going to conquer peace and then sleep on its laurels and roses; for the instincts of war are ever alive and so the instincts of peace must be ever awatch. A point which peace-bleaters usually forget, for they see peace as a rest between men, while peace is the organization of men to fight against the evils of war by co-operative means. No institutions, no co-operation; no co-operation, no peace. . . . Easier, ten times easier to drift into war than to defend peace against the ever-recurring attacks of the war disease. The world must know that if it wants peace it must work hard for it."

Mr Madariaga will, no doubt, forgive, and will, I believe, approve an adaptation of his saying. If the world wants peace, it must educate for peace. [74–83]

Appendix

A Note on the Foundation of a Federal School

1 *Purpose.*—The conception of the school is based on the principle of federalism between the races of the civilized world and the cultural and social groups which it comprehends; its aim is the education of a citizen of the world able to take a world view without ceasing to be a patriotic citizen of the State or the group to which he belongs. To this end it is believed that federalism and not internationalism is the true means. That belief is based on the following convictions:

- *a* That a world view of politics and a capacity for co-operation between racial, cultural and social groups is now essential to the safeguarding and progress of civilization;
- *b* That the present generation, however individuals may realize and admit this necessity, is fettered in instinct and volition by separatist habits of mind; that these habits are due to separatist systems of education (since man is governed more by habit than by reason and our habit of mind is radically separatist), and that this generation is therefore commonly unable to overcome the inhibitions and provincialisms of mind which menace our peace;
- *c* That such habits of mind, together with the separatist reactions (founded upon fear and the craving for security) which have characterized post-War history, can only end, sooner or later, in another European conflict in which civilization will probably collapse;
- *d* That 'Internationalism' is unhistoric and unsound in so far as it tends to suppress national characteristics, traditions and patriotisms and to intensify a reaction towards nationalism;
- *e* That the next step in the evolution of a world-polity is federalism—*i.e.,* a combination of nations and groups for the protection and furtherance of their common civilization and interests;
- *f* That the over-nationalist, separatist, and dominating habit of mind can be corrected, and a mentality more favourable to a world-view of world-problems and to the principles of federalism and co-operation can be produced, by a

system of education based on the federal principle together with a greater concentration upon the teaching of those subjects which are the common heritage of mankind, which shall be universal in scope and free from separatist bias;

2 *Site*.—For the purpose stated it is believed that Switzerland is the most suitable centre owing to:

a Its central position in Europe, facilitating contact with most of the chief groups of the modern world;
b Its linguistic facilities;
c The presence of the League of Nations, with which such a school should be in close contact;
d Its system of federal government, providing a valuable political object lesson in the principles and possibilities of federation.

3 *Organization*.—It is proposed that the school should be organized in separate groups according to nationality, each group being housed either in separate houses or in semi-detached blocks of the main building, with central and common schoolrooms, lecture-hall, dining-hall, recreation-rooms, library, chapel, etc. Each group will have its own group-staff under a Group Headmaster (of the same nationality as the group), while the group staffs will be correlated and controlled by a central Directorate. Staffs of groups will be interchangeable for the purpose of modern languages and those subjects and activities for which a foreign language is not detrimental to efficiency. The size of the staff will be governed by the principle that there should be one teaching master to every ten boys.

4 *Education*.—The education given will be governed by the principles stated above, but will otherwise conform to the educational requirements of the respective countries concerned, with a view to giving each boy as effective and practical an education as he would receive in the best schools in his own country. Heads of groups, working under the general control and supervision of the Directorate, will be responsible for the curricula of their groups, which will be arranged to meet the requirements of the various examination systems. History, languages, science, sociology and allied subjects will be taught with a special view to the objects of the school, and correlation of subjects will be practised wherever possible. Each boy will be required to study Universal History, at least one modern language and literature other than his own, and to take a course of general science during his school career. The group staffs will teach in the group languages since it is believed that an efficient education cannot be given in a foreign tongue. For purposes of general intercourse English and French will be the official languages.

5 *Athletics*—It is believed that team games have a definite and desirable educative value for a school of the type contemplated, but that in the English public schools in particular they have often been allowed to usurp too large a place in the system. Such team games will therefore be encouraged without being made a fetish, and will be supervised by a games master and a central sports

committee. A healthy rivalry between groups will be encouraged, provided that school teams, chosen impartially from all groups, have the first claim on each boy's patriotic feeling.

At the same time the desirability of allowing each boy to develop resource and self-confidence by the individual employment of his leisure time (under proper supervision) will not be lost sight of, and country and mountain expeditions, camping, climbing and winter sports will be encouraged with this end in view.

6 *Discipline.*—The aim of the school is to cultivate a type, with not only breadth but dedication of mind and a sense of duty towards the traditions and civilization which he has inherited as, firstly, a citizen of the world, and secondly, a citizen of that nation or a member of that cultural or social group to which, by the accident of birth, he belongs. He will be trained to regard the achievements of humanity in its politics, arts, and sciences as a warfare against chaos, ignorance, and disorder, and himself as a soldier in that enduring conflict in which, by reason of his education, he is especially equipped to lead. He will be taught to obey in order that in the future he may learn to lead, and that his highest freedom and self-expression consist in service to the cause of civilization. The aim of the school will be to train a boy in the exercise of self-control and later of leadership. With this end in view discipline will be maintained on the lines of the English public-school system with 'prefects' elected by a Council on which both staff and boys will be represented. There will be certain essential school regulations, but these will be made as few as possible, and the reason for them will always be made plain.

7 *Co-operation.*—While the school will be organized on the group principle which has been described, and within bounds, a group *esprit de corps* will be encouraged (as a house-spirit is encouraged in English public schools without detriment to the *esprit de corps* of the school), every possible encouragement and stimulus will be given to the promotion of a corporate feeling and loyalty for the school as a whole. With this purpose there will be a school magazine common to all groups, school prefects (as distinct from group prefects) elected from the whole school impartially, a school council on which all groups will be represented, and school societies (debating, literary and scientific) in which all groups will co-operate.

8 *Religion.*—It is believed that a true and whole education should be based on religion, and that religion should not be confined to chapel services on Sundays. Although boys of other faiths will not be excluded, and facilities for the practice of their faith will be provided, the religion of the school will be Christian and limited in doctrine to the broad elements of Christianity common to all Christian Churches. Doctrinal religious teaching, when required, will be supplementary to this. As far as it is possible to do so, in view of the fact that a boy cannot be expected to worship sincerely in a foreign language, services will be arranged according to groups. In addition to such services it will be the aim of the school to convey a sane and sincere religious tone into all its activities, to emphasize the religious nature of the unbiased weighing of evidence

and thorough scrutiny of facts, of a candid and courageous pursuit of truth and the understanding and appreciation of beauty, of a clean and hard austerity of living and thinking, of school friendships and loyalties, and particularly of that dedication and service to civilization which it is the first aim of the school to inculcate.

9 *Touring.*—A special feature will be made of yearly tours in European countries (and later, it is hoped, in the New World also), with a view, not to mere sight-seeing, but to an acquirement of knowledge of social, industrial and political conditions in foreign countries and an understanding of the mentalities of other races. Such tours will be annual and progressive.

10 *Endowment Fund.*—In accordance with the aim of the foundation, the school will be adapted with a view to those classes in the countries of Europe and the New World most likely, from their culture, heredity, wealth, or position, to influence the course of world-politics in the future, i.e., from the monied classes. Every possible advantage as regards premises and modern equipment will be provided, and it is considered essential that the staff should be of the highest quality and so salaried and provided for that no monetary anxiety for present or future shall lower the quality of their work. The fees will therefore be high, but reducible at the discretion of the Directorate.

At the same time in order to provide for boys of the right type, the situation of whose parents forbids the payment of high fees, an endowment fund will be instituted and a proportion of free or assisted places will be reserved.

From: *Aphrodite; or, the Future of Sexual Relationships*

Ralph de Pomerai (1931)

Headnote

Ralph de Pomerai (born Walter Theodore Tolley; 1892–1977) ran away to Africa in his teens and joined the Rhodesian Mounted Police. He sailed home in 1914 with seven or eight friends, but was the only one who survived the First World War. In the 1920s he worked for the Colonial Service and traveled widely in the east.

He came back to the UK in the 1930s and was involved in the theater in Richmond. He again saw active service in World War II—both in North Africa and in the Allied D-Day Invasion. He is thought to have been one of the first Allied troops to enter the concentration camp at Belsen (a theme which crops up in one of his unpublished plays). He moved to Bideford in North Devon in the mid-1950s, where he acted as a political agent for the Liberal candidate Jeremy Thorpe in the local constituency.

Before *Aphrodite* he had published *Marriage: Past, Present and Future: An Outline of the History and Development of Human Sexual Relationships* (1930).

******

To MAY, my wife, companion, co-worker and kindliest critic

Preface

In this little book I have endeavoured, in the limited space at my disposal, to set forth as clearly and concisely as possible the real facts about the history, physiology and psychology of sex, its relation to life, and to deduce therefrom a positive and rational code of sexual morality. That such a code is considerably overdue, few thoughtful readers will dispute. The old sex morality is disintegrating rapidly, and few will regret its passing. It failed largely because it was negative; because mere prohibitions can never hope to be an incentive to right action. In indicating what I believe will constitute the code of the future, therefore, I have attempted to outline a positive and vitalizing ideal, being fully convinced that man needs must love the highest when he sees it, and that the present ills we suffer from are due primarily to the pursuit of false values and to the ingrained habits that issue therefrom.

DOI: 10.4324/9781032625522-22

Incidentally, I have made no attempt to deal adequately with the reproductive or family aspect of sex and marriage. The omission is due not to any under-valuation of this side of the problem but to the fact that it has received more than its share of attention of late, and to the further fact that other writers have dealt specifically and exclusively with the question in the Series to which the present volume forms a modest contribution. R. de P. February, 1931.

Chapter I: Forbidden Fruit

Aphrodite, the Greek Goddess of Love and Beauty, or Venus, her Roman counterpart, is undoubtedly the most widely known of all the old pagan deities. Her lovely form, either entirely nude or draped only in a scanty embroidered girdle or cestus, has been made familiar by innumerable exquisite paintings and statues, and her amorous exploits and intrigues have served as themes to generation after generation of poets and writers. In the pagan world, of course, she was even better known, and it is safe to say that she was the most popular and beloved of all the inhabitants of Olympus. Nor was her fame and popularity confined only to Greece and Rome, for she had her counterpart in the celestial hierarchy of practically every ancient civilization. Thus, to give only a few examples, the Assyrian Ishtar, the Phœnecian Astarte, the Babylonian Mylitta, the Armenian Anaitis, the Syrian Antargatis, the Arabian Ilat and the Cyprian Aphrodite were all virtually one and the same person, and she was openly revered in each and all of these countries.

Gradually, however, for reasons that will be discussed later, her cult became suppressed, and at long last the lovely and amorous goddess achieved the reputation of being the most notorious and dissolute of all the ancient deities. Secretly her beauty and seductiveness still enthral us, but openly we affect to repudiate her and all her ways. Her beauty, according to conventional modern standards, is debased by wanton exhibitionism: her ardour profaned by its demand for illegitimate satisfaction. No matter how much we may sympathize with her misfortune in being married to the misshapen Vulcan who, by the way, was no moral paragon himself, we at least simulate disapproval of her intrigues with Mars, Adonis, Anchises and a score of others.

To her influence, too, we attribute the naughty intrigues of innumerable other classical deities, semi-deities and mortals. Thus we conjure up visions of Diana rapturously visiting Endymion in her remote and secret cave; of the fiery ardour of Jupiter for the lovely Danæ; of the enamoured intimacy of Eros and Psyche in their enchanted island palace; of Hero's passionate nocturnal dalliance with Leander in her lonely sea-girt tower, and, in more recent times, of the voluptuous abandon of Tannhäuser in the seductive grottos of the Venusberg. Approaching yet nearer to our own age, we picture the exquisite faery islands of the South Pacific as they were prior to the advent of the stern, repressive code of the white invader. We see enchanted lagoons, embowered grottos and limpid sylvan pools wherein disport the last surviving votaries of the old voluptuous goddess. Aided by imagination we see again their superb, golden-brown figures, their lovely, nude limbs, the garlands of flowers and the bright hibiscus blossoms gleaming in dusky tresses, the flashing, amorous eyes of youths and maidens and their unrestrained abandon to the delights of beauty and love.

At times, at any rate, we envy them, and in such moments find ourselves in profound sympathy with our Germanic forefathers who, according to Grimm, portrayed in the myth of the Venusberg their regret at the death of paganism and their instinctive dislike of the stern morality of the new religion. Being the victims of Puritanical convention, however, we are usually rather ashamed of this unadmitted envy and, lacking both the courage and opportunity to revive the ancient and discredited cult of Aphrodite, we console ourselves with the belief that all forms of physical gratification – not enjoyed by ourselves – are heinous, and that virtue and self-denial, at least theoretically, are alone worthy of the dignity of highly civilized beings. The yoke of Aphrodite, we reflect, is nearly always hard in the long run; and her votaries, like the beautiful Helen and Cleopatra, almost invariably come to a sticky end. It is regrettable, of course, but there it is, and there is not much sense in crying over spilt milk. The gods are a jealous lot and ever grudge their favours, and if ever they give with one hand it is only to take away with the other. This being the case, it saves ultimate disappointment if we make it a rule to take as little as possible and, following the example of the fox in the fable, endeavour to deceive ourselves into believing that the grapes which are beyond our reach are necessarily sour. In actual fact we do not believe it. But we pretend that we do, even when we privily share the conviction of the modern flapper that the forbidden fruits of sexual delights, although unquestionably "naughty," are nevertheless indubitably "nice."[1] The old cult, in fact, is still in our blood, and a few of us, somewhat more rash or courageous than our fellows, openly or surreptitiously continue to pluck the fruit and to damn the consequences. The majority, however, owing either to timidity or vitiated natural passions, avoid the tree like the plague and console themselves with the convention that all fruit-stealers are eventually caught, and that the penalty then exacted is such as to render the game not worth the candle.

This traditional belief arises from our curious conception of the nature of God – a God, by the way, still substantially identical to the ruthless deity revered by a primitive Semitic tribe. This will be apparent when we remind ourselves that, according to a tradition accepted by European civilization for nearly two thousand years, this deity is reputed to have placed Adam and Eve in the Garden of Eden and, in order to tempt them, to have planted a most attractive tree in the garden while expressly forbidding them to touch its fruit. The fact that Adam and Eve succumbed to temptation is not surprising; the only surprising thing being that innumerable men and women still continue to revere the perpetrator of this despicable trick. And despicable it surely was, for in this matter the action of the deity resembled nothing so much as that of a class-prejudiced mistress who deliberately leaves money about, hoping that her servant may discover it and yield to temptation, and so offer a sop to her own moral superiority and afford her an opportunity for the exercise of tyrannical vindictiveness. Strangely enough, most sane people are capable of

1 Allusion to the music-hall song 'It's Naughty but It's Nice' (1871).

realizing that the action of the mistress is far more reprehensible than that of her maid, though they seem quite unable to appreciate that the reputed action of the deity is even more repugnant than that of the mistress.

In point of actual fact this despotic deity is feared rather than revered, and is merely accepted because the majority of us are, or were until recently, apprehensive of his power to inflict eternal torment upon rebels and unbelievers. Nowadays he is increasingly regarded as a sort of super-bogy, but being officially recognized by the bulk of our spiritual and temporal rulers, who are invariably half a century in arrears of their fellows in thought, we still yield nominal allegiance to his archaic taboos. And of all these taboos the cult of Aphrodite is the most outstanding. The pleasure of sex is the pleasure of sin: to the ecclesiastical mind it is the one forbidden fruit; the essential fruit of the originally forbidden tree, and Gibbon reminds us that the early Christian Fathers were convinced that if Adam and Eve had preserved their virgin purity, some mode of harmless vegetation would have peopled Paradise with a race of innocent and immortal beings. Nor was this belief an invention of the third and fourth-century theologians, for traces of it are implicit and even explicit in the original Hebrew myth. Thus we are informed in the book of Genesis that the first and immediate consequence of the theft of the forbidden fruit was a sense of nakedness and an accompanying consciousness of shame. Prior to the Fall, the nudity of the first human pair had no more significance than it has to new-born babies, and it was only on the morning of the Fall that its significance leapt into being. The inference is inescapable. Something had happened to them; something that had "sexually conditioned" human nudity, and what that something was does not call for the assistance of a Sherlock Holmes to discover. Sex, then, is the supreme sin of the religion enshrining this naive and ludicrous myth of the Fall, and inevitably it has become the supreme taboo of all communities yielding allegiance to the faith. From a racial point of view, official Christianity regards it as a regrettable necessity – at least it is a reproductive necessity in a sin-conditioned world – but even so it should be indulged in only as a duty and never as a pleasure. Furthermore, since strict monogamy was reputedly enforced upon Adam and Eve, it is necessarily the only legitimate form of marriage open to their descendants, and non-procreative sexual activity, or procreative activity in any form outside the bond of holy monogamic matrimony, is utterly reprehensible. Similarly, human nudity or anything serving to stimulate sexual pleasure or desire is rigorously taboo. Thus we suppress, or endeavour to suppress, any painting, statue or photograph that attempts to depict the beauty of the unadorned human form, and establish a censorship designed to deter any rash author from seeking to portray the alluring pleasures of sexual intimacy in drama or novel. Sex, in short, as Professor J. Huxley reminds us, is the last of the great taboos,[2] and the survival of this taboo has perverted our whole sense of moral values and has been responsible for some of our most irrational and inhuman actions.

2 Taboo is the subject of another ingenious To-day and To-morrow volume: *It isn't Done; or, the Future of Taboo among the British Islanders* (1930) by Archibald Lyall.

A husband, for example, may be a depraved, drunken sot and an inhuman bully, but society, although it may afford his wife the grudging protection of a judicial separation, will not, unless the degenerate has consented to commit the arch offence of adultery, grant her the necessary freedom to seek happiness elsewhere. A wife, again, may be nagging, lazy, slatternly, inefficient and shrewish, and the husband is granted no redress; but no matter how perfect or affectionate a wife she may be, her husband, if once she is rash enough to snatch an unauthorized sexual pleasure, is entitled and even encouraged to turn wife and mercy out of doors. A man may, in the name of commerce, torture and oppress little children, stunting their minds and bodies by sweating their parents, and provided he keeps within the letter of the law no anathema will be hurled at him, but if one of his female victims yields the use of her body in return for food, then, and then only, were it better that a millstone were hanged about her neck and she were cast into the depths of the sea. Similarly, a wealthy and powerful man or corporation may, for mere personal gain, establish an unfair and anti-social monopoly, juggle with world markets, or establish a lockout that may occasion privation and misery to countless thousands and it will be accounted good business; such an individual or group may also, if influential enough, corrupt and debase their fellows by manufacturing and purveying drugs or narcotics and, provided their market be as distantly situated as, say, China, their virtuous countrymen may even find it expedient to wage a war in defence of their unholy monopoly, but should a public man be unfortunate enough to be cited as co-respondent in a divorce suit, then, no matter how valuable his services to his country and fellow-men may have been, nothing short of the complete deprivation of the offender's office and usefulness will assuage our shocked and outraged virtue.

And the prejudices that apply to the sex act itself apply also to anything that may be construed as "aphrodisiacal" or conducive to sexual desire. Thus an author may describe murder, suicide, disease or crime in horrible and revolting detail and be applauded as a realist, but should he be unwise enough to portray, no matter how delicately, the beautiful intimacies of married lovers, or, perish the thought, of unmarried lovers, then the guardians of conventional morality will hurl their anathemas at him; he will be dubbed lustful, debasing, wanton, lascivious, prurient, pornographic and sensual; his work will be suppressed, and he may even be apprehended by the police. Similarly, it is quite legitimate for a member of either sex to appear on the stage in such a guise as to render the human body more grotesque, distorted or repulsive than that of Caliban, but should a beautiful girl or a perfectly proportioned youth dare to reveal the beauty of human nudity, in no matter how appropriate or lovely a setting, the censor would immediately intervene and Mrs. Grundy and all her legions would unite in pillorying the daring actor or actress.

Examples of like nature could be multiplied *ad lib*. Indeed, it sometimes seems as if society, having but a limited amount of moral bias at its disposal, tends to expend so much in the suppression of so-called sexual delinquency that it has little or none left over wherewith to combat a host of far more insidious evils. Nor is this the worst of the matter. Our misplaced energy, in the vast majority of cases, is being expended not honestly but hypocritically, and hypocrisy is of all things the most debasing. To the majority, the dogma of the evil of all physical desire is a convention not a belief; in their heart of hearts they do not believe it, but the tradition is so old and so firmly rooted that there is always a vague fear there may be a grain of truth in it; and there exists also the more tangible fear of the consequences of running counter to accepted standards. Such being the case, the modern commandment is not "Thou shalt not covet thy neighbour's wife or daughter," but "Thou shalt not covet openly." Hence there is an ever-widening gulf between profession and practice, and the letter of the taboo is more or less maintained while its spirit is flagrantly outraged.

Never before has there been such a vogue for the *risqu[é]*, for insidious suggestiveness, and *double entendre* in fiction. And what applies to literature applies also to the theatre. The public flock in their thousands to see the pretty and ever-popular Miss X in her cami-knickers, and hordes of sexually-starved spinsters and convention-abiding citizens nightly besiege the playhouse where Mr. Y and Miss Z are playing in the naughty French farce containing a bedroom and semi-seduction scene. Thus stimulated imagination is substituted for personal intrigue, and vicarious satisfaction for actual experience; and while reluctantly conceding that the forbidden fruit is "naughty," we are more than ever convinced that it is certainly "nice." The net psychological result is that a totally exaggerated or disproportionate interest is centred upon sexual excitement, and this is due primarily to the fact that the "forbidden fruit" is *forbidden*, and that we are ashamed of our own desires. Nor are we even rational in the matter of vicarious satisfaction, for we stimulate our sexual imaginations and desires only to thwart them at the last moment. Nor, although half-convinced that such desires are immoral, will we permit them to be sublimated. Human nudity is a case in point. Your average Parisian, like the average person elsewhere, having but scant artistic or æsthetic appreciation, rarely patronizes the Folies Bergères, where nude and lovely maidens may sometimes be seen in lovely and artistic settings, but both Parisian and provincial audiences will go into semi-hysterics over the ever-popular act in which an acrobat appears on the stage, apologises for the unavoidable absence of his partner, invites a young lady from the audience – his actual partner – to assist him, induces her to undress, which she does with provocative and expostulating slowness and, finally, when the last diaphanous under-garment is literally torn from her by the hectoring acrobat, reveals herself clad in orthodox tights or conventional trunks and brassière.

The audience, despite the fact it knows perfectly well the girl will not completely undress, never fails in its enthusiastic applause, the applause being far greater than that ever granted to beauty naked and unadorned at the Folies Bergères. Complete nakedness, as our savage forbears discovered thousands of years ago, is neither exciting nor sexually stimulating when familiar: its appeal being primarily to the æsthetic sense. But dress, or rather partial undress, adds a fictitious glamour to its wearer, concentrates attention upon the sexual characteristics which alone are concealed, stimulates the imagination and deliberately arouses desire. Thus the girl in the acrobatic act would be far less intriguing to the average person if she disrobed in a perfectly normal, unself-conscious and ingenuous manner. It is her simulated protests and reluctance that whet the appetite of audiences by reminding them that they are to see, or by deluding them into the hope that they will see, not something innocent and natural, but something forbidden and exciting – something "naughty but nice." And the same psychology applies to our more restrained English and American bedroom scenes in popular plays, and to the scanty attire of chorus girls in most of our revues. Individually we are each and all far more erotic than were our primitive forefathers. Nor, as will be shown later, is this to our discredit, but we seem quite incapable of appreciating the fact and facing the consequences. As a result eroticism tends to develop into sexual neuroticism, and our current civilization has lost its mental balance. Each and all of us, if normal and healthy, are conscious of sexual needs, but nine-tenths of us, thanks to accepted conventions, are afraid to gratify them. Instead, we half-gratify them, and this automatically increases their urgency. Consequently we become literal bundles of repressions, our outlook is perverted, a sense of false shame is engendered, things naturally clean and beautiful are rendered unclean and sordid, energy is dissipated, humiliations multiply, and sex becomes a grotesque ogre that dogs our steps, perverts our imagination and vitiates our utility. Thus the thing we fear and yet desire battens upon us, assumes unreal and gigantic proportions and refuses to give us any peace. Sexual repression, in brief, is the greatest menace of the age, and unless we can either rid ourselves of our desires or grant them more rational satisfaction, our eroticism will undoubtedly prove our undoing. [11–23]

[. . . . Chapter II speculating on the evolution of 'Sexual Taboos' omitted]

Chapter III: The Truth About Sex

Our conventional sexual morality, as has been shown, is directly due to the survival of a number of taboos which have long since outlived their original utility. The old virtue has gone out of them, and the urge of compulsion has decayed with the passing of ancient exigencies. Similarly, the once widespread, though by no means primitive, notion of the inherent impurity of copula is no longer tenable, since the hopeless inadequacy of the theories upon which it was based is now becoming increasingly recognized. What, then, is the truth about sex?

Quite obviously we can approach this problem in a number of different ways. We might show, by reviewing the ascertained facts of biological development, that there is a definite trend in sexual evolution and that, following what the late Professor Moore called the Law of Complexity,[3] sexual activity now subserves much higher and more complicated ends than it did in the early days of sexual dimorphism. Similarly, by reviewing the consensus of medical and psychological opinion, we might deduce the fact that sex, since it produces admittedly beneficial physical and psychical effects, must be both necessary and legitimate; and, finally, by following the methods of abnormal psychology, we might discover what

3 British biochemist Benjamin Moore, FRS (1867–1922), *The Origin and Nature of Life* (London: Williams and Norgate, 1913), 188–189: the law he states being 'that matter so far as its energy environment will permit tends to assume more and more complex forms in labile equilibrium.'

consequences, if any, result from sexual repression and deduce therefrom the true function of a normal and healthy sex-life. This last method, yielding as it does singularly impressive and universally recognizable results, may conveniently be examined first.

The peculiarity and acidity of aged spinsters and the irritability and crustiness of old bachelors has long been a by-word, and although exceptions naturally exist it is nevertheless apparent that the complete suppression of sexual activity does lead to abnormality. This is especially obvious among women, partly perhaps because women are organically more specialized for sex purposes than men, but more especially owing to the fact that the percentage of aged virgins is far higher than that of male celibates. Very few men of mature years, other than priests, have entirely repressed their sexual impulses, while in priests the results of repression are possibly less obvious because training and habit are conducive to the production of a stereotyped character. But with true spinsters the case is different, and in most instances, even to the casual observer, it is obvious that the individual character has been considerably warped. Thus, among the mental characteristics we find gaucherie, lack of poise, nervousness, unconscious sadism, narrowmindedness and bigotry. In other cases, the resultant characteristics, being "reaction formations," may be strongly contrasted to each other. Nervous self-consciousness, for example, may be replaced by sheer aggressiveness, or strong acidity of temperament by semi-maudlin sentimentality. Unpleasant bodily consequences are sometimes also apparent and include lack of physical tone, chronic anæemia and susceptibility to severe and recurrent headaches, while acute disease is by no means uncommon.

The foregoing consequences of sexual repression, which are apparent to all and sundry, are trivial compared with those revealed by skilled investigators. As long ago as 1859, Dr. Adolphe Bertillon published, as a result of his researches, a monograph in which he showed that the celibate third of the French population was infinitely inferior to the married two-thirds. This celibate third, he pointed out, showed twice as many deaths, twice as many cases of insanity, twice as many suicides, and twice as many criminals as the married majority. His deductions, as shown by Professor Letourneau,[4] are subject to certain modifications, but his main thesis remains absolutely unshaken, and it is significant to note that the last census of the County of London proved that the number of females in homes for lunatics nearly doubled that of the males – which fact may be accounted for by the excess of spinsters over male celibates.

Twenty years after the publication of Dr. Bertillon's monograph, Sir Francis Galton drew attention to similar facts in his classical Inquiries into Human Faculty. "Among the morbid conditions which accompany the show of excessive piety and religious rapture," he wrote, "none are so frequent as disorders of the sexual organization"; and again, "by enforcing celibacy, fasting and solitude, they (the great teachers of all creeds) have done their best towards making men mad." Even more emphatic was the verdict of Professor S. Freud, who, as a result of his unequalled

4 Charles Letourneau, *The Evolution of Marriage and of the Family* (London: Walter Scott, 1900), 350.

researches in the field of morbid and pathological psychology, laid it down that sexual repression or sex-conflict is always the cause of neuroses, hysteria, and most forms of abnormal mentality. Freudian psychology, as is well known, has been subjected to a very considerable amount of adverse criticism, but the general truth of this statement has never been disproved by any competent authority. Thus Dr. [C. G.] Jung, who refuses to admit the validity of the Freudian theory of libido, when asked why it was that the erotic conflict, rather than any other, caused neuroses, replied: "There is but one answer to this. No one asserts that this ought necessarily to be the case, but as a simple matter of fact, it is always found to be so." Again, the late Dr. J. J. Putman emphatically declared that "some abnormality of the sexual life is always present as the cause of especially insistent emotions and repressions."[5] The late Dr. W. H. R. Rivers, one of Freud's most competent critics, similarly maintained that "the great majority of the neuroses of civil practice depend on the failure of balance between. . . the sexual instinct and the very complex forces by which this instinct is normally controlled";[6] while Dr. William Brown, who also disagrees with much of Freud's psychology, frankly admits that "sex is very frequently the determining factor in the formation of pathological complexes."

In view of these and numerous other statements by highly competent specialists there can be no shadow of doubt but that a healthy sex-life, quite apart from procreation, is a human necessity, and that repression is productive of mental abnormality. Such being the case, it is quite impossible to maintain that copula, apart from or outside marriage, is necessarily impure and displeasing to God, unless we are also prepared to admit that the deity has played a despicable practical joke upon humanity. The sneer of the eminent Churchman who, at a recent Modernist Conference,[7] remarked that in the old days people refrained from sexual sin from fear of hell, but now indulged in it from fear of complexes, came with peculiar inappropriateness from the lips of a self-styled Modernist, while his attempt to belittle the proved evil results of sexual repression reflected very adversely upon his knowledge and integrity. The fact that a few people may escape contagion despite contact with the victims of a highly contagious disease is no proof whatever that the disease is non-contagious, and similarly it is ludicrous to argue that sexual repression is harmless simply because a few celibates *apparently* suffer no evil consequences – mental and emotional poverty, for example, are often conspicuously present even when actual abnormality is seemingly absent.

It is now known, furthermore, that the bodies of celibates suffer no less than do their minds. It has been proved, for instance, that insomnia, neurasthenia, uterine ailments, and even malignant growths are among the commonest consequences of

5 Putnam, introduction to the translation of Freud's *Three Contributions to the Sexual Theory* (New York: The Journal of Nervous and Mental Disease Publishing Company, 1910), vii.

6 Rivers, *Instinct and the Unconscious* (Cambridge: Cambridge University Press, 1920), 120.

7 Probably the Dean of St Paul's, William Inge's presidential address to the Modern Churchmen's Conference in Oxford, August 1930, on 'The Modern Outlook in Morals.' The 1930 Lambeth Conference had earlier that month passed the landmark resolution 15, reversing the Anglican Church's position on birth-control and giving qualified assent to the use of artificial contraception.

sexual repression, and a modern doctor has gravely declared that "the chaste morality of women not infrequently revenges itself in the cruellest forms of disease."[8] Nor is it merely absolute chastity that exacts such penalties, for an incomplete sex-life is, if anything, even more dangerous than no sex-life at all. This is proved quite conclusively by many of the diseases suffered by married women, as witness a writer in the *British Medical Journal*, who detailed a number of serious cases of nervous diseases in married women attributable to this cause. Similarly, Dr. Havelock Ellis quotes an eminent Austrian gynecologist who asserts that "out of every hundred women who come to him with uterine troubles, seventy suffer from congestion of the womb, which he regarded as due to incomplete coitus."[9] The commonest cause of this is the absence in women of the nervous and physiological climax to copula known as the orgasm, and Dr. Marie Stopes has estimated that "70 or 80 per cent. of our married women (in the middle classes) are deprived of the full orgasm through the excessive speed of the husband's reactions, or through some maladjustment."[10] The absence of adequate preliminary courtship is unquestionably also a contributory factor, but may conveniently be considered later.

Mental and physical pathology, then, prove quite conclusively that sexual repression gives rise to mental and physical disease and abnormality, and it is significant to note that it is the absence of copula and not of procreation that is the determining factor. It inevitably follows, therefore, that a healthy sex-life is essential to human beings, and that copula is valuable in itself and plays just as important a rôle in individual, as reproduction does in racial, well-being. This latter fact cannot be too strongly stressed for it is fashionable among certain biologists to ignore all the aspects of sexual dimorphism other than the strictly reproductive. Such an attitude, needless to say, is both prejudiced and unscientific.

Sex, at least originally, was not a reproductive necessity, and was not evolved until comparatively late in the history of living organisms. The whole class of Protozoa, which embraces the most primitive forms of life, is asexual and reproduces by simple fission. With this type of reproduction there is necessarily no such thing as parent or child in the ordinary sense of the words; each half or sub-division is absolutely identical: each is as old as the "parent" and as the species itself, and incidentally natural death is unknown. Apart from accidents, these lowly creatures are organically immortal, the development of sex among the Metazoa being alone responsible for the introduction of physical death. Such immunity, however, exacted its own penalty. The absence of sex involved the absence of variation, and this in turn largely precluded the development of higher or more complex forms of life. But complexity of structure is not an end in itself, and is valuable only in so far as it is conducive to the better adaptation of the organism to its environment or, in other words, to the enjoyment of a fuller and more complete life. Nature, in short, seemed to agree with Byron in believing that "a crowded hour of glorious

8 Dr Hegerisch, quoted in the Appendix to Edward Carpenter's *Love's Coming of Age* (Manchester: Labour Press, 1896), 162.

9 Ellis, *Studies in the Psychology of Sex*, vol. 6 (Philadelphia: F.A. Davis, 1921), 551.

10 Stopes, *Married Love* (London: A. C. Fifield, 1918), 50.

life is worth an age without a name,"[11] and in any case it is quite certain that primeval sex was as much concerned with individual well-being as with reproductive necessity.[12]

This becomes increasingly apparent as we ascend the biological scale. Rut, as already pointed out, is not limited among the higher species to the minimum demands of reproduction, but by an absence of surplus bionic[13] energy resultant from a precarious or inadequate food supply. This is completely proved by the fact that man has emancipated both himself and many of the animals he has domesticated from the limitations of primitive rut. It is significant, moreover, that sexual enjoyment and indulgence progressively increase as we ascend the biological scale, and that this increase is accompanied by a paralleled decrease in the resultant number of offspring. Letourneau has already been quoted as stressing the "veritable frenzy" which characterises copula among certain species, and it is a well-known fact that man is even more erotic than are animals, as also that modern civilized man is even more erotic than were his savage ancestors. And just as primitive man produced rather less offspring than did most of the higher animals, so civilized man, in turn, tends to have far smaller families (quite apart from birth control) than did primitive man. Biological history thus renders it apparent that sexual indulgence and its emotional content have developed in an inverse ratio to the demands of procreation, and that there must be some explanation of this other than that of mere compensation for the increasing disability of gestation and parturition.

This explanation, as previously indicated, lies in the fact that sex tends increasingly to subserve individual as well as racial ends, and that it promotes, as Mr. H. G. Wells suggests, "not only a renewal of life in the species," but also "a renewal of energy in the individual."[14] This is noticeable among animals no less than among human beings. "Each period of rut is for the animals a sort of puberty," wrote the late Professor Letourneau. "The hair, plumage, and the scales often assume rich tints which afterwards disappear. . . . The psychic faculties of the animal, whether great or small, are then over-excited, and rise above their ordinary level." In rut, then, life is lived with far greater zest than at any other time and reveals its maximal possibilities, hence it is not surprising that the season of love is also the season of joy. And what applies to the lower species applies even more significantly to man. "Human rut," we are reminded by Letourneau, "Is essentially rut in an intelligent being. It exalts all those vital forces of the man just as rut over-excites those of the animal. If it seems to differ extremely from it, this is simply because . . . in radiating

11 Actually from 'The Call' by Thomas Osbert Mordaunt (1730–1809); though it was often attributed to Sir Walter Scott, who quotes it in *Old Mortality*.

12 [Author's note] Reproduction, both among the Protozoa and Metazoa, constitutes the organisms' escape from the nemesis of continuous growth, and is, in the case of the former, an individual necessity.

13 De Pomerai appears to use this term in a sense not recorded in the *OED*, to mean biologically-produced energy. In the context of sexuality, it anticipates Wilhelm Reich's claims of the later 1930s (not confirmed by other scientists) to have identified 'bions,' or rudimentary structures between inert and living organisms.

14 Wells, *The Secret Places of the Heart* (London: Cassell, 1922), Chapter 4, 80.

from highly developed nervous centres, [it] awakens and sets in commotion an entire psychical life unknown to the animal."[15]

The vital factor in awakening this psychical life, both in animals and in man, is courtship. In all cases the female must be won, and in order to be won she must be stimulated. This is accomplished by two methods, the one unconscious and the other conscious. At the onset of the season of love, the animal (by virtue of its organic endowment) unconsciously develops or accentuates its secondary sexual characteristics, and the animal world becomes rich with unwonted odours, magnificent coats of hair, brilliant tints, gorgeous plumage and innumerable other manifestations of exuberant vitality, while in the field of conscious activity, animals vie with each other in vociferation and song, in antic and dexterity, and in feats of prowess and daring.

It is apparent, therefore, that sexual activity is the parent of creativeness, art, and of the mastery of life, and especially is this true in the case of man. The original function of dress, as demonstrated by Westermarck and numerous other authorities, was to stimulate sexual desire, and Mr. Gerald Heard has pointed out, in *Narcissus*, that the elaboration of dress by man, which is a conscious task, is simply an extension of nature's activity in the elaboration of secondary sexual characteristics among the lower species.[16] Primitive man decorated himself in order that he might court more successfully, and he probably also developed his powers of declamation and song for the same purpose, hence it is obvious that art, oratory (including written speech or literature) and music were all fostered by sex. Mr. Bertrand Russell has particularly stressed the dependence of art upon the tender emotion, in his *Marriage and Morals*, and the reader may also be reminded that it has been long recognized that a creative genius is always more highly sexed than is an ordinary man. This fact was admirably recognized by the late M. Romain Rolland. Passion, he wrote, "Is a necessity of the nature of men of genius. Even the most chaste, like Beethoven and [Bruckner], must always be in love; every human capacity is raised to a higher degree in them, and as, in them, every human capacity is seized on by the imagination, their minds are a prey to a continuous succession of passions."[17] Similarly, it has been suggested by Mr. Bertrand Russell that if a musician and a mathematician were deprived of sexual experience, the effect on the life-work of the former would be tremendous, but practically negligible on that of the latter.

Sex, then, is a primary requisite of creativeness, but it is also essential to the attainment of mastery of life – to the maximal development of human potentialities. Nature, no less than human society, demands specialization, and males and females are psychically as well as physically specialized. This specialization, moreover, is essential to the full realization of life, for since each sex is literally the complement of the other, perfect psychical development depends upon mutual co-ordination.

15 Letourneau, *The Evolution of Marriage and of the Family*, 8, 9 (though he has 'human love,' not 'human rut').

16 Heard's To-day and To-morrow volume, *Narcissus: An Anatomy of Clothes* (1924).

17 *John Christopher,* IV: *Journey's End: The Burning Bush*, 318. Gilbert Cannan's translation has 'Bürchner,' but Rolland wrote 'Bruckner.'

It is here that courtship plays a stupendous rôle, for in no other sphere of activity are specifically male and female characteristics so stimulated and developed. Men, thanks chiefly to predominant katabolism[18] and bionic endowment, are naturally the leaders, initiators and inventors in life. Man's rôle is essentially active, whereas woman's is comparatively passive, and it is significant to note that masculine hegemony is stimulated by and largely dependent upon the existence of feminine sequaciousness. The development of these complementary qualities is peculiarly fostered by the respective rôles of the sexes in courtship. "The female," writes Mr. William Thomas, "Will not submit to seizure except in a high state of nervous excitation (as is seen especially well in the wooing of birds), while the male must conduct himself in such a way as to manipulate the female; and, as the more active agent, he develops a marvellous display of technique for this purpose. This is offset by the coyness and coquetry of the female, by which she equally attracts and fascinates the male, and practises upon him to induce a corresponding state of nervous excitation."[19] This play of the sexes leads naturally and inevitably to the consummating sex act and this, presuming all has been what it should be, gives rise to profound psychological consequences. Thus, the complete and final surrender of the female invariably inaugurates a period of ardent discipleship, while the successful exercise of masculine hegemony in courtship invariably stimulates the male to equally successful initiation and leadership in the ordinary affairs of life.

Mr. Thomas, however, seems rather afraid of this play of the sexes and stresses the danger of sexual life "becoming a play interest and a dissipation." The danger of dissipation naturally exists, but it is entirely erroneous to suppose that play is necessarily an idle diversion. "Play-activity," declares Professor T. P. Nunn, "is subject to the general law that spontaneous activity, when not baffled or obstructed by unfavourable circumstances, tends always towards increasing perfection of form, to more complete expressiveness."[20] Similarly, Professor Schrödinger informs us that, "When man had to some extent conquered his environment, so that his whole life was no longer a struggle for bare existence, he began to play. He invented sports, he invented the arts, he invented science. All these things are forms of play. They exist to give man pleasure."[21]

It is here that we come to the crux of the matter. Play-activity is not only utilitarian but also, as Mr. Shand[22] reminds us, inseparably connected with joy, and the crowning justification of sex lies in the fact that it is capable of producing the most perfect and supreme form of human happiness. Man is social through and through; of all living creatures he can the least endure solitude; the more human he

18 The metabolic process in which complex molecules are broken down to simpler ones to release energy.

19 Thomas 'The Sexual Element in Sensibility', *Psychological Review*, Vol. 11 (1904), 61–67, 63.

20 Percy Nunn, *Education: its Data and First Principles* (London: Edward Arnold, 1920), 69.

21 'Interviews with Great Scientists: IV: Prof. Schrödinger, *Observer* (January 11, 1931), 15–16, 16. Interviewed by J. W. N. Sullivan, author of *Gallio; or, the Tyranny of Science* (1927).

22 Alexander Faulkner Shand (1858–1936) was a founding member of the British Psychological Society and wrote *The Foundations of Character* (1914).

is the more he needs companionship, sympathy and affection, and the more highly civilized he becomes the more ardently he longs to experience and enjoy his felt identity with his fellow-creatures. And nothing is more conducive to this necessary experience than the love of the sexes, for passionate love breaks down the barriers of egotism and strives ever for closer and more intimate union. And in love, as in most activities, mind and body co-operate. "Expression," declares Professor A. N. Whitehead, in his *Religion in the Making*, "is the one fundamental sacrament,"[23] and this truth applies to sexual love no less than to religion. The emotions of lovers demand tangible expression, and in kissing, fondling and embracing they experience far more than mere sensual satisfaction. Shelley, in *Epipsychidion*, makes a passionate lover protest, "I am not thine: I am a part of *thee*";[24] and the outward and physical demonstrations of lovers are chiefly valuable because they enhance and give expression to this inner yearning for complete unity. This is especially true of copula, which, as Dr. Marie Stopes rightly maintains, "symbolizes, and at the same time actually enhances, the spiritual union."[25] The sex act, in brief, can and should be the supreme sacrament of love, and love itself is the fulfilment of life.

Love, moreover, is a principle of growth and expansion. It strives ever for richer and fuller intercourse with the beloved object and increasingly endeavours to comprehend and sympathize with the thoughts, aspirations and desires of the beloved. For this reason a passionate lover inevitably enriches and fosters his individual mental and spiritual capacities. He learns new interests, new viewpoints and new truths, and as his outlook is widened, so also is the scope of his sympathies extended. Sexual love, therefore, possesses a distinct social utility, for since civilization and society cannot exist without co-operation, and co-operation, if it is to be effective, pre-supposes sympathy and mutual understanding, it will generally happen that a lover will possess wider and warmer sympathies than will one to whom this enriching experience has been denied. Nor is this all, since if we are prepared to admit that the love of a person for another leads to an enrichment of experience and a wider range of sympathy, we are inevitably forced to the further conclusion that the love of an individual for more than one, supposing this to be possible, must yield even greater benefits. And here we must categorically deny the popular fiction that the majority of human beings are incapable of loving more than one. Proof to the contrary could be piled up *ad nauseam*, the simple truth being that although some individuals are satisfied with a single love, the vast majority are as capable of a plurality of loves as they are of a plurality of friendships. To sum up, then, the truth about sexual love is that it is wholly good and one of the most potent and beneficial factors in human life. Sexual love, qua sexual love, can never be immoral in itself, although like every other good thing, it can be, and very frequently is, both degraded and abused.

23 (Cambridge: Cambridge University Press, 1926), 117.
24 Line 52.
25 Stopes, *Married Love*, 75.

Chapter IV: New Wine and Old Bottles

It is apparent to all possessing eyes to see and ears to hear that modern sexual relationships are in a chaotic condition. Feminine chastity is no longer a necessary virtue; pre-marital and extra-marital relationships are surreptitiously or openly indulged in; sex-interest, and even sex-perversion, constitutes a major interest in modern fiction and drama, while even the institution of marriage is openly attacked. The old sexual code is disintegrating rapidly; the spirit of revolt against ancient taboos is rampant, and respect for authority has largely become a thing of the past. The current revolt, however, is far from being due to enlightenment. Old taboos are attacked not because they are understood and perceived to be inimical to social progress, but because they restrict individual freedom, and sexual indulgence is defended not so much because its psychical and physical benefits are comprehended as because sexual freedom offers transient excitement and a means of escape from personal ennui. What, then, is responsible for this sudden *volte-face*?

As is usual in such cases, no one cause is adequate to account for the situation, but rather a number of causes all of which have mutually contributed. Some, of course, have exercised greater influence than others, and conspicuous among these may be mentioned the decay of orthodox religious beliefs; the progressive economic emancipation of women; the increasing dissemination of a knowledge of birth control; the effects of the late War, and the general reaction to the standardization imposed by modern industrialism. The decline in religious beliefs may conveniently be considered first.

Scepticism in religion and declining clerical authority has been increasingly apparent throughout the present century, and especially during the past two decades. This is primarily due to the gradual popularization of the great scientific discoveries of the past seventy years, the effects of which are often underestimated. The man in the street, in brief, although he may possess scant scientific and even scantier theological knowledge, nevertheless has an acquaintance with the general tenets of both sufficient to enable him to appreciate a broad incompatibility. Popular astronomy has given him a dim appreciation of the colossal magnitude of the cosmos, and this ill accords with the earlier religious conception of a three-storied universe comprised of heaven, earth, and hell. Heaven and hell, in fact, lost much of their credibility when they lost their cosmographical location. Similarly, as soon as it was generally realized that man had evolved from the lower species instead of being created a perfect and sinless being, the Adam and Eve story was once and for all exploded, and men and women, having discovered that at least a part of Holy Writ was false, tended to jump to the conclusion that what characterized the part more than likely applied to the whole.

The progressive economic independence of women has played an equally important rôle. As soon as women entered the commercial world and succeeded in earning their own living, chaperonage necessarily became a thing of the past. Girls, prior to this, were virtuous not because they lacked any inclination to be otherwise, but because they lacked the opportunity and were fearful of the consequences of the discovery of unchastity. Economic independence brought comparative immunity from authority, either parental or connubial. Apart from the bond of affection, a woman capable of earning her own living was enabled to indulge her sexual impulses with little or no risk, since even if husband or parent turned her out of doors she was no longer faced with virtual starvation and social degradation. Real defiance became possible for the first time, and parents and husbands were compelled to rely upon the restraining influences of affection and conventional morality rather than upon authoritative prohibitions. For similar reasons, wives have tended to become rather less apprehensive about the alleged infidelities of husbands. In point of fact, they rarely were apprehensive when the "other woman" was merely a prostitute, since there was then little likelihood of the liaison leading to the desertion or loss of livelihood of the wife. If the woman was of the man's own class, however, the case was different. Woman, under the old regime, had only one means of earning a livelihood – the exploitation of her sex. She could sell her body to one man in marriage or to a number of men in prostitution, but no other course was open to her unless she happened to be an heiress. Sexual indulgence by another woman of her own class, therefore, was regarded as an actual or potential act of rivalry, it being assumed that the other woman, being likewise economically dependent, must necessarily be attempting the permanent monopoly of the husband. The fear of loss, of course, always exists in cases where there is genuine affection and where neither husband nor wife is sure of the other's fidelity (to the marriage bond more than to sexual monopoly), but it is even more accentuated where the wife is not economically independent. When both the wife and the other woman are economically independent, however, this natural fear tends to be minimized.

The spread of the knowledge of effective contraceptive methods has been responsible for a still greater tendency to sexual indulgence, particularly among women. The "pleasures of sin" have unquestionably been envied for countless generations, even if only unconsciously, but until comparatively recently these pleasures have always carried with them the undesirable risk of pregnancy. And the pregnancy of an unmarried woman, quite apart from the fact that it branded the child with the stigma of illegitimacy and probably sentenced it to inadequate nourishment and attention, has hitherto involved loss of social status and practical condemnation to the rank of prostitute. Pregnancy, moreover, was a visible sign of "sin"; there was no concealing it, whereas the modern perfection of contraceptives has made delinquency attractive by eliminating or minimizing the risk of discovery. Sexual indulgence, as in the old days, is still largely regarded as being "naughty but nice," the naughtiness tending to add a superficial spice to the niceness now that naughtiness is no longer necessarily punished by conception. The effectiveness of modern contraceptives, furthermore, has affected the conduct of men almost, though not quite, as much as women. Most men hitherto, apart from promiscuous commerce with harlots, have been somewhat chary of sexual intercourse with respectable girls owing to the risk that they might become mothers; motherhood in such cases often being used by a girl as a weapon to force the man into matrimony or to saddle him – as was only right and legitimate – with a part of the burden of the financial support of the child. And what applied to single men applied with even greater force to the married. The risk of conception, moreover, unquestionably induced men to take strong views about the extra-marital indulgences of wives. Children, in highly civilized societies, are a financial handicap rather than an asset, and to be unwittingly saddled with the support of another man's child was a misfortune no man voluntarily cared to risk. The possibility of a wife enjoying extra-marital pleasure with little or no risk of inflicting disability on the husband, however, has tended to remove one of man's strongest motives in the imposition of strict marital fidelity, and the complete elimination in the future of the surviving residuum of this fear will doubtless be productive of still wider toleration.

Less important but nevertheless potent factors in the current revolt against ancient sexual taboos are the influences due to the late War and to the standardization imposed by modern industrialism. The War involved the whole of Western civilization. For several years men and women lived a life of unnatural strain; privation and death haunted them, and from sheer reaction they attempted to find relief from uncertainty and gloom in the most ephemeral and superficial of pleasures. Men on leave from the front, anticipating an immediate return to the Hell of Flanders and the possibility of a premature and horrible death, sought to crowd as much pleasure and excitement as possible into the few days of life that might remain. The quest for amusement and pleasurable diversion became urgent and all-absorbing; compensations were demanded for untold misery, suffering and privation; time was short, and inevitably the moral standards and conventions of normal life were swept irrevocably to one side. The fate of civilization seemed to hang in the balance; permanence of values was scarcely considered in a world where permanence seemed but a wild dream: each and all lived but for the moment, and the habit once acquired survived the cataclysm responsible for its formation.

This reaction against the abnormality of war is curiously paralleled by that against the cramping atmosphere of modern industrialism. The present age is one of mass production and standardization; everything seems turned out to a pattern, even human beings. Personal initiative has been largely replaced by machine-like routine; individualism in public and commercial life tends increasingly to be at a discount, and from sheer reaction men and women seek to assert their thwarted individuality in private life. And the same dissatisfaction that permeates our work characterizes also our leisure. The rapid multiplication of mechanical contrivances has thrown us all into a condition of chronic passivity and dependence. We no longer, if we possess musical inclinations, seek enjoyment or self-expression in amateur singing or instrument playing, but switch on a gramophone, the wireless or a player-piano. Instead of walking and obtaining enhanced health and æsthetic enjoyment in physical exercise and a leisurely appreciation of the beauties of the landscape, we jump into a car and race through a desecrated countryside enveloped in a cloud of dust. The art of talking has become almost obsolete; the "talkies" give us all the conversation we want, and a multiplicity of cheap and superficial amusements leaves us with no time whatever for contemplation or thought. True happiness and contentment are but rarely experienced; the art of being "passing rich on forty pounds a year,"[26] or its modern equivalent, is something that passes modern understanding, and in consequence we are superficial, disillusioned and increasingly dissatisfied. Nor is this to be wondered at. The creative impulse is part of the very stuff of human nature. Ordinary men and women, no less than artists, sculptors, poets, composers and skilled craftsmen, feel the imperative urge to some form of self-expression. In the old days there was ample scope for this. People were thrown more upon their own resources. The man satisfied his creative impulse in his garden, in amateur carpentry or some other hobby; the woman in creating a home that in some measure expressed her personality, or in needlework, or some analogous occupation. But nowadays the home is little more than a dormitory and restaurant annexe to a garage, or a depository for mechanical devices, and dress is a mere product of mass production. Inevitably human nature rebels against this passivity, and this rebellion sometimes expresses itself in a craving to do something startling; in the cult of the bizarre and the bohemian. It explains, too, a good deal of the alleged scandalous behaviour of our modern bright young things; their apparently wanton defiance of convention, and their pre-disposition to explore the possible excitements of "free love."

But beneath the surface there is, I believe, a vague striving for the satisfaction of a felt want. Amours, even of a transitory type, do offer some momentary escape from the sense of loneliness that oppresses most men and women throughout the greater part of their lives, and they do also afford some means of self-expression in the captivation of a fellow-being. Real escape and maximal satisfaction, however, are rarely achieved in the vast majority of pre-marital and extra-marital relationships. Often they are sordid and mean; indulged in under the cloak of deceit and subterfuge, or in an atmosphere that is inimical to genuine happiness. More often still, they are hedged about with repressions and reservations as though the participants feared the consequences of unleashing real affection and emotion. Seriousness, indeed, seems to be the one genuinely modern taboo, hence lust is far more prevalent than love. And the difference between the two is primarily this, that whereas lust seeks principally its own physical gratification, love is essentially reciprocal and is never content to seek its own satisfaction at the expense of the desires or interests of the beloved.

The word lust, however, is as unpopular today as it is unfashionable. What our forefathers called lust the modern generation prefers to designate free love. This is a misnomer. Free passion may exist, but free love never, for the moment love intervenes in a relationship it forges its own emotional, psychical and spiritual chains. It is quite impossible to love anyone and remain mentally and emotionally free, and the greater the love the greater is its bondage, so that if love is withdrawn or suffers violence it involves real deprivation and misery. The person seeking merely sensuous enjoyment, therefore, must necessarily deny his capacity for true affection and the deeper satisfactions of love. And swayed by vague fears of human

26 Oliver Goldsmith (1728–1774), 'The Deserted Village,' in lines thought to describe his father the Rev. Charles Goldsmith: 'A man he was to all the country dear,/And passing rich with forty pounds a year' (lines 141–142).

caprice, the uncertainty and insecurity of life, and by the dictates of bohemian fashion, innumerable men and women deliberately do endeavour to exclude affection from their sexual relationships nowadays. Walter Bidlake and Lucy Tantamount, in Mr. Aldous Huxley's novel, *Point Counter Point*, are excellent examples of this, and also of the consequences of the attempt. Bidlake enters upon the affaire as light-heartedly as Lucy, fully determined that the liaison shall make no real demands upon him. But it does, and when he at last experienced tenderness and emotion "love produced a desire to be loved." But Lucy had no desire to experience that deep tenderness that would involve a surrender of her individuality. "She wanted to be herself, Lucy Tantamount, in full command of the situation, enjoying herself consciously to the last limit, ruthlessly having her fun; free, not only financially and legally, but emotionally too –emotionally free to have him or not to have him."[27] The result, as far as Bidlake was concerned, was inevitable. The relationship, stripped of response, emotional content and true companionship, was reduced to stark sensualism, to crude physical satisfaction; and since no man is a mere physical entity but a complex being compounded of body and mind, he was soon left disillusioned and dissatisfied. Possessing a more ephemeral character and disposition, Lucy's disillusionment was possibly a slower process, but its final realization was inescapable, as it is to all who pursue false values. The Lucys of modern life, and their male counterparts, are no more inherently wanton or vicious than were their forebears; if anything they are possibly more honest and courageous, but their tragedy lies in the fact that they are utterly lacking in a sense of true values. Like children who prefer bright new pennies to golden sovereigns, they are swayed by the superficial attractions of quantity rather than quality. Their new-found liberty, too, has dazed them by its suddenness, and they have the misfortune of living in a transitional period in which the old standards have fallen into disrepute and new ones have not yet been evolved capable of winning approbation and acceptance.

But if the rebels are pursuing phantoms, the diehards are in no better case, and the inner psychology of the modern opponents of sexual freedom would probably surprise even themselves. The case of the average man who insists upon marrying a virgin and in regarding physical infidelity as the arch marital crime is illuminating. More often than not, such a man has had previous sexual experiences; experience with a prostitute or with some little servant girl or with a social inferior, and such relationships having been predominantly mean, sordid, and crudely sensual, he is predisposed to imagine that all unauthorized intercourse must necessarily be equally degrading. His motives, however, are frequently even more selfish. A man may, and very often does, object to his wife earning her own living or being economically independent for the simple reason that it reduces his sense of proprietorship. Such a man is not infrequently conscious of mental or physical disability, of lack of hegemony and initiativeness, and experiences a conscious or unconscious inferiority complex, hence he finds compensation in the possession of a door-mat wife. The wife who dutifully hands her husband his hat and coat every morning, warms his

27 *Point Counter Point* (London: Chatto & Windus, 1928), Chapter 15, 278.

slippers at night, defers to his every wish and subordinates her own personality, ministers to his stunted and under-developed ego, whereas his union to an equal or superior and virtually independent partner would add materially to his discomfort and sense of inferiority. Other men, again, demand fidelity in a wife because they are inexpert lovers and seek a maximum of physical satisfaction at a minimum of effort. This type is extremely common, and Mr. Bernard Shaw reminds us that the popularity of marriage lies in the fact that it affords a maximum of temptation with a maximum of opportunity. There are countless numbers of men who, either from ignorance or blunted emotional susceptibilities, demand no more from sex-life than crude, physical sensationalism. To such a man a wife is little more than a legalized harlot, a passive instrument of sensuous satisfaction, and he would no more think of patiently stimulating her desire, fondly wooing her and securing for her a reciprocal satisfaction than he would dream of appearing at his city office in his pyjamas. He has bought his wife by virtue of his support, and the least she can do is to earn her keep by conceding his full conjugal rights. And here it is useless for ingenious feminists to argue that the term "conjugal rights" is a mere mis-translation of "conjugal rites." Conjugal rights are a natural corollary to the economic dependence of women, and the diminution in the size of the modern harem has scarcely detracted one iota from the proprietary attitude of innumerable modern husbands. The exaggerated value they attach to exclusive proprietorship, moreover, effectually blinds them to the existence of the superior values which they sacrifice, and to the misery they inflict upon their partners, hence there is even more degradation of sex within the marital bond than outside it. The kept woman, in fact, be she wife or mistress, is more often than not a sacrifice on the altar of masculine egotism, and the egotist, by purchasing that which nature decrees he should win by personal effort, deprives himself of all that is most satisfying, valuable and ennobling in sexual relationships.

Nor is the possessive instinct in matrimony confined merely to men. Women, too, seek to enslave their husbands and convert them into conjugal possessions. As already explained, this is partly due to a fear of losing their sole means of livelihood, but even more is it due to sheer acquisitiveness. Women, no less than men, have yet to learn that exclusive possessions constitute limitations. The lust for possession erects an artificial barrier between oneself and one's fellows; it destroys the sense of human homogeneity, and leads to a segregation and isolation that is only imperfectly compensated by the intimacy between the two partners. Nor is it adequate to maintain that this limited form of happiness is comparatively satisfying. One ought no more to be satisfied with but a minimum experience of love than one should with a minimum experience of knowledge or of beauty. The riches of life are man's heritage, and to monopolize them or fear to partake of them is to fear life. Nor is this all. Possessiveness defeats its own aim. The man who feels himself possessed instinctively feels his freedom curtailed and his capacity for happiness impaired. From sheer affection, or unwillingness to inflict sorrow, or from a sense of duty, he may voluntarily deny the full scope of his capacity for extraneous friendships or affections, but the fact remains that he will feel himself cramped and tied. And the more unreasonable he feels the restriction to be the more will he inwardly resent it, and such sentiments inevitably blight and stunt the full development of spontaneous love. There can be no true love without trust, and although trust necessarily involves risk, it does not in the end entail nearly so much as does distrust. Many of the supposed risks of conjugal trust, moreover, are entirely fictitious. Among sane, normal and decent people a second friendship rarely, if ever, necessitates disloyalty to an earlier one, and what applies to friendship should apply also to affection, for both are inestimably valuable and enriching, and no man or woman desiring to live fully and fearlessly can enjoy too much of either. What is required by the world today, then, is not the reinstatement of the old and cramping sexual code, but emergence of a new morality adequate to secure the maximum benefits to the individual and to humanity as a whole. The principles and requirements of this long overdue ethic I shall now proceed to examine. [45–74]

From Chapter V: Ethics, Religion and Sexual Morality

[. . . . The section omitted here turns on a contrast between Kantian and Aristotelian concepts of the good life.]

Man, as distinct from the lower species, is characterized by his intellectual and æsthetic capabilities, and by his capacity for conscious social co-operation, hence his maximal satisfaction must ever be found in the pursuit of Truth, Beauty and Goodness. This is not to maintain that mere transient and superficial pleasures or excitements should never be experienced; far from it, but it is to maintain that we should endeavour to apply the Aristotelian doctrine of the Mean and to guard against the ever-present possibility of over-developing certain tendencies and desires at the expense or atrophy of others even more valuable and satisfying.

Applying this ethic to sexual relationships, it becomes at once apparent that the old standards of conduct were hopelessly inadequate. Love, as already shown, constitutes one of the most essential factors in the realization of a full and perfect life for, in so far as it is the expression of our sense of felt unity with the deity, with nature and with our fellow human beings, it belongs to the very stuff of our existence. But love is obviously capable of varying degrees of intensity and its graduations are ultimately dependent upon the existence of varying degrees of affinity. Love of Nature is usually vague and diffuse, as also is our love for the deity, until we have learned to love him through our fellows. Love of humanity is invariably less intense than that felt for a loved individual, while even among friends or kinsfolk the greatest intensity of love is always directed towards those in whom we are conscious of the greatest measure of sympathy. Of all the degrees of love, however, sexual love is undoubtedly the most intense and most peculiarly personal. Nor is this surprising. The sexes are essentially complementary, both psychically and physically, and the act of sexual union, when it is genuinely sacramental, is capable of bringing about such a concentration of affection as to elevate love to its highest intensity. No two persons can experience the supreme ecstasy of the perfect physical consummation of a deep and passionate love without being profoundly affected, and no matter how much they may wrong or injure each other at some future date, neither can ever completely eradicate the feeling of tenderness and affinity that was then engendered.

The physical factor in sexual love, moreover, operates in a twofold manner: it can both evoke tender emotion or an other-regarding sentiment by drawing out what Freud calls the libido in its sensory form towards the other person, and it can even more give expression to, and actually raise, an already existing love to its highest and most intense concentration. As an example of the out-drawing of sentiment by physical activity, Dr. William Brown instances the fondling and nourishing of a child by its mother, and declares that "if a child is denied that and brought up on the bottle, it tends to seek sensory satisfaction in its own person, whence arises the danger of auto-eroticism."[28] The effects of physical fondling, and especially of fondling involving a sexual element, is also demonstrated by the well-known fact that, among healthy and normal children, a greater intensity of affection invariably exists between mother and son or father and daughter, than between mother and daughter or father and son. The son may respect the father even more than the mother, and the daughter the mother more than the father, but as long as real affection exists and has been accorded physical expression, the intensest form of emotion will inevitably be directed towards the member of the opposite sex. This also constitutes an example of the sacramental aspect of fondling, in so far as it issues in a fuller concentration of already existing affection, and needless to say such a concentration is even more profound where the supreme intimacy of the sex act consummates the affection of true lovers.

Such being the case, it would seem apparent that the sex act, at its highest, should never be consummated except as a means to securing a mutual intensification and realization of existent love; and that, even at its lowest, it should never take place without strong mutual desire and attraction, or in the absence of a full determination to explore its maximal benefits – physical, emotional, psychical and spiritual. The lower experimental type of sex-life is undoubtedly legitimate among the young and inexperienced, but only so if it is implicitly understood that the goal of such experiments is the realization of true love and not the mere gratification of stark, sensuous impulses. Similarly, with regard to the higher or sacramental form of sexual intercourse, which presupposes the existence of a considerable degree of mutual love, it must always be remembered that civilization and culture place distinct limitations upon the number of persons towards whom any one individual is capable of experiencing both intense sympathy and strong physical attraction. Some individuals, indeed, are so unfortunate as to go through life without coming into contact with a single person capable of meeting these exacting demands, while even the most fortunate can never hope to meet more than a very small number.

These guiding principles, I think, must equally apply to all the relationships of the sexes, whether pre-nuptial, marital or extra-marital. Additional principles and considerations must, of course, be taken into account in particular cases, but these need no more than be indicated. A husband or wife, for example, has no ethical right to repudiate a tacit understanding or to practise deceit. If they cannot agree as to the degree of liberty to be conceded to each other they must either honour their

28 Brown, *Mind and Personality* (London: University of London Press, 1926), 189–190.

original contract or consent to terminate it; any other solution would be intolerable. Similarly, youths who are dependent upon their parents for their economic support have no moral right whatever, while still accepting maintenance, to outrage parental wishes or injunctions, no matter how unreasonable they may seem. Such persons are not free agents, and until they have achieved economic freedom, they must remain bound by their obligations. I am firmly convinced, however, that wise parents in the future, while always endeavouring to instil the highest ideals and to proffer friendly and disinterested guidance, will nevertheless concede to their offspring who have reached years of discretion – say, university age – the widest possible freedom of action. The ensuing gain to morality, idealism and fullness of life will be immense, since when such young people eventually find a suitable life-partner the criterion of choice will not be the subordinate one of physical immaculateness, but rather the superior one of "moral excellence" in its Aristotelian significance.

And what applies to the single will also apply to the married, for marriage will unquestionably persist. A permanent and intimate association of two persons of the opposite sex must always yield a fuller, richer and deeper content than any transient association, and the slow, progressive growth throughout the years of intense comradeship and mutual dependence, which is the essential characteristic of the matrimonial bond, will doubtless be even more fully appreciated by our descendants than by ourselves. But I think that the very real benefits contributory to a fuller and richer life conferred by ideal extra-marital friendships and sexual relationships will also be recognized and appreciated and that, granted the safeguards of contraceptives and the existence of idealism and trustworthiness in partners, no restraints will be placed upon such relationships other than those self-imposed by the moral, æsthetic and intellectual standards of the individual. [91–96]

Part III

Culture, Media, and the Arts

Complete Texts

Proteus; or, the Future of Intelligence

Vernon Lee (1925)

Headnote

The British writer Violet Paget (1856–1935) published under the pseudonym Vernon Lee. She is best known for her supernatural fiction, writings about Italy (where she spent much of her life), and her work on psychological aesthetics.

Lee was from an older generation than most of the To-day and To-morrow writers. She was a follower of Walter Pater, involved in the Aesthetic Movement, met Oscar Wilde as well as Pater, and was a friend of Henry James. The idiom of *Proteus* is occasionally more old-fashioned than that of most of its companion volumes. But its intellectual commitment to modernity—to using intelligence to respond to modern challenges—and Lee's restless imagination of new possibilities, ways in which thing could be other, makes it effectively a manifesto for the whole series.

******

I:

Intelligence and Proteus

There seems not to exist a word—for words are old while meanings may be new—which answers exactly to what I shall speak of as *Intelligence*. But space being short for what has to be said, I will not waste any in preliminary definitions. That which I mean by Intelligence will become evident by what I expect from its presence and attribute to its absence. I start from the assumption that it already exists, however insufficiently; and I deduce from what it has done that its nature is to intensify and extend. Whether this will be witnessed in the near future, or whether it may be checked by adverse circumstances, is no concern of mine. Writers of this series, and several others besides, have enlarged on the political and economic contingencies to which Intelligence, or persons presumed to have it, seem likely to be exposed. Whether Intelligence may become the weapon of a dominant caste, as was the hope of Comte, of Renan, and, at one moment, of Mr. H. G. Wells; or

DOI: 10.4324/9781032625522-25

whether, as proposed by M. Charles Maurras,[1] Intelligence shall be honoured with a subordinate function in some sort of Fascist State, I am inadequate to judge. Nor do I even feel certain that history has shown, or economic theory demonstrated, that Intelligence can be bullied or starved out of existence. Meanwhile let me confess that what I have to say about the Future of Intelligence is the expression as much of my hopes as of my convictions, both, however, arising from a longish experience of changes already brought about, and changes beginning to be brought about, by the particular, and perhaps rather modern, something I mean by *Intelligence*. What *I* mean, and what, under restriction to that meaning, appears *to me* likely or desirable. By underlining these personal pronouns, I am able to forestall the mention of one great change which Intelligence is already initiating, namely, the recognition and avowal that what one thinks (as distinguished from what primers, manuals and other authorities have taught one to believe) is—well, *just what one does think*, and neither the consensus of human opinion nor the revelation of the Deity's irrefragable truth.

Returning to the word *Intelligence*, the meaning I attach to it will become sooner obvious by clearing away some misconceptions thereof which may occur to my reader. And first: The Intelligence whose future interests me is not the same thing as the *Intelligentsia*. Those of us who belong to that class presumably possess *Intelligence*, since we live, or try to live, by its exercise. But it is no monopoly of ours, nor do we always employ it in the manner which answers to my meaning. For living on or by its employment may, as is often seen among men of science and philosophers, result in their capital of natural Intelligence being sunk in a few enterprises of especial value, leaving them, as in the notorious case of Dr. Faust,[2] but a scanty balance for current use and pleasure. I have brought in the word *pleasure* because the pleasantness of its varied exercise is one of the chief characteristics of what I mean by *Intelligence*, fostering that nimbleness, elasticity, hence also pervasiveness, which makes it a chief factor of human progress, as well as one of progressive mankind's indisputable marks and unalienable rewards. Now these same pleasant properties, so often sacrificed by very studious persons, turn Intelligence into the stock-in-trade (eked out with plentiful surrogates) of that other branch of the *Intelligentsia*, those who make a livelihood by living down to their readers, relieving their boredom, lapping their thick skins in sentimentality, and keeping up the sooty flame of their collective passions; for alas, the Man of Letters is tempted to serve his public not merely as an unconsidered jester but as a respected moral guide.

1 French author Charles Maurras (1868–1952) was the chief philosopher and organizer of the right-wing political movement Action Française, committed to monarchy and opposed to parliamentarianism and revolution.

2 The scholar and magician makes a pact with Mephistophilis that the demon will serve him for the rest of his life (in Goethe's two-part play *Faust*) or 24 years (in Marlowe's play *Doctor Faustus*), after which the Devil will claim him body and soul. As his time on earth runs out, the pleasures and entertainments Mephistopheles offers come to seem less of a bargain.

Thus it comes about that we of the *Intelligentsia* cannot stand as faultless specimens of Intelligence. Besides, our facility for self-expression and our habit of holding forth unchecked combine to exaggerate, stereotype and warp our best ideas: only think of Carlyle and Ruskin, let alone Tolstoy and Nietzsche!

Having so far established what I do *not* mean by Intelligence, and before entering on discussion of Intelligence's future achievements, it seems fitting to say a word or two about Proteus,[3] to whom this little treatise is consecrated. It is so, I confess, partly because I am attracted by the classical titles, *Daedalus*, *Icarus*, *Tantalus*, of my predecessors,[4] and then because, as described by Virgil, Proteus is to me one of the most engaging figures of mythology: … "Ille, suae contra non immemor artis, omnia transformat sese in miracula rerum, ignemque, horribilem feram, fluviumque liquentem…"[5]

But here again I must forestall another wrong identification likely to jump into the reader's mind: to wit, of Proteus with Intelligence. On the contrary: Proteus, multiform and ever-elusive, represents that which Intelligence (lighter equipped than specialized Intellect for such rapid hunts) can sometimes catch sight of and, for however brief a contact, sometimes even clutch. Proteus, in my mythology, is the mysterious whole which we know must exist, but know not how to descry: Reality. For, whatever else we may believe it to be, Reality when thus partially revealed, is never twice the same. Nor merely because of what we call waxing and waning, growth and decay, and whatever other phases of individual and racial transformation biology has made us superficially familiar with. There may well be some πάντα ῥεῖ[6] outside and irrespective of our thoughts; indeed, it may have been in miming the universal flux that our thoughts themselves have grown *protean*. Look, for instance, at that strange (well named *auxiliary*) verb whereby we testify belief in reality, *esse, to be*; which holds in its emptiness the possibility of all qualities and happenings and implies in its assertion of mere blank existence the assurance of continual change: a future and past. For, whenever we speak of what we call a *thing*, its mere name, like the name of Virgil's Proteus, is a spell making us witness aspect after aspect, take stock of relation after relation, admit likelihood after likelihood. And our belief in that thing's reality, in its being *that thing* and no other, means that it has had a certain, however unknown, past, and will have a more or less certain future. In this sense *Reality*, the fact of aspects perceived, remembered and expected in regularly connected sequences and combinations, *that* is what I mean by Proteus. Maybe that Proteus does not change at all except in our

3 In Greek mythology Proteus was a sea god who could foretell the future, but shifted his shape to avoid doing so. He would only prophesy to those who could stop him slipping out of their grasp.

4 *Daedalus* and *Icarus*, the first two volumes in the series, are included here in full. F. C. S. Schiller's *Tantalus; or, The Future of Man* (1924), was the sixth.

5 Virgil, *Georgics*, Book 4, 440–442: 'he nathless,/All unforgetful of his ancient craft, /Transforms himself to every wondrous thing,/Fire and a fearful beast, and flowing stream.' *Bucolics, Aeneid, and Georgics of Vergil*. J. B. Greenough (Boston: Ginn & Co. 1900).

6 Greek: 'everything flows'; phrase attributed to the philosopher Heraclitus, who probably didn't say it.

narrow, and shifting, field of vision. Maybe that the multiform Virgilian Proteus might turn out to undergo but one first and last transformation, into that great auxiliary *esse*, *to be*, holding in its stark emptiness all that, for us, is things and happenings... Such a transcendent and sole real Reality I leave to metaphysicians, not without wondering secretly whence, save from occasional experience of this (to them) unreal Proteus, they ever got to think about Reality at all.

So, dealing in this shallow treatise solely with such (even if spurious) Reality as Proteus represents, I need now only justify my outrageous claim that mere Intelligence can have any privileged intercourse therewith. My ground for saying so is that specialized intellect screws its marvellous lenses down on only a single, and *singled out*, aspect of Reality; employs subtle reagents revealing only the properties for which they have been devised. Moreover, that the world of regular, foretellable sequences which science constructs is a map teaching us why to turn to the right or the left, but not a moving slice of the landscape we are moving in. Instead of which mere Intelligence, with its rule-of-thumb logic and wellnigh automatic movements, may be fairly fitted, not indeed to inventory and schedule separate items of Proteus' multifold embodiments, but to keep us aware that Proteus is there at that eternal game of his: changing his aspects perpetually, whether you watch him or not, nay, changing aspect by the very fact of your watching him.

This may suggest that Intelligence is never at rest; and no more it is. But its movements being responsive to what strikes it from outside, are, just as the outside's own ways, orderly, and such as organize themselves into regular rhythms of sameness and diversity. For Proteus is absolutely unexpected only to persons like Virgil's Aristæus who, you must remember, was so hidebound in his business of *h*oney-making (alter one letter, you won't alter my meaning!) as to be wholly unaware that his own caddish behaviour had occasioned the death of Eurydice and so remarkable an event as the Descent of Orpheus into Hell.[7] Practical people like that are nearly always astonished and dismayed when confronted with Proteus; "they had forgotten..." Now Intelligence is as much memory as perception; and for it there is always in the transformations it is watching something familiar which carries it back to what it has already witnessed, and forwards, expectantly, to something it may be going to witness. Hence to Intelligence there is never mere repetition, just as there is never utter novelty. And its frequent doubts are always conditioned by its habitual beliefs. That explains why Intelligence is so chock-full of prejudices, as all those are aware who have ever asked it to accept miracles and ghosts on their testimony or on someone else's authority. Such people exclaim at the sceptic's blindness to evidence, because they do not know that doubting and even denying are part of Intelligence's active rhythm of grasping and acquiescing; a process of assimilation and elimination in which the already experienced and accepted selects that which shall be accepted or rejected. Moreover, such selective action often expresses itself in the most impertinent (because most pertinent)

7 In the passage from the *Georgics* quoted above, Aristæus' mother Cyrene tells him how to get Proteus to explain to him why his bees are dying. Lee's joke is that '*m*oney-making' is the modern equivalent of a preoccupation which blinkers intelligence.

queries, as: “Now how would you explain that?” “In what sense are you using that word?” etc., etc., etc. Queries, all of them, which in their exasperating amateurishness have probably done more than the elaborate arguments of specialized Intellect to shoo away some of the many Chimæras, Entities, and Essences, which, as Rabelais already remarked, had gone on *bombinating in vacuo* through the resounding spaciousness of philosophy and science, leaving behind only the fainter buzz of *Historical and Economic Laws, Entelechies, Teleologies* and *Vital Elans*.[8] It was, I take it, Intelligence which first scoffed in Molière's play at opium's *Virtus Dormitiva*...[9]

At this point a parenthesis must be opened on account of a reader asking, not impertinently, whether what I have been talking of under the name of Intelligence is not plain *Common Sense*. Yes; but also *No*. Since, on behalf of practicality, Common Sense usually warns us off from just such questions as Intelligence should deal with. So one might say that Intelligence is a kind of Common Sense, but applied to uncommon (not common or garden!) subjects, and as yet, alas, only by rather uncommon people.

If Proteus be taken to represent that Reality which all save metaphysicians believe to be real, he represents especially that half of it which I have (elsewhere) called *Otherness*, that is to say, whatever is not *ourself*.[10] And just as the essential, unshareable *ourself* is what we feel, to wit: moods, passions, efforts, hope and fear, liking and disliking; so the *not-ourself* (other persons as well as other things, and even our own personalities when viewed as if they were not our own)—the *Otherness* in short, is, on the contrary, *seen*, because it is outside us. Seen by the mental eye of Intelligence, which, like the bodily one, moves in every direction and focuses to all distances, thereby informing us of the proportions and relations of whatever is not ourself, and following step by step the actions which are not our own. And though it must borrow the lenses of Science (which centuries of thinkers cut and polished) before it can know things in their microscopic detail or their astronomical remoteness, yet with no aid save everyday experience, Intelligence suffices to teach us the most important and most overlooked fact concerning that Reality which is *Otherness*: namely, that it has ways of its own and does not exist merely to suit our likings.

8 François Rabelais' *Pantagruel* includes an imaginary Latin book title which translates as: ‘A subtle question, whether a Chimera buzzing in a vacuum can devour secondary intentions.’ ‘Entelechy’ is the supposed vital principle that guides the development and functioning of an organism. ‘Teleology’ has the theological sense that the material world is designed for divine purposes. ‘*Elan vital,*’ or the life force, is particularly associated with Lee's contemporary and pre-eminent philosopher of flux, Henri Bergson (see *Archimedes*, 102, n. 33). Lee suggests all these terms are scholastic fictions (or *Vital Lies*, as she described them in her 1912 book on Pragmatism), rather than corresponding to anything in reality.

9 In Molière's *The Imaginary Invalid*, physicians are mocked for their obscurantist tautologies: opium induces sleep because of its ‘dormitive principle.’

10 Lee had written of ‘the great otherness of contemplation and understanding’ in her preceding book: *The Handling of Words and Other Studies in Literary Psychology* (London: John Lane: The Bodley Head, 1923), 315

The habit of taking "otherness" into account, and a wider and wider circle thereof, might serve as a rough test of Intelligence and of its progress: young children, as is notorious, referring everything to themselves; and "uneducated" people, from the narrowness of their knowledge, rarely conceiving anything beyond their own personal experience. At the risk of incurring the same criticism, I hazard my own impression that the dominance of possessive pronouns, the restriction of interest to one's own history and circle of acquaintances, has become less usual among "educated" persons.

Similarly, that there is getting to be something rather old-fashioned about settling general questions on the strength of single personal experiences. Except where strong likes and dislikes come into play, it is rarer than formerly to hear (shall we say?) divorce condemned because of the sad case of Mrs. Blank; or the eight-hours day rejected on account of last harvest having been soaked; or the practical utility of a classical education justified by the career of Mr. Gladstone. Modes of thought like this seem to be (slowly!) disappearing in the wake of the anecdote-mongering and epigram-and-joke button-holing of ancient bores who may have been brilliant conversationalists at Meredithian dinner-tables.[11] And when one thinks of it: was not such the substance of much of our grandparents' wit and wisdom? Nay, a little further back did not "gentlemen" ask the ladies riddles after themselves exchanging smutty *Joe Millers* over their wine?[12] And behold! there opens up a vista of euphuism, of pedantic discussions, of "sonnet, c'est un sonnet," of "deliciæ eruditorum," and "facetiæ"; boredom incalculable back through Hôtels de Rambouillets and Medicean academies to Courts of Love and the stale scurrilities of Shakespearian clowns…[13] Nay, was not Shakespeare himself ready to adorn with supremest poetry and philosophy stories often preposterously cock-and-bull? Which makes one suspect that Intelligence, in the sense in which I have been using the word, is of amazingly recent growth; and that the people of the past, superior though they may have been in genius, wit, humour, and even wisdom, would have struck us (and we shall doubtless strike future generations) as decidedly stupid.

For instance (returning to Proteus!), in their capacity of *thinking in terms of change*. This seems an *intrinsic part of thinking in terms of otherness*; yet, as, a

11 The novelist George Meredith (1828–1909) was notorious for his convoluted syntax. Oscar Wilde said his: 'style is chaos illumined by brilliant flashes of lightning. As a writer he has mastered everything, except language; as a novelist he can do everything, except tell a story'. *A Critic in Pall Mall* (London: Methuen, 1919), 204–205.

12 Stale jokes; from the eighteenth-century joke book *Joe Miller's Jests*.

13 Euphuism is a highly mannered literary style, named after John Lyly's romances of *Euphues* (1578 and 1580). In Molière's *The Misanthrope*, Act 1, Scene 1, Oronte reads his love sonnet but keeps interrupting himself. *Deliciae eruditorum* (Latin: 'delights of the learned') was an 18-volume collection of antiquarian lore compiled in the eighteenth century by Giovanni Lami. *Facetiæ* are 'witticisms.' The *Hôtel de Rambouillet* was where Catherine de Vivonne, marquise de Rambouillet, held a famous literary salon (ridiculed by Molière in *Les Precieuses ridicules*). Cosimo de' Medici founded the Platonic Academy in Florence in the 1460s; his grandson Lorenzo kept up the tradition. The Courts of Love were the – probably mythical – tribunals in the medieval courts of Aquitaine, Provence, and elsewhere, presided over by women adjudicating on matters of love. These phenomena were all culturally significant, but Lee objects to their verbosity and pedantry.

fact, it dates only from the days of Montesquieu, Voltaire, Gibbon and Condorcet. This last name brings home that until the eighteenth century the only Future which people thought about was the Future in Heaven or Hell. The importance of the latter alternative explains quite sufficiently why no interest was left over for any other after-life, to wit, that of unborn generations. Indeed, the sway of religious conceptions accounts also for our ancestors having been no less cut off from the Past and replacing it (as their painters dressed Abraham or Cæsar in Renaissance costume) by the Present. For all religion tends to think *sub specie æternitatis*,[14] as of the god who is sacrificed afresh at every celebration, and who consecrates the routine of the seasons and the seasonal monotony of agriculture and pastoral life; whence, no doubt, the persistence of the amazing fallacy *that there is nothing new under the sun*,[15] with its corollary that there ought to be nothing not *old*. Whence also the double superstition (till Science broke in with something different!) of chewing and rechewing the cud of Scripture and the Classics. With, in turn, the practical results that Milton's Puritans modelled themselves on Joshua and Gideon; and frilled and waistcoated French Revolutionaries postured as heroes of Plutarch. Why, at this very moment do we not see the rods and axes of antique hangmen[16] figuring (not merely in figurative manner!) as emblems of post-war Italy, itself identified (to the neglect of schools and irrigation works) with a particularly high and palmy Rome? Rome! to rule which squalid mediæval village Dante called on a Cæsar who was a *Kaiser* elected by German feudatories;[17] Rome, which we may take as a *reductio ad absurdum* of the refusal to realize that Past is Past and Present is Present. Which is perhaps the only "Lesson of History"; and whose application would dissolve many mythical alloys of conflicting nations welded together by the passionate white-heat of a name: England, France, America, Christianity, and nowadays, I fear, also Socialism, nations, and creeds concerning which, when asked for our allegiance, we have need to inquire: In which of its phases, which of its characteristics and embodiments?

For Intelligence warns us that we are dealing with Proteus, with him of ceaseless change. Not with the eternal, immutable divinities to whom our forebears brought their sometimes quaint and lovely, but, quite as often, obscene and grisly oblations.

But while ignoring distinctions between Past and Present, even our nearer ancestors conducted much of their thinking in elaborate water-tight compartments; for they conceived of "Truth" as a battleship, continually exposed to the murderous broadsides of "Error." Of these hermetic partitions, say, between *Faith* and *Reason*, *Body* and *Soul*, or *Good* and *Evil*, Intelligence has already rammed in a number, without drowning us. *Error* itself has lost its capital E, being usually called *Mistake*. And, what is more important, we have begun to notice that it and *Truth* are not

14 Latin: 'under the aspect of eternity'; coined by Spinoza in his *Ethics*.

15 *Ecclesiastes* 1:9.

16 The *fasces*, the ancient Etruscan and Roman symbol of state power, was adopted by Mussolini's Fascist party when Lee was living in Italy.

17 In the *Purgatorio*, Canto VI, Dante beseeches the German emperor Albert I to return to Rome. A 'feudatory' owes feudal allegiance to another.

at all irreconcilable, but cradled originally together, and sometimes intermarrying, with mixed or alternating generations, as by Mendelian rules;[18] but very rarely, either Truth or Error, affording us a pure breed.

These examples will have justified, I trust, my contention that Intelligence is specially fitted to deal with Change. Not to praise or blame it after mature deliberation, like solemn and sedentary Reason; still less to filter concrete realities into immutable, because purely abstract, entities, which is the business of scientific thought; but just to perceive change on its passage and in so far help us to make the best of its coming.

Need one add that Intelligence is far more liable to mistakes than either "Reason" or "Logic"? But its mistakes, though so much more numerous, are, methinks, less massively enthroned and less likely to block the way than theirs, for there is something self-satisfied and without appeal about "Reason" and "Logic": does not the one issue "dictates" and the other enunciate "laws"? Whereas the mistakes which Intelligence commits to-day, it will, in its light-hearted way, correct to-morrow, being as little ashamed of revokes as its disconcerting friend Proteus is of transformations. Of course, Intelligence *is* rather irresponsible and, one might add, cannot help being so because it is essentially *responsive*. Like the human eye (to which I have compared it) Intelligence turns to whichever side the light comes from, adjusting itself, in discursive, often desultory fashion, to all manner of directions and distances, comparing and measuring with unabashed slovenliness, extracting the qualities which strike it and hastening on to connect them with something it was struck by before. Being thus rapidly responsive, Intelligence may often, I admit, seem *on the pounce*, and more so than politeness warrants. But it can also take its time, poise circling round and round, and reverse its movement, because it is never motionless and always able to readjust its balance.

Such do I see Intelligence in those who possess it; such do I feel it, on some delightful occasions, in myself. Such also I frequently notice it failing to make itself agreeable to some kinds of persons. Those who take a just pride in Reason or Logic are often a little ruffled; or else, as Wagner said of Mozart, they find Intelligence just a little *frivol*. But in the long run they recognize an ally; and their conscious superiority makes them indulgent. Not so with people—I might have said Peoples—who happen to be indulging in the glorious unimpeded violence of collective passions, specially those which are magnanimous and cruel, as, for instance, in war time, when a conscientious objector may come off better than an intelligent one.

In like fashion Intelligence's passionate pleasure in dealing with *Otherness* and in looking out for Proteus, Intelligence's frequent indifference to *here* and *now*, disrespect to *self* and refusal to regard *means* as *ends*—all this renders it unpopular with those practical-minded men who are bent on personal advantage and on outstripping competitors in the great race to *Nowhere*. These acute persons are quite aware that Intelligence might make an invaluable slave, only you cannot keep its nose, with any regularity, to the grindstone. In default of such practical usefulness

18 See *Icarus*, pp. 47–8, n. 5.

it may be worth hiring, as one buys a yacht or an old master, for a mark of wealth and being in the know. But let us have none of your *whys* and *wherefores*! Besides, the Rulers of Men have by this time mostly recognized that Intelligence is harder to deal with than any number of High Principles, for you cannot hope to bamboozle it into serving you unawares.

But Intelligence, though thus in some quarters deservedly unpopular is adored by all who have it; and that is the reason why, once it has got a footing in the world, it is bound to increase and multiply and eventually conquer its promised land.

II: Proteus and Ethics

I have just come across a passage from Huxley's famous *Romanes Lecture*,[19] read thirty years ago and long since forgotten; and which has brought home to me all our elusive Proteus has been doing in the domain of ethics; moreover the share of Intelligence in confirming those changes. Huxley is pointing out a fact which he finds disconcerting, namely, "that ethical nature, while born of cosmic nature, is necessarily at enmity with its parent."[20] The allusion to the harmony reigning in Victorian families may make one smile, like some well-bred Du Maurier illustration.[21] But how those words bring back what some of us are old enough to have suffered in days when Free Thought drew a terrifying line between religious dogmas and moral ones, clinging to these to steady itself after jettisoning the others! One's youthful deistic anguish (as cruel, perhaps, as any believer's sense of God's forsaking him) at discarding God for insufficient morality, was merely transferred to one's terms with Huxley's ogre Cosmos, devouring the moral instincts itself had begotten. Occupied as my studies then were with art-history, I can remember wrestling with the horrid inconsistency of the art of Michael Angelo and Rafael having arisen in a civilization described by Taine as partaking of the brothel and the cut-throat's den. And I remember the heavenly relief of hitting on the notion that, since such art is not born in a day, it must have been begotten and incubated during the Franciscan Age, immune from all Borgian infections.[22] Of course, the generation immediately younger than mine was taught by Nietzsche that Michael Angelo's greatness was, on the contrary, due to presiding Renaissance villainy; but that pseudo-Nietzschian generation is, in its turn, superannuated, and the cult of *immoralism* along with it. Not only because paradoxes do not bear repetition, but for another reason which that quotation from Huxley has made me realize. Namely, that we have left off thinking of art as either moral or immoral, simply because

19 The biologist T. H. Huxley (grandfather of the novelist Aldous and biologist Julian Huxley), gave the 1893 Romanes lecture in Oxford on 'Evolution and Ethics.'

20 The quotation comes not from the lecture but from Huxley's Preface to *Evolution and Ethics and other Essays* (London: Macmillan, 1895), viii.

21 George du Maurier (1834–1896), author of the novel *Trilby*, was also well-known as a cartoonist for *Punch*.

22 The Franciscan order was founded in the thirteenth century; the Borgias became prominent in the fifteenth.

morality no longer holds the same place in our thoughts as, say, in those of Ruskin, George Eliot, or, as that quotation shows, even in those of Huxley. Not the same (if one may say so, *ubiquitous*) place; a place more clearly defined, but only the more important, ever since Intelligence, ferreting about among *Golden Boughs*, *Religion of the Semites*, and similar books, has quietly stripped from our moral valuations that half-supernatural, half-æsthetic halo which is but the shrunken religious involucrum wherein they came into the world.[23] The "problem of evil" has already become the problem not of its toleration by God, but of its diminution by Man. *That* is the great change we are still witnessing; a change, I cannot but think, greater than any brought about by the material applications of science, and implying a deliverance from individual suffering not less than that we owe to Pasteur and to Lister.

Whether we notice it or not, Morality is already taking a new status, independent alike of an absentee (or absent) Deity, and of an indifferent Cosmos. But its new domain, narrow and self-governing, essentially *sui generis*, has sanctions and imperatives only the stronger for being man-made and man-regarding. And, one may add, only the more austerely binding on the present that we shall recognize them as different from those of the Past and different, no doubt, from those into which the Future will transform them.

Thus we are already conceiving of punishment only as a mechanism, successful or not, for social defence. And we scarcely ever hear more than the last echo of those incentives to virtue and deterrents from vice of the *Sandford and Merton* type of my own childhood's copybooks.[24] Still less of the Stoical, and (alas!) Platonic mendacities about remorse torturing evil-doers, and the unhappy life of Browning's *Instans Tyrannus* with his "Then *I* was afraid."[25] Neither do we talk any longer of the virtuous glows which (failing the increase of flocks and herds!) used to reward the virtuous acts of the generation adorned by Butler's Mr. Pontifex.[26] We are getting to think of our own virtues, supposing we have any, as conducive not to our own advantage but to that of other folk.

Consonantly with the psychologist's recognition that, of the two polar feelings determining human action, the (positive) attraction of pleasure is far less potent than the (negative) repulsion of pain, it seems as if our future ethics would emphasize not good actions but bad ones. That will be following up the rule-of-thumb wisdom of the Commandments, of which the surviving ones are all "thou *shalt not*"; the positive ones about loving God and honouring Father and Mother having become either difficult to enforce or optional. I am glad of that word *optional*, because it leads to the remark that Intelligence is surely abolishing that neutral

23 J. G. Fraser's *The Golden Bough* (1890–1915) and William Robertson Smith's *Lectures on the Religion of the Semites* (1894) were both influential studies of comparative religion. An involucrum is a layer of new bone growth outside existing (and sometimes dead) bone.

24 Thomas Day's *The History of Sandford and Merton* (1783–1789) was a set of moralistic children's tales.

25 A dramatic monolog in Browning's *Men and Women* (1855), in which a tyrant realizes the limits of his power when confronted with a righteous victim.

26 The hypocritical and sanctimonious father of the protagonist Ernest in Samuel Butler's *The Way of All Flesh* (published 1903, but written between 1873 and 1884).

territory whence "good actions" can issue at "good" people's good pleasure and as an expression of their goodness, but which no one has a right to insist upon; indeed, which they have a perfect right to withhold, since they are patted on the back for doing them, or have their hands kissed, as children were taught to kiss those of the "revered author of their being." I expect that before so very long Intelligence may bluntly suggest that if the action, whatever it happen to be, is *really good*, that must mean that it is *really needed*; and if it is really needed, your fellow men can claim it and oblige you to claim it from your unwilling self. And to dishonour that claim may become in their eyes (mirrored in your own), mean, disgraceful, dirty. In the language of contemporary youth, it will not be *decent*.[27] That substitution of the word *decency* for the word *virtue* gives, methinks, the clue to the future revaluation of our moral standards. It implies, as I have suggested, a more intelligent and, in some ways, more indulgent, morality; but a morality on the whole more austere, a stark notion of duty armed with the relentless imperative which nowadays makes us abashed at the revelation in ourself of physical cowardice or bodily dirtiness. A morality, I venture to add, eventually able to do without the adornments coming under the head of "Moral Beauty."

And, speaking of a future standard of "decency," there will necessarily come sundry revaluations quite intolerable to our present morality. I will not speak (since far too much is nowadays being spoken concerning what, after all, is but a small part of conduct) about such revaluations of sexual morals as *Dædalus* prognosticates from transplantation of ovaries. *That*, and coming facilitations for changing one's sex, cannot, indeed, fail to modify family arrangements; although I have greater belief in the effects of future methods of producing and exchanging, not offspring, but other commodities, and the consequent alteration in our tenure and conception of property. Indissoluble marriage, which already strikes some of us as scarcely decent, will lose its practical utility once inheritance is more or less abolished, and the subsistence and education of children no longer a charge on parents. Nor is this all: a more restricted practice and therefore habitual notion of ownership may at some distant day educate men and women, parents and children, lovers and friends, nay, masters and disciples, to admit Proteus even into the impregnable stronghold and inviolable sanctuary of human selfishness called *Love*. The "marriage of true minds" may, like the other one, come to be supplemented by honourable divorce. Exclusive reciprocal attachment, surely of all spiritual essences the most delicate, if not most volatile, may cease to be regarded as an inalienable piece of property, guaranteed by honour more terrible than law; and which, while all else (and ourselves most!) alters and shifts, cannot be altered and shifted without guilt of theft. There may come an end to the ideal of such fidelity as implies the claim of him or her once preferred to be preferred for ever; the duty also of continuing to prefer once having begun. Like much of the morality of a more intelligent age, "decent" behaviour in matters of sentiment will be based less upon an *ought* than

27 [Author's note.] "No, we may not be as moral as they (*i.e.* the older generation) are, but we are fifty times decenter."—G. B. Stern, *Tents of Israel*, 1924; pp. 244.

an *is*. And I can conceive that such a change may make love's tenure less insecure and less routinish and perfunctory. It will, at least, save one of the finest kinds of happiness (and the multiplying factor of many other ones) not indeed from the passing misery of change, but from the ignominy of claimed or accepted sacrifice, and the cruel pollution of jealousy, not between lovers only, but between all who love. And when there shall be applied to love the solemn saying "the Lord giveth and the Lord taketh away" we may learn to temper our loss by the intenser gratitude for whatever, even if only for a time, has been ours.

As with fidelity in love, so also with "loyalty" to persons, even to causes and ideas. But such, too often degrading, loyalty will, I imagine, be more than compensated for by the condemnation of a new sin against the Holy Ghost, and by insistence on a minimum (at least!) of consistency in one's own ideas and a minimum of conformity between one's judgment of others and one's judgment of oneself: the mote in one's brother's eye awakening the suspicion of the beam in one's own.[28]

At the same time (which is not our time!), and as Intelligence takes on a leading part in morals, there will come the indulgent recognition that such a "decency" as we may exact (or try to exact) from ourselves, cannot, any more than personal cleanliness in our own day, be exacted from all our neighbours. It may take a good many transformations of Proteus before the mote can always be removed from our brother's eye, even supposing the beam to have been taken away from our own. It is no easy matter to be always clean inside and out, especially when, like the little boy in Stevenson's rhyme, "your dear Papa is poor,"[29] poor in spirit, perchance one of a long line of moral paupers. Neither is "decency" always attainable where there has been no past charwoman to prepare your easy tidiness at expense of previous dirty hands. Still less when, as nowadays, wallowing in excess or in cruelty is the only excitement many people can get out of life. Hence it may be a long while yet before the bare decencies of the spirit, even if recognized for such, can lose the value of rarity and the status of virtues. For, let us remember that, the fouler mankind's surroundings and sores, the greater the need for incense and myrrh and even for the questionable odour of sanctity. Is not early Christianity's, say St. Paul's, insistence on chastity and mansuetude the expression of the otherwise inexpressible bestiality, cruelty, and vaingloriousness of decadent Rome? And what is the foolish Franciscan laudation of beggary save the measure of mediæval rapine and simony?[30]

So, for the time being and the world as it may, alas! long continue, mankind will need something besides a taste for moral decency, to wit, an admiration for generous, nay, quixotic impulses and for tender sensibilities. These Intelligence, respectful towards the need for them, can neither create, nor, except by negative

28 *Matthew*, 7:3: 'And why beholdest thou the mote that is in thy brother's eye, but considerest not the beam that is in thine own eye?'

29 Robert Louis Stevenson, 'System,' from *The Child's Garden of Verses* (London: Longman's, 1885), 23: 'The child that is not clean and neat,/With lots of toys and things to eat,/He is a naughty child, I'm sure – /Or else his dear Papa is poor.'

30 Mansuetude: gentleness. Simony: buying pardons, relics, or preferment.

measures, even increase. But it can do something as necessary. Intelligence, and only Intelligence, can see to it that, even to-day, such rare and precious impulses and sensibilities be not diverted to evil results, wasted in barren self-sacrifice, or the fostering of hide-bound selfishness. Wasted, above all, in hecatombs to the Molochs of collective superstition, like the one which is only just over, and may begin again to-morrow.[31]

III: Proteus and Æsthetics

"If there be any truth in these forecasts of what your fine Intelligence may bring about or justify in the domain of ethics, then" (it's a certain kind of reader interrupting), "Then may I never enter, nay, cast a glance into this detestable world of rationalized righteousness! Human do you call it, because you have made it godless? What is this Intelligence of yours worth if it fails to perceive that God exists because Man has need of Him; and that the true mission of virtue, of truth, of heroism, is not to make the world more endurable, but to satisfy our deepest human craving, that for greater harmony and loveliness, for deeper, steadier passion than otherwise life affords?"

My answer is that such a craving will become stronger, or at least wider-spread, in proportion as mankind grows more intelligent, therefore less exhausted by struggling against adverse circumstances, inbred defects and inherited superstitions. In proportion, likewise, as it will have learned to value its own virtues as they can minister to man's prosperity and betterment. Nay, more: the time will come when we shall turn with disgust and wrath at their cultivation for any other purpose; and when pleasure in virtue and in heroism for its own sake may come to be accounted so much æsthetic dilettantism, questionable and well-nigh obsolete. For with the orientation of morals towards human usefulness, towards dutifulness conceived as decency, people will get to understand that what man craves for as consolation and enhancement, the passion deeper and steadier, the harmony more complete than real life furnishes, Man creates for himself in poetry and art, and in the things of reality seen as poetry and art. In all this he has made himself a realm where truth is never betrayed, because in its sheer existence *true* and *false* become words without a meaning; moreover, where the deepest and highest passions are satisfied without being misapplied or wasted, because satisfied by their mere expression. Art and whatever the poor word *Art* may stand for—is the man-made sanctuary of the legitimate, the innocent, the immaculately *decent*, because it is closed to the shifting *needs*, the partial truths, above all, the *mine* and *thine*, which trouble real life. This, in its way, is also a realm of *otherness*, inasmuch as it transcends the self with its *here* and its *now*. Yet an *otherwise* not merely recognized by Intelligence, but made by the heart's desire out of the heart's own substance and in desire's own shape; for of such are the forms of the painter and sculptor, no less than the counterfeit

31 In ancient Greece or Rome, a hecatomb was a great public sacrifice. Moloch was traditionally thought to be a Canaanite god demanding child sacrifice. Lee is referring to the First World War. Her *Satan the Waster: A Philosophic War Trilogy* had appeared in 1920.

presentments of the poet. Above all, in the twin arts of architecture and of music do we already meet the clarified embodiment of the longing and clinging, the solemn appeasement and victorious stress and fulfilment of human passion. Here, in art's interludes of life, we can obtain what religious creeds lay open to the reproach of being false because they give it for true; and what love seeks to make unchanging, only to taste the bitterness of change. So that many as have been and will be the successive responses to our æsthetic cravings, the manifold satisfactions thereof, embodying as they do the purified essence of our feelings and activities, will, in the endless shifting of our valuations, perhaps constitute the one region where we need not be watching for Proteus.

IV: Proteus and Intellectual Manners

Though many other causes will bring about such moral revaluations as I have mentioned, Intelligence will play its part in justifying them. But Intelligence will itself effectuate great changes, methinks, in the minor moral realm of intellectual manners. For instance, proscribing *perfunctoriness*; making us ashamed, which we are not, of offering in the guise of opinion much which we know to be stop-gap and shoddy. The condemnation of perfunctoriness will lead to discarding heckler's tricks and dialectic pit-falls like that concerning the future status of the widow of seven successive husbands, wherewith the Sadducees, though disbelieving in any after-life at all, tried to trip up Jesus. His answer: that there would be neither marrying nor giving in marriage, may stand as the typical silencer to many queries with which debaters embarrass each other without advancing a step in the inquiry. For instance: "then, what do you propose to do?" Well! we may be intelligent enough to know that in nine cases out of ten there *is* nothing to propose. Since the more habitually we get to regard the Future as resulting from the Present and the Present from the Past, the more often we must admit that we know too little of the hidden Past, and less of the fleeting Present, to make sure what new combinations, including reciprocal neutralizations, are preparing to arise. The oftener we have watched the Old Man Proteus, the less, perhaps, our cocksureness about his next metamorphosis. Or, take another question which intellectual good breeding will refrain from because Intelligence foresees no answer: "In that case, what will you put in its place?" *It* being, let us say, indissoluble marriage, exclusive private ownership, war (you remember William James was, shortly before 1914, looking out for a *surrogate*!) or even what used to be called *God*, but may now be thankful when philosophers (like Mr. Lloyd Morgan)[32] allow it merely adjectival rank as "*Deity*." How do you know that *it*, whatever it is, *will leave a place*? Do we not daily see that, when things vanish, their place (place in the world, in our thoughts, and also, in our hearts!) are apt to vanish along with them and be forgotten? And among such things as may some day vanish and be forgotten there will, I trust, be the dialectic ju-jitsu which makes many pages even of Plato such dreary reading.

32 See *Scheherazade*, p. 402 n.3.

Lacking, as the world then did, all discipline of experimental science, such acrobatics may have afforded an indispensable training to logical thought, and a preparation for that latest-comer of all intellectual habits: care for the exact sense in which a word is being employed. Apart, however, from this, controversy of this kind has but a personal value, adding nothing to knowledge, just as nothing is added to wealth when one gamester loses money to another: the personal value of downing an adversary and magnifying oneself by mere comparison, which may be reckoned in some distant Future an intellectual entertainment fit for cads.

Such are a few of the improvements one might foresee in our intellectual manners. Allied with these is one which appertains to our intellectual morals. In another book[33] I have written at some length against the survival of the lawyer's and politician's arts of *Persuasion*, and of the priestly arts of *Exhortation* and *Denunciation*, both sets of them intended to influence men to think, feel and act differently from how they would otherwise do, but in compliance with the persuasive person's wishes. Some day or other such attempts may be accounted impertinent where they fail, and dishonest where they succeed; they and the sway of words should constitute a chapter of intellectual morals.

Returning to mere intellectual *manners*, I think intellectual prize fights, duels and *vendettas*, such as wasted half the life of the greatest intellects from Abélard to Samuel Butler, are a little going out of fashion, like the quarrels for precedence we read about in *Herbert of Cherbury* and such-like: they stop the traffic, make a noise and, after a minute, bore us! Moreover, I fancy I see a reason why, let alone mere spectators of such frays, even those who might have been principals in them will refrain and call them ill-mannered. I mean that, as people grow more intelligent, or more people grow intelligent at all, we shall discover other opportunities for exercising intellectual energy and for obtaining the thrill and uplift of intellectual prowess. There will appear other adversaries to wrestle with and circumvent: *Things, Reasons why*, the Universe's riddles; not any longer mere other people trying to make us write ourselves down asses as we try to make them. The finest sport in all the world is hunting Proteus…

Nor will I let myself be heckled with the objection that joys like these are reserved for minds like yours and mine, dear reader, minds *Creative*… As if books, pictures, policies, opinions, etc., etc., were created *ex nihilo*, obeying that august *fiat* which evolutional philosophy has filched from the old Creator of all things to bestow on every member of the Intelligentsia. No, no, the joys I speak of are unprofessional. And the chief creative joy is that of understanding and appreciating; say the joy of every deserving reader outrunning the straight path of the writer in circles like those of a dog pleased to be taken a walk.

Amateurishness! I can hear those of you *pshah*! and *tosh*! who believe in training the (involuntary) attention and who value work less by results than by efforts. Amateurish? Why, of course, that's just the fun and the good of it. Also the unsought moral gain. For, are we not made more "decent" by these private, irresponsible

33 [Author's note.] *The Handling of Words*.

dealings with the Unknown (at least to us); for instance, those secret inaccurate guesses at geological and historical riddles which make up half the pleasure of a journey? Since these amateurish stalkings of Proteus attain an attractiveness such that personal controversy seems insipid or odious by comparison.

I have had the good fortune once or twice, even in an old-fashioned lifetime, to witness the full flowering of such selfless intellectual happiness: to watch a mind so passionately interested in certain subjects as to care nothing whether the enchanting new ideas were its own or other folks'; nay, whether its own were by them confirmed or utterly demolished. I have seen that unusual spectacle, but once or twice only. For Intelligence has yet to establish its claim to such generous happiness. Once or twice only. But never more clearly have I seen it than in you, Mario Calderoni, dear dead young friend, who have embodied my hopes for the Intelligence of a distant Future, when you will not have received a posthumous recognition and I may be entirely forgotten.[34]

V: Uses and Abuses

Considering the great pleasantness, let alone the various uses, of the mental habits I have summed up as *Intelligence*, it is surprising there should not yet be more of it forthcoming. It has, at present, a way of giving out suddenly in individuals and nations, just when its mixture of light-hearted scepticism and steady hopefulness would seem most needed. By which inadequacy of its supply I am confirmed in the suspicion that Intelligence is a much more recent human accomplishment than the Past's other achievements in art and poetry and wit and humour would lead us to expect. Indeed, the deserved prestige of that Past, and the consequent survival of its educational and religious traditions and institutions, is very likely what has so strangely delayed the advent of modern Intelligence. And it is their dwindling, itself partly attributable to nascent Intelligence, which has delivered our intellectual activities from sundry blind alleys and sloughs of despond, like those presided over by the terrible, and most unintelligent, word *Salvation*. For Intelligence, one of whose virtues is abolishing Fear, is itself stifled by the obsession of danger in this world or the next: are we not seeing the most naturally intelligent of all countries fallen into incredible self-defeating stupidity through its present mania for "Security"? It is, of course, evident that, apart from the decay of religious and classical superstition, the growth of Intelligence in our own days has been enormously fostered by increase both of scientific knowledge and of civic liberty, and also, as far as it goes, of well-being; also of opportunities for variety of impression at least for the well-to-do classes. And one hopes there may be other kinds of yet unforeseen novelties coming to Intelligence's assistance in the Future. Yet, the chief obstacles once removed, my hope is chiefly in what one might call Intelligence's own natural proliferation. One intelligent mode of thought inevitably leads to another, and puts out of action an unintelligent one. Every intelligent book adds, let us hope, to the

34 Italian philosopher noted for his advocacy of Peircian pragmatism.

intelligence of at least one reader; so that we could almost do without the tremendous launchings-forth of the great challengers, Ruskin, Tolstoi, Ibsen, Nietzsche; even of the more purely beneficent (because lighter-hearted) stirrers-up of thought like Bernard Shaw.

Believing, therefore, in such spontaneous multiplication of Intelligence, I do not find much use for the methods, whereof that one is but an extreme example which is attributed to the late Dr. Metchnikoff, proposing (it is said) to increase the output of genius by judicious doses of syphilitic virus.[35] Surely, the supply of raw genius would be fairly adequate if only we could put it to the best use? And the best use of genius is not, in my opinion, the practical application of science to methods of reciprocal slaughter and devastation. Nor even its application to easier locomotion, intercourse and the cheapening of food, heat and light, except to the extent which would secure more health, more leisure and more opportunity all round; certainly at present not the case. And, speaking for myself, the best use of genius and the most necessary application of science, seems to be teaching people what Descartes called (whatever his precise meaning) the *méthode de bien conduire son esprit*;[36] at least to the slight degree of not letting obsolete shibboleths and new-fangled catch-words carry us, as they did ten years ago, headlong into the disasters we were trying to avoid.

The next best thing to be done with our existing supply of genius might be to train it to check, by application of common sense and a little modesty, certain ailments inherent in its own constitution, namely: exaggeration, contrariness and, of course, megalomania. Such self-purification, on the part of persons of genius or what passes for such, would save the rest of us the disheartening task of throwing half of their sayings onto the scrap-heap, and of picking out of the scrap-heap some of the sayings of rival persons of genius consigned thereunto at their bidding. There is quite enough to be done in the way of selection, assimilation and elimination (since all understanding means that) without setting such Intelligence as we have to play scavenger to wasteful or slatternly genius… And yet, and yet… May not those trashinesses of genius, and the scavengering entailed thereby, be that which secures to Intelligence its highest activities, and in so far fulfils one of Genius's chief missions? For, after all, Intelligence is the living, changing mass of unprofessional thought, the averaged, habitual thought of the majority of us. And is not the chief use of all such genius as is not set aside in science or in art, rather to make the rest of us think, than to furnish us with ready-made thoughts, however true or sublime? Nay, I would hazard the supposition that it was because the men of the Past

35 Ilya or Élie Metchnikoff (1845–1916) was a zoologist known for his pioneering work on immunology (for which he shared a Nobel Prize), aging, and probiotics.

36 René Descartes published his *Discours de la Méthode Pour bien conduire sa raison, et chercher la vérité dans les sciences* ('Discourse on the Method of Rightly Conducting One's Reason and of Seeking Truth in the Sciences') in 1637. But it was Buffon who argued sceptically in the first volume of his *Histoire Naturelle* (1749), 52, that '*la méthode de bien conduire son esprit dans les sciences est encore à trouver*' ('the method of rightly conducting ones mind in the sciences is still to be found').

were presented with such a mass of ready-made thoughts, creeds, philosophies, and moral formulae (think of *Deuteronomy*!) all given for perfect and definitively valid, that there did not appear till so late in the day just what I have called *Intelligence*, which alone could give that without which the greatest genius is solitary and barren: an audience, a reader, a mind able to carry on the thinking and, in so far, able to eliminate the deciduous, the rubbishy elements of the thought already offered to it.

The consideration of this loss which mankind may have suffered through the notion that thoughts must be accepted ready-made rather than transformed into one's own, and that, consequently, men must be set apart to do the thinking (as the priests once did the praying and sacrificing) for others; this consideration is at the bottom of my present hatred of the idea (which I shared in my youth) that exceptionally gifted persons should consecrate themselves into a caste, ministers, maybe, of a new religion, chosen vessels for a new-fangled deity.

And since we are discussing the reciprocal uses of Genius and of mere Intelligence (Genius ever stimulating Intelligence, Intelligence ever keeping Genius within the bounds of sense and of decency), allow me to set my face against all those oligarchies of genius and virtue which every Utopian philosopher from Plato to Comte, from Renan to H. G. Wells, has wanted to foist on dull, driven Mankind. And let us take to heart the *reductio ad absurdum* of all such schemes in the latest and hugest joke of our one great laughing philosopher, when he shows us the world governed by bald, toothless and passionless sages, who, even without having reached the years of Methuselah, have long survived the age when every decent person should retire to his or her coffin.[37]

And before leaving the subject of the services (sometimes scarcely desired) which Genius may need at the humble hands of Intelligence, let me point out how our men of genius or thereabouts ("Creative Intellects" is the official expression) have latterly taken over one of the most remunerative and mischievous employments of all obsolete priesthoods, to wit: of frightening believers with bogies of their manufacture. For, just as the Torments of Hell and the Pitfalls of Sin formerly supplied matter for all the learning and eloquence of centuries of Divines, so nowadays encyclopædic science and journalistic emphasis are being applied to making our flesh creep with prophecies of Perils. There is Peril from black, brown, yellow races; from Semites, Mongols, Latins (in "Nordic" countries), Teutons (in Latin countries), Celts all over Anglo-Saxondom, Jews throughout the globe; Bolsheviks, Fascists and Junkers, International Communism and International Finance, Militarism which was put an end to by the War, besides our old friends Jesuits and Freemasons. There is Peril from the multiplication of Idiots and the multiplication of Supermen; Peril from depopulation and Peril from overpopulation, from unsexed women and over-sexed women; Peril from over-much altruism, and Peril from insufficient altruism. Perils which I cannot even remember, but by whose side those of War, Pestilence and Famine are, of course, too familiar to be noticed. Indeed, it is characteristic of this latter-day apocalypse that none of the prophets

37 See *The World, the Flesh and the Devil*, p. 119, n.11.

of disaster prophesied the War and the Fruits of Victory, except, if I remember correctly, Mr. H. G. Wells, who, however, once the War had been declared, enlisted at once for the Fleet Street Front and bid us unsheath the Sword of Peace for the final extermination of Militarism…[38]

Therefore, it strikes me that in view of this multifold reincarnation of the spirit of prophecy in our Men of Science and of Letters, some increase of Intelligence may well be needed to steady our nerves and allow us to recognize the real dangers of which, heaven knows, there are plenty requiring to be faced with… well, the far-too-little Intelligence already at our disposal. For, possibly because it is not "creative" (and *creation* usually implies chaos and refuse-heaps), Intelligence is especially preservative and sheltering. It is the natural purifier and tidier-up where Genius and Stupidity, disrupting and corrupting by turns, have between them played the deuce with our poor mortal heritage. And in the face of the millionfold sacrifices of self and others which Ideals and Heroisms have once again presented to our foolish admiration, I would go so far as to add that Intelligence is often more humane than Sentiment, and, oftener still, more beneficent than what we call Virtue.

From the misapplications of our Science, the exaggerations and lunacies of our Genius, and the havoc wrought by our higher instincts, we therefore need to be saved, not by Reason, which is always too long in getting under weigh, but by Intelligence, active, alacritous and ubiquitous, afraid neither of being laughed at nor of laughing at others… But even as I stammer out this old-fashioned demand for *Salvation*, the name of Proteus sounds suddenly in the ear of my spirit. How can I tell what Proteus may next be—perhaps already is—changing into? And after prating about Intelligence being one-half light-hearted scepticism and one-half steady confidence, am I going to join the mixed choir, ecstatic or growling, of prophesying optimists and pessimists?

But, though restrained by that thought of Proteus and his frequently inconsiderate metamorphoses, I should like to add a word on one question regarding the near future of Intelligence, but with the understanding that I do so not as a prophet of what may happen, so much as a witness of half a century's already accomplished changes. The question, or rather query, has doubtless occurred to some of my readers, and is as follows: Granted that Intelligence rids us of dangerous superstitions, and rids us, moreover, of the habit of superstition, which is a matter less of *what* than of *how* one believes; granted that along with lucidity Intelligence brings also intellectual equity, cleanness and dignity; granted all this, may not such gains be paid for in disproportionate loss? And may not Intelligence itself constitute a danger? Has it not already begun despoiling life of many of the shelters built by the Ages with unknowing or inspired hands? Worse still: will it not replace with its narrow and wavering lucidity those dark unquestioning instincts and aspirations, lurking ever ready in the obscure organization and the mysterious formula inherited from our remotest ancestors?

38 Soon after the war was declared Wells started writing *The War That Will End War* (1914).

To this I would answer that, so far as my observation tells me, the soul will always find some shape and some material wherein to build, or to restore, the shelters needed in its moments of weariness and sorrow, there to await the consolation which no creed seems to bring without the supreme aid of time. On the other hand, that the instinctive part of our nature, when it is truly instinctive, can surely be trusted to keep itself alive in the face of the (alas!) inevitably feeble imperatives of such new-comers as Reason and Intelligence. Moreover, that all the sciences dealing with man point to the fact that traditional commandments and, even more, physiologically transmitted tendencies, have constituted themselves as responses to changing environments and needs, so that their transformation may be expected as a result of the very movement of things which has produced them.

And, finally, I would add that, even if all this were doubtful, we must accept the risks which the coming of Intelligence may entail upon us, because (so at least appears evident to me) whatever sets-back and temporary overwhelmings it may suffer in the future, Intelligence is of such nature that, once come, it must develop, or at least bide its time and revive in as yet unforeseeable manner. And Intelligence itself must prepare us to expect that every change may mean a loss, but likewise mean an opportunity. Perhaps it may even sometimes show us how the one can turn into the other; for does not Intelligence keep an eye on Proteus?

Timotheus; or, the Future of the Theatre

Bonamy Dobrée (1925)

Headnote

The British academic Bonamy Dobrée (1891–1974) was an expert on Early Modern English Literature. He served in France and the Middle East during World War I. In 1920 he took advantage of a tuition discount offered to veterans, taking his BA from Christ's College, Cambridge in 1921. In 1925 he was appointed lecturer in London, and in 1926 Professor of English at the Egyptian University, Cairo, where he remained until 1929. In 1936 he became Professor of English Literature at the University of Leeds, where he remained until his retirement in 1955. He published books on Restoration comedy and Oxford History of English Literature on the early Eighteenth Century. He delivered the Clark Lectures at Cambridge in 1953. After retiring from Leeds he edited the Writers and their Work series of pamphlets for the British Council and the National Book League.

In *Timotheus* he borrows H. G. Wells' Time Machinery, but operates it in the manner of a satire by Jonathan Swift. Nevertheless, it is one of the most suggestive volumes about possible developments in media and the arts. Dobrée's speculation about the emotions of theater audiences being controlled by psychotropic drugs may have influenced Aldous Huxley's presentation of the immersive 'feelies' in *Brave New World*.

******

"Even the powerful mind of Dr. Johnson seemed foiled by futurity"
—BOSWELL

Prefatory Notice

This booklet is Chapter Twenty-three of a work already largely in being, but of which very little will be published in the reader's lifetime; for though the author has none of that false respect for the wishes of the dead and the privacies of contemporaries which still causes so much avoidable inconvenience in social life, that feeling of delicacy towards posterity, now so active an influence as sometimes to shrink from exposing its members even to existence, hinders his speaking fully.

DOI: 10.4324/9781032625522-26

Being obnoxious[1] to the sufferings of others, he had, in 1915, the good fortune to acquire Mr H. G. Well[s]'s Time Machine. Choosing a remote corner of our island, and building due safeguards against possible bumps in time—I will not forestall the account given in Chapter Two of his work—he arrived at the year 2,100 (O.S.[2]) with no further damage than a slight bruise on his knee caused by the shovel of an archaeologist in search of human remains thought to be of the same period as the Cro-Magnon man.

The details of how the author was greeted, conducted, honoured, and spied upon are not here to be told—much is left to the reader to infer; *verb. sap.*,[3] as we say. But there are two points the latter must bear in mind: the first that there are things the author has bound himself not to divulge, and which will never be known until they occur: the second, that he has had to rely on his memory alone. A little thought will make the reason plain: although he took copious notes, and these are in existence, they are so only in future time, and will not become available until the year 2,100. This last statement, I fear, bristles with issues, and opens up deep scientific and philosophical questions, involving on the one hand relativity and on the other vitalism, which I have neither the space nor the ability to dispose of. But very vivid in the author's mind is the remembrance of his emotion on first seeing the writing fade backwards out of his notebooks, and becoming bitterly aware that time was a reversible flux. He returned to 1920, a study of history having informed him, as far as he could unravel the evidence, that his feelings would be less lacerated in that year than seemed likely in 1915. He then settled down to write his great work, which he will give to the world piecemeal as discretion permits: (the reader has only to glance at our law reports to see that were futurity displayed, the enjoyment of life and that nice adjustment of personal desires to social duties which our time has perfected, could not exist), and I have persuaded him to allow me to make known this chapter on the theatre, which can break no bones, or even abrade the most delicate skin.

I have, indeed, taken the liberty of making some omissions in order to brevity,[4] and, I freely admit, for decency's sake; for I do not hold with the modern fashion of protesting that nothing is withheld, and forthwith teasing the reader with a series of dots or stars.

Why the book is called 'Timotheus' will be evident to those who bear in mind the name of the 'Mighty Master,' the Wagner of Alexander the Great's day, who

"Cou'd swell the Soul to Rage, or kindle soft Desire."[5]

1 Dobrée uses the word in the first sense given by the *Oxford English Dictionary*: 'Liable, subject, exposed, or open to a thing (esp. something actually or possibly harmful). (The usual sense before the 19th cent.) Now rare.'

2 [Author's note] Our Style.

3 Abbreviation of the Latin phrase *verbum sapienti sat est*: 'a word to the wise is enough.'

4 Another archaic usage, 'in order to' here meaning 'for the sake of.'

5 From Dryden's poem 'Alexander's Feast; or, the Power of Musique. An Ode, in Honour of St. Cecilia's Day. 1697', VII, lines 4–6.

Timotheus, to his breathing Flute,
And sounding Lyre,
Cou'd swell the Soul to rage, or kindle soft Desire.

Timotheus was one of Alexander's court musicians.

I: The National Theatre

Our air-taxi landed us at what I took to be the nineteenth floor, and we walked almost at once into a huge hyperboloid pit, the walls of which consisted of tiers of seats. It would hold, I gathered, some twenty thousand people, and much resembled a Roman theatre, except for the peculiar curve of the walls, and the seats continuing to the very bottom of the funnel. There was no sign of any stage, and on my questioning Fabian,[6] he pointed to the saucer-like dome which formed the roof, or lid of the building. I was afraid that to keep my eyes fixed upon this airy stage would mean ricking my neck, but I was reassured on being shown the shape of our seats. Not only were they well slanted back, but they were also provided with rests for the head, such as we are familiar with at our barbers' and dentists'; and I was told that with the body in the position proper to the chairs, our emotional apparatus lent itself most readily to suggestion.

I then asked him if the performance was to be a good one, and he replied that "The clutch was officially ranked as A2 for efficiency, but that he did not know what it was for." I was much puzzled as to his meaning until I learnt that 'clutch' was the name given to a drama of the kind about to take place, where everything was under the control of one man, the 'fairfusser'[7] as he is called, who designs the movement, the emotional sequences, the voices, and whatever else is needed. I laid myself open to much banter on the part of Ierne[8] by asking whether it was to be a tragedy or a comedy: such a crude distinction, she said, was typical of the muddle-headedness of our age, on a level with the antitheses classical-romantic, conservative-liberal, matter-mind, and even intellect-emotion we were so fond of making, and which for absurdity were only equalled by our craze for dressing men and women in different sorts of clothes. The object of a drama, Fabian enlightened me, was to summon up a given state of being, pure or complex; and once the fairfusser knew what the clutch was for, it was his business to produce the right emotion. I began to speak of emotion for its own sake, but Ierne hurriedly checked me, saying that I would shock anyone who might overhear, for there was no biological value in emotion for its own sake. This made me think less agreeably of her kindness to me on the last evening.

I was therefore still confused by their speaking of what a clutch was 'for', as though it might be a sort of charity matinee, and was about to put the question, when the theatre became pitch dark: the clutch was beginning.

At first I was aware only that the roof, or 'stage' had become luminous, the light varying in strength, as it does on the ceiling of a room when clouds travel across the sun. Soon it became more steadily bright, and vague human figures began to take shape on it, shadows at first, some of enormous size, advancing and retreating,

6 [Author's note] The author's general guide—Vergil to his Dante.

7 This obscure word is possibly Dobrée's coinage. It may refer to a hustler trying to rustle up excitement at a fair; with perhaps a suggestion of the German *Vierfüssler*, or 'quadruped.'

8 Ierne is a rendition in Ancient Greek of the mother goddess of Ireland, Ériu. [Author's note] The author's guide in the more intimate social relationships.

making wide gestures of an import I could not grasp. Sometimes the shadows would assume solid shape and stand up as live beings, seeming to detach themselves from the dome so as not to appear in the least like those extravagant persons who populate the ceilings of many of our own theatres: and among them was one singularly graceful form which seemed to dominate the rest, and whose motions I could not help following, so great was the pleasure they gave me. Soon I became conscious that the air of the theatre was pulsating in a manner which never quite became sound, and in a definite rhythm, which varied occasionally, but yet seemed to conform to the original beat, much as a poet will modulate his verse. Now a faint perfume hit the sense, while an uneasy feeling stole over me, as if something had been done I did not want. Then, from the body of the theatre, as from a member of the audience, a voice spoke, in the tones of a man resigned to grief.

> No means at all to hide
> Man from himself can find:
> No way to start aside.
> Out of the hell of mind.[9]

And I felt myself sinking into such an agony of despair as I can remember having gone through only in dreams, or under the influence of supernatural fear. Struggle as I might against the weight of oppression, I was forced to abandon myself to the flow of dire tribulation, in which remorse succeeded terror, and all the passions of the world were black. And from all around the theatre, now from here, now from there, above me and below me, sometimes in front and sometimes at my back, I could hear voices and the noise of approaching events. Once I thought a voice cried out:

> Desolate, as she is desolate, in ruined cities, and when the sun has gone down to his rest.[10]

And in the midst of a tumult of pulsations and perfumes and shadowy occurrences, a woman whispered, it seemed close by my ear:

> And Pity, like a naked, new-born babe.[11]

At that the sense of intolerable woe lightened; the rhythm changed, the figures appeared human and brave, while joy seemed to issue from the very walls of the theatre with the words:

> Love's banners on the battlements of song,[12]

9 The first four lines of the second stanza of the Chorus at the end of Act I of Samuel Daniel's play *The Tragedy of Cleopatra* (1594); a source for Shakespeare's *Antony and Cleopatra.*

10 In 'Levana and Our Ladies of Sorrow,' De Quincey describes three sisters. This is said of the second, 'Our Lady of Sighs': *Suspiria de Profundis* (1845).

11 Shakespeare, *Macbeth*, I.vii.21.

12 Keats, *Endymion*, Book II, line 41, where it reads: 'Love's standard on the battlements of song.'

Which trickled from every side. At last, without warning, in a triumphant burst of sudden glory such as makes us laugh with active lungs, a loud but harmonious cry resounded from the very middle of the theatre, where there was nothing visible but empty air, calling:

Where are the eagles and the trumpets?[13]

And I remembered no more till we found ourselves perched on the outer landing of the theatre waiting for our taxi to take us home.

It was then that I found myself prey to strange and mingled, but insistent emotions, partly of kindly generosity, and partly of self-sacrifice. Looking at the men and women around me I could see that they too were strongly moved, making gestures foreign to the occasion, such as taking out their pocket-books, searching in them feverishly, and doing sums on slips of paper. Some whom I could see were giving themselves up to despair, and others were arguing with their wives. Fabian then pointed out to me that most of the carriages taking people away from the theatre, instead of flying in all directions, made for a building upon which was written large

---

SUBSCRIBE HERE FOR THE EURBANK LOAN. OFFICES OPEN

That, he said, accounted for the clutch. There was a crisis, he continued, in the bank upon which the credit of the League of Europe was founded, and the governments were anxious to sell the scrip of the new loan. The clutch we had seen had, no doubt, been performed that afternoon in the larger towns all over the continent, the language alone being suitably varied; and by this means the bank would be placed on a firm footing once more. My emotion was damped on learning this, for after all, I could have little interest in the finances of a country in which I had no stake: but enough of my feeling was left to make me give a foolishly large tip to the driver of our machine.

I was naturally curious to know by what means the frame of mind had been aroused, and in the evening Fabian was kind enough to enlighten me, going very learnedly into the origins of the form which met with such success. I was very surprised, and not a little proud, to find that a large part of the science had had its starting point in our own day, as he showed from several old books: but he on his part seemed inclined to think we had been wanting in genius to have had so much knowledge to hand, and yet not have been able to use it.

The shape of the theatre had been chosen for acoustic reasons, on account of certain properties of the hyperbola, which I had not the mathematics to understand, but which, Fabian said, had been utilised in the third (1914–1918) of the five great wars of European settlement, for finding out by their report, the exact post of hidden guns.[14] It was this which had enabled the fairfusser to make the last

13 T. S. Eliot, 'A Cooking Egg' (1919), line 29.

14 See *Wireless Possibilities*, 63–4 and *Automaton*, 159, n.7, on 'sound ranging.'

cry seem to come from the void; the other speeches had merely been delivered by variously placed loud-speakers, connected in due turn with a wireless gramophone. I may here say that the phrases I have remembered and written down are only a very small number of those used in the performance, and which, for some reason, seemed familiar. The other words spoken in the clutch were of like great emotive power, chosen or invented by the fairfusser for this reason alone: and though they may seem to have no logical thread, or connection in real life, their place in the scheme was very carefully thought out. The reasons, and the terminology, for all this were too far advanced for me to be able to hold them in my head, but I have since traced some passages Fabian showed me as early sources of the form, and which will give the reader some idea of the great cleverness of the design.

> Thus the indirect methods of hypnotising, like many of the technical procedures used in making jokes, have the effect of checking certain distributions of mental energy which would interfere with the course of events in the unconscious, and they lead eventually to the same result as the direct methods of influence by means of staring or stroking.[15]

From there the high road is plain to see; the phrases of the clutch check or loosen 'certain distributions of mental energy,' for art is only a kind of hypnotism: but the perfection which I had 'felt' had not been arrived at without much arduous trial. At one time jumbled up words had been tried, or single ones, but even the most striking, such as *death,* or *beauty,* or *ruin,* had not had an effect at all to be put beside that of the shortest sentence. Familiar quotations had also been made use of, but they were put by for two reasons. The first was that all men did not respond in the same way, since all men are not equally noble, some even finding risible "Tears, idle tears, I know what they mean."[16] The second was that hardly any quotations were familiar enough to be known by everybody: for example, the words "Till the conversion of the Jews" moved people quite unevenly, some connecting them with religion, others with their pass-books, and a few with an old obscure poem.[17] This, besides preventing many from entering into the proper mood, destroyed that singleness in the audience without which the highest suggestible state cannot be reached: for an emotion is infectious only if the units of the crowd are ready to agree together, as I have often noticed on first nights when the friends of an author try to sweep the critics away on a tide of noisy enthusiasm. And further, emotion caught on the wing is always stronger than when it is the result of deliberate thought.

As to the shadows on the 'stage,' these were for fixing the attention of all upon the same thing; and I discovered that every member of the audience had been

15 [Author's note] Freud, *Group Psychology and the Analysis of the Ego*, page 97.

16 A song in Tennyson's *The Princess* (1847), nearly. The laughter perhaps because the 'not' has dropped out of 'I know not,' meaning the speaker seems to know not what they mean.

17 It is from Andrew Marvell's 'To his Coy Mistress.'

greatly drawn towards the figure which had seized upon my imagination, and had to some extent made himself one with it, as we now do sometimes with the hero of a play. This had served to transform the loose 'herd' into a unified and thus suggestible 'horde,' if I do not mistake the terms.

The air being made to throb was merely to create a rhythm, the effects of which had been keenly studied. Again to copy a passage I have traced:

> Among the results of rhythm susceptibility and vivacity of emotion, limitations of the field of attention, marked differences in the incidence of belief feelings closely anal[o]gous to those which alcohol and nitrous oxide can induce . . . may be noted.[18]

and I can willingly believe this, for I have myself often felt very curiously stirred when listening to the jazz-band at young people's parties.

The naming of nitrous oxide, or laughing gas as it used to be called, brings me to the perfumes, which, I learnt, were led along each row of seats by what I had taken for hot-water pipes. This again, Fabian said, was a legacy of the third (1914–1918) Great War of European Settlement, and he gave me to read an account I have since recovered of a gas which caused "the most appalling mental distress and misery."[19] Of course the means had been much refined, and the fairfusser could at will set free gases which brought about sorrow, fear, joy, shame, the love of glory or of animals, and indeed any emotion, all without the least risk of harm; though it is true that some serious mishaps, especially in the early stages, had unluckily happened.

The combined result was that almost any feeling, and any required degree of that feeling, could be produced by the fairfusser, and this the government found of the greatest use at times of political or European crisis, when wars were to be declared or averted, or any controversial measure passed.

I was bound to utter my high admiration of the lengths to which the art of the drama had been carried, and made so salutary an influence, though I could not help doubting whether such a tool in the hands of rulers might not be a little dangerous: but I was assured that this had already been foreseen, and that the national theatres were closed during the period of a general election, and of debates of high moment, such as those on the budget.

I asked if there were no theatres in which human beings came upon the stage and strutted and talked after the manner of common life, as they do today, and I was told that there were many kinds: but that before going to see them I would be taken to the Dramatic Academy, which had been handsomely endowed by an Anglo-Caucasian millionaire. I thought I should learn more of the trend of the art by going there than by attendance at a number of theatres, and gladly consented to the proposal.

18 [Author's note] I. A. Richards, *The Principles of Literary Criticism,* page 143.
19 [Author's note] J. B. S. Haldane, *Callinicus.*

II: The Dramatic Academy

This academy is not an entire single building, but a continuation of several houses on both sides of a street, which, growing waste (owing to changes in fashion), was purchased and applied to that use. I was received very kindly by the warden, and went for many days to the academy.

This was composed in three parts; one for research professors, another for play-makers, or fairfussers, and the third for students; the first being trained psychologists, the last, young men and women remarkable for beauty, fine feeling and intelligence, as they are in our own day. The notion was for the professors to find out facts, for the fairfussers to apply them, and for the students to carry them out: and if it may occur to the reader that the workers in one side paid small heed to the discoveries of the rest, a little consideration will show that this is all to the good; for slavishly to accept the opinions of others can never lead to clarity of thought, and the warden was anxious to maintain in every inmate that active spirit of self-reliance without which no advance can be made in any of the sciences. I saw many of each division, but to write of them all would be to take up inordinate space in these memorials, since the arts are not of large importance to the state or to the public. I shall, therefore, confine myself to describing one or two of each kind, choosing those which throw most light on the methods of those days, and the great progress we can shortly expect.

My first visit was to a small, active professor, with a tiny clean-shaven chin, and unusually bright weasel eyes, who, speaking very fast, and with vivid gesture, easily convinced me of the usefulness of his discoveries. He had applied his mind for fifteen years to proving that there was no such thing as thought, for, he said, the conversation he had to listen to, or the acts he was able to observe, could be accounted for without supposing such a thing existed.[20] Speech, he explained, was merely a habit, like that of scratching when something causes our skin to itch; and, moreover, all our deeds were like scratching, which we do thoughtlessly, however much we may flatter ourselves that we prepare great things far ahead. He did not go to the lengths of some of his colleagues, who denied the existence of consciousness: for, so as not to be too positive, he preferred to regard 'awareness of the emotions' as a fiction convenient to his purpose, or, to put it differently, as words merely to describe a sequence of events. By his system, when impulses are set up, something occurs, such as eating, and we are 'satisfied,' as we say: or we are 'disappointed,' as when prevented from a kindly action.

I paid close heed to this part of his discourse, because he begged me not to confuse his theory with that of a rival professor, who believed that allusive gestures gave rise to a corresponding emotion. This, he pointed out, was merely the out-of-date heresy that "A dog does not wag his tail because he is happy; but that he is happy because his tail wags." This rival believed that if he made the gestures which usually go with certain emotions, he would undergo these feelings; and the spectators would, by 'in-feeling' or 'empathy' (such were his barbarous terms) put

20 Dobrée's satire of Behaviorists, who, following J. B. Watson, rejected the study of consciousness.

their muscles in readiness to go through these movements, and so, in their turn, experience these emotions.[21] For his part, when he had been to see this professor acting woe, far from feeling unhappy, he had barely been able to master his mirth. But, so as to give the theory a fair trial, he had wished with one of his pupils to observe the result of moving the tail of a dog quickly from side to side, but that unluckily, the pupil had been bitten before the experiment had reached a stage from which anything could be learnt.

His own idea, he informed me, was to set up a known order of stresses and strains in the watchers' nerves, and this could be done by cunning movements performed in front of them. I was not, however, to tax him with inconsistency, for his motto was 'Not miming but movement,' and though an actor had to use gestures, they must by no means be after the naif manner of his rival. He himself had spent over four years in the abstract study of the movements proper to the passion of benevolence—abstract, because nothing is more misleading than what people relate of their own feelings: cruelty, for instance, they often describe as a wish to better their neighbours. Indeed, another of his mottoes, he declared chuckling, was 'No motion, no emotion'; and I could not but agree with him when I considered that after all, to lie in one's bed all day and simply 'think' as we stupidly call it, is no life at all.

Yet I could not help trying to argue, feebly enough, that the showing of dreadful acts would call forth feelings of horror, as we may judge from the murder of Desdemona, and that fitting words would arouse pity in us, as they do in the same play. But, smiling at the clumsiness of my example, he remarked that just there lay the error; for if we were truly to see a Moor, however splendid, plunging a knife into the body of a beautiful lady, our emotions, if we knew her to be guiltless, would be very different from those we feel in a theatre; and that, instead of sitting still, we should most certainly interpose, or run to fetch the police. Again, while we were children, before our minds are distorted by what we erroneously call thought (but which is only an idle luxury-habit), Mr Punch beating his wife causes us to laugh very loudly. This instance, he said, would also prove how much more useful movements were than words, since for ten who would laugh at a Punch and Judy show, hardly one would smile at the wittiest things in Pascal. It was only on reflection that I could altogether make his views my own, for it is not easy for us to give up opinions we have held ever since we can remember.

To test the movements appropriate to various emotions, this professor had invented a machine, which, by reason of the changes in electrical resistance a body

21 The German philosopher Theodor Lipps famously based his theory of aesthetics on the concept of *Einfühlung* ('in-feeling'), which was translated as 'empathy' by Edward Titchener, *Lectures on the Experimental Psychology of Thought-Processes* (New York: MacMillan, 1909), 21: 'Not only do I see gravity and modesty and pride and courtesy and stateliness, but I feel or act them in the mind's muscles. This is, I suppose, a simple case of empathy, if we may coin that term as a rendering of *Einfühlung*.' See Remy Debes, 'From *Einfühlung* to Empathy: Sympathy in Early Phenomenology and Psychology', in Eric Schliesser, ed., *Sympathy: A History* (Oxford: Oxford University Press, 2017), Chapter 9.

undergoes under the action of the passions, recorded the feelings of any person subjected to it. This machine he had just brought to perfection, and to give me a demonstration, he sat me in a chair made of some amalgam unknown to me, and fitted with sockets into which my head, hands and feet were clamped. A piece of wireless apparatus, supplied with a diaphragm such as make a part of our telephone receivers, was placed over my heart, and, my loins being bared, my lumbar vertebrae were played upon by a peculiar ray. Above my head, where I could not see it, was placed a marker, much like our telegraph morse-code dials, but corrugated and rayed after the manner of a fan. The professor could watch this while evolving before me the strange movements I could connect with nothing I had ever seen, and so could vary his gestures according to the results shown. After about a quarter of an hour of erudite passes, the professor, wiping the sweat from his brow, triumphantly announced the successful issue of his experiment, and asked me what emotion I felt. At that moment, having overcome my awe, I was filled with a profound sense of pity, and on my confessing this, the professor danced with glee. Crying "Typical! Typical!" he pointed at the dial; but as the needle showed 'Lust tempered with Sentimentality' I could not but feel that his wonderful invention needed alteration in a few details to perfect it. Nevertheless, while doing up my braces, I framed a few remarks to make known my pleasure at seeing the drama make such strides in his hands; and promising to meet at a later day we parted with many expressions of esteem. I must also add this tribute to his ability: when I did visit him again, very fast, on my backward journey through time, even when all the motions were reversed I once more felt very deeply the compassion his gestures had provoked.

I next went to see a fairfusser, though not one in the service of the government, and was much impressed by his freedom from doubt as to the way in which the best result was to be reached. He said, with justice, that the artist's desire was to communicate with his fellow creatures, and that the object of an actor was to place his soul in touch with that of others. Man, being in each case a unique individual, was fitted for work more noble than that of a mere interpreter, or conduit pipe from an author to an audience, and the contrary view had been the grand error of all producers from the time of Shakespeare almost to his own day. The aim of an actor was to express himself (as a part of universal nature) and reveal a cup overbrimming with passions. Any ideas introduced by an author were to be deprecated, for his business was strictly to provide the raw material; and so the teacher's main efforts were to be directed towards training his pupils to rid the authors' words of any meaning they might contain, simply by the manner of speaking them. In this way nothing was allowed to come between the actor and the audience. This he claimed to be the especial discovery of his age, and one of which he could not help regarding with more than a little pride.

I made bold to tell him that his notion was not so new as he imagined, and that we too had actors who disbelieved that words had any plausible meaning apart from the emotion the actor could register through them: men (as a rule well set up, or even bulky, since these are always the most passionate) who by a clever alteration in stress, or an abrupt cleavage of a sentence in the middle, could effectively cancel any extraneous idea the words of an author might interpose between the feelings of the player and the minds of the audience. This constructor was good enough to say that he was quite sure our age had not been so dark as was commonly supposed, but that, at least in our classical plays, which had been in verse, a form which compels a certain manner of speech, he thought the 'pure' actor must have met with difficulties hardly to be overcome. I was able to assure him that this was not so, and that, indeed, it was just in these very plays many of our actors had shown their highest genius; that one might know *Hamlet,* for example, quite well by the book, and yet go to two or three versions of the play and hardly recognise any of the speeches, so much were they heightened and made subtle in the speaking of them.

I was also much taken by an investigator who had made a highly diverting play simply with scenery, and a few mutes who now and again varied their place. It was his view that we had always been astray in making people the centre of our dramas: it was their surroundings that mattered—for who, he said, given the choice of seeing Brown eat his dinner, or a thunderstorm on Mount Everest, would not prefer to look at Mount Everest? A modern producer could not help laughing at the remark of Aristotle, if he ever read it, that 'the spectacle was the least artistic part of a drama.'[22] A comely staircase, he averred, or even a rickety ladder if it was tall enough, had more significance than a tale of hopeless love; and he was about to design a series of scenes in a logical order of forms and colours, green following pink, which would make a spectator sadder than even a play by Sophocles. This I could well believe: but I found it hard to understand how it was right to allude to pink as though it were a premiss, for after all, nobody dreams of calling a rainbow a syllogism, any more than they do of saying 'paradox' when they mean a hill.

On another day I was taken to see a fairfusser at work on an old-fashioned play which was to be 'acted,' in our sense of the word, by students. He made his actors rehearse a scene; and then all sat down on chairs and took up stereoscopic glasses. Immediately, at the other end of the room, two coloured films appeared, exactly reproducing the movements of the actors, while at the same time a gramophone repeated the words they had uttered, in such a manner as to seem to come from the mouth of each actor who spoke. With the glasses the illusion was complete, and I could hardly believe I was not re-dreaming the scene I had just witnessed, except that the producer could stop the play at will, or even go back to a phrase or gesture to point out the errors of voice or movement of which the actors had been guilty. He could also show how a gesture would be more effective if performed at a greater or less speed; and how admirable this method was I could judge from the looks of pleasure or mortification on the actors' faces as they saw themselves displayed.

There was one handsome young actor who seemed by his vehemence and assurance to be more talented than the rest, and to him I asked to be introduced, that I might learn his views from him. He led me aside, and with great earnestness explained to me how experience had shown that one could not take for granted the least intelligence in an audience. Words, he said, conveyed nothing to them unless

22 Aristotle, *Poetics*, VI, 6 .1450 ^{b} 16–20. trans. S. H. Butcher (London: Macmillan, 1895): 'The Spectacle, though an attraction, is the least artistic of all the parts, and has least to do with the art of poetry.'

accompanied with appropriate action: and this he ascribed to the fact that an audience was a crowd, and therefore followed the normal law of mass psychology in being much stupider and more primitive than a single person. The actor, therefore, had to deal with the simplest objects or ideas, indicating them by a kind of airy drawing. The connection between them, the grammar or the syntax as it were (so he was kind enough to phrase it for my understanding) was portrayed by the actors' emotion as expressed in gesture or tone. That was why plays with few words were better than plays with many words, as in the latter case the number of gestures became very tiring both to the eyes of the audience and the muscles of the actors. He had a noble, yet reasoned, scorn for any player who stood still and with hardly a movement allowed sentences merely to trundle out of his mouth, and he considered his place could very well be taken by a gramophone.

To illustrate the stages of his art, he took me to a room to see a young actress practise an easy passage, and I was much gratified by the manner in which she expressed by gesture the meaning of the words she was uttering; and I could not but admire the subtle difference she made in pointing to the floor when she said in one case 'down 'meaning merely downstairs, in another the infernal regions. The same variation was introduced in her rendering of 'up,' and I did not fail to note that each gesture emphasised a new beauty in her arms. She also practised some 'tone-work' as they call it, and for my benefit declaimed an old-fashioned line "To lie in cold obstruction and to rot,"[23] and it is hard to imagine, as it is impossible to describe, the frigidity she put into the word 'cold,' or the horror and loathing with which she vivified the word 'rot,' so making their meaning quite clear to any audience.

My actor friend afterwards told me that she had sadly bungled the word 'obstruction' because she had not yet reached the year in which abstract terms were studied; and being himself nearly at the conclusion of that period, he gave me a finished version of the line "The quality of mercy is not strained,"[24] which quite transported me, and which I should not know how to praise sufficiently. The gesture for 'quality,' initially too simple, was raised to a high state of complexity by the young man's genius, and I should no doubt have understood it perfectly had I been more used to the method.

I later asked the young woman if such interpretation did not involve work almost too arduous, since nothing is more tiring than to bring one's ideas to the level of common minds; and she told me that though the intellectual labour was harder than in any other profession, their task was lessened by the fact that so few authors had any idea of what they really meant that the actors could substitute such phrases as lent themselves more readily to their temperament.

There was another actor, with a mobile mouth and masterful manner, whom I saw practising for his thirty-third performance of a part, and therefore engaged in working out a thirty-third reading. I was amazed at this, which is so contrary to our own method, but was soon persuaded of its rightness. For a work of art, this actor

23 Shakespeare, *Measure for Measure*, III.i.134.

24 Shakespeare, *The Merchant of Venice*, IV.i.184.

said, did not exist, apart f[ro]m the observer—it was a collaboration; and as no observer was ever twice in the same mood, he could never experience the same sensation from the identical thing. One might say, to adapt the words of an old Greek sophist, "No man can go to the same play twice."[25] He then went on to argue very brilliantly that since the actor made the work of art, was indeed himself a piece of it, his share of the collaboration was to make it on each occasion as different as possible from the last, so as to help any observer who might come more than once to any play. (He knew several ladies who had been to see him no less than seventeen times in the same character). This also had the extra advantage of avoiding that dull monotony—for what is art without an element of surprise?—so often to be seen in our actors, who think they have achieved a final rendering, and attempt day after day to repeat a thing which can never really occur even twice.

I was much satisfied at what I had seen and learnt at the Academy, but was made slightly melancholy by the thought that if ever I should return to my own time, I should find our actors and actresses much below the level of what I had come to expect from their calling.

III: Other Theatres

The National theatre which I have described was not, of course the only kind, though it had many imitators; and I shall now pass to some others, beginning with that which I think will most interest readers of the present day, but which was rarely mentioned in the polite society of the age. This was the Cathartic Theatre, where people went to be cured of the passion of love. In our day, as has been for many generations past, we often refer to love-sickness, but it is half in jest, and there are few of us who do not think the undoubted pains of the state amply repaid with its joys. Its dangers, however, now only beginning to be recognised, were fully taken into account in 2,100, for it was seen that the claims of society were incompatible with an emotion then relegated to the songs of derivative poets. Already we know that the battle between the self and the ideal social self gives rise to the most frightful diseases, but in our day we only try to cure the unhealthy symptom instead of going to the root of the matter and abolishing the cause. At this time, though the malady was well in hand, its approaches were so insidious that patients going to a doctor for what they thought was one sickness would often be surprised by a diagnosis which convicted them of love, and would later be seen entering the Cathartic Theatre with shamed faces or that air of studied indifference we assume when we do not wish to be noticed, thinking that by this method we shall appear to have no face at all.

The theory of this theatre was very simple; and this too, I was proud to think, had been foreshadowed by writers of our own time, one of whom had written, "We have lost the orgy, but in its place we have art,"[26] and another "Poetry acts

25 The pre-Socratic philosopher Heraclitus is thought to have said that 'No man ever steps in the same river twice, for it's not the same river and he's not the same man.'

26 [Author's note] Havelock Ellis, *Affirmations.*

as a physician."[27] The aim of the performance was to break down the obstacles we wrongly oppose to our thoughts in rude attempts to fit ourselves to social life, and so allow to drain away those impulses which in any really harmonious nature should never be set up. I must warn the reader of this age not to confuse the method with one of 'sublimation' as we say, for this involves the 'will,' a fiction of which the futility had long been exposed. It is true that I did not very clearly grasp these matters, which are too far beyond our time—just as Dryden would, perhaps, have found it hard to grasp the true subtleties of Expressionism—but I hope I do not err in saying that an element of vicarious fulfilment also entered into these dramas, on the ground, which no one will contradict, that it is the function of art to provide what everyday life denies us. The name of the theatre, I found indeed, had arisen from that reading of Aristotle which confuses the meaning of Catharsis with that of 'purgation by excess': for even in those days the Faculty of Med[i]cine was not always happy in the names it chose for the ills and remedies it invented.[28]

The authorities were, of course, fully aware of the risks run in using such a specific, and its abuse was hindered in the same way as we curtail the buying of opium and other drugs, which wrongly applied prove harmful; and a man was only able to buy a ticket if he showed a paper signed by a doctor to declare him a fitting subject. Medical men were themselves allowed to buy as many tickets as they liked, and a large percentage of the audience was always composed of them, because they wished to observe the effects of the cure upon their patients. It was not without difficulty that Fabian was able to get us the needful pass, and it was only after representing to the authorities that any account of their age which omitted so beneficial a device would be very faulty, that I was granted it.

Before entering the theatre Fabian told me to abandon myself freely to any impulse to laugh, as that was a condition of perfect purging: and of this I was very glad, for I have often, not only before, but since my visit to this time, felt the pain of constraint at light plays, especially if I was with relatives, or else friends with whom I did not wish to become too intimate. But what was my surprise to find, instead of the fescennine[29] jesting I was prepared for, a play I had already seen in Manchester, a drama of the most correct sentiment by one of our notably respectable, even titled playwrights, which might call forth smiles and tears alternately, but not those crude outbursts of mirth I now heard on every side. So little was I able to enter into the spirit of the thing that after the first act Fabian took me out, whispering to me that my callous behaviour might have the worst effect upon the patients seated near me. He seemed to think my bearing had been of set purpose, and only grudgingly gave the explanation I longed for, which was that the people in these times saw a wealth of allusion lurking beneath the innocent phrases; and that what the audience so much relished and admired in our author was the simplicity with which he had hidden the 'latent content' under the 'manifest.'[30] When I pro-

27 [Author's note] Robert Graves, *Poetic Unreason.*

28 Aristotle, in his *Poetics*, 1449b, says the effect of tragedy on its audience is a catharsis (purgation) of the emotions, especially pity and terror.

29 Scurrilous, obscene.

30 Freud's terms in *The Interpretation of Dreams*.

tested my belief that nothing had been more remote from the writer's mind, Fabian looked coldly at me, as though he were sure I was trying to dupe him.

It was in vain that I pleaded with him to be allowed to attend another of these plays, one of later date, for Fabian would by no means recommend me for a pass, saying that the sense would certainly be beyond me. There was, however, another level of play of the same nature, designed for the stupider sort of people, such as members of Parliament, wardens of libraries, teachers in science or religion at the Public Schools, municipal architects and so on, for which, not without a contemptuous word, he recommended me. He excused his own absence however; for even, he said, if he could obtain permission, which he doubted, he would not care to go. And indeed, after I had seen the piece, I could not blame him. For determined as I was, this time, to behave with propriety, I allowed myself the license of which we are all guilty sometimes, when our conscience, or 'censor' to use the modern term, is off its guard; and laughing very heartily, I have since felt ashamed at my acquiescence in a remedy that must prove so greatly worse than the disease. The least things there were such as make us keep some of our Callot[31] etchings locked in a drawer, and to leave certain portions of Mr Loeb's excellent library in the original tongue,[32] though some I am told, regard this only as an ingenious device to outwit the laziness of students. At all events, since I was not ill, the performance, I fear, did me no good: whether it would have done so in less happy conditions I am unable to tell, and I dwell upon it no longer, as in any case the experiment will not, I think, be tried in our day, even in the hospitals. Nor would I be convinced of the wisdom of the venture if it were.

In going about the streets I had often noticed, especially in the business quarters, what appeared to be shops or booths, not unlike those places one may see abroad, where men sit in rows to have their boots polished. Above them was displayed a sign on which was written "Two Minutes," or "Thirty Seconds," or some like period of time. I had seen the backs of men standing in lines, with pads clamped over their ears and their faces pushed forward into a sort of camera, and had supposed these retreats to be telephone boxes. I was much surprised when Ierne told me they were "Hurry Theatres," and invited me to accompany her to one of them.

They were erected, as the name implies, for those without the leisure to attend longer performances, and were found very beneficial to brokers and such, who, hurrying from their offices to snatch their midday meal, could pause for the declared number of seconds, and gain, without waste of time, a modicum of 'organised emotion'[33] as Ierne called it; and this was often a great relief to them. For when we are too much troubled by our affairs we may usefully go to art for escape or refreshment. There were a few booths in the more fashionable parts, for errand

31 The French master printmaker Jacques Callot (1592–1635), known especially for his series *The Great Miseries of War*.

32 The Loeb Classical Library, then published by Heinemann in the UK, now by Harvard University Press, provides parallel texts of ancient Greek and Latin authors.

33 The phrase is used by the French psychologist Théodule-Armand Ribot (1839–1916) in his *Essai sur les passions* (1906). 'Organisation' is also a key term in I. A. Richards' *Principles of Literary Criticism* (1929). Richards was lecturing in Cambridge when Dobrée was a student.

boys, journalists, and taxi-drivers, while it was found that those in the dentists' quarter were much patronised: for in going to have our teeth seen to, if we do not like to be late, yet we shrink from entering the place until the last moment, although the waiting rooms are made homely and cheerful with time-tables, comic papers, and copies of Academy pictures. These theatres were also agreeable to those who had arranged to meet friends at a certain spot and were kept waiting, and some had even, in the early days, gained a notoriety as rendezvous.

The camera through which one looked was simply a stereoscopic glass directed on a double film screen, and the pads were the telephone receivers through which one heard the voices of the actors, which seemed to come from their mouths. The plays themselves were most dramatic in character, since their object was to endue the spectator with a highly disturbing emotion in a minimum of time. They were therefore very allusive, and I should have found it hard to understand many of them if the gestures and tones of the actors had not been profoundly striking. I grew to be fond of them and, indeed, with weakly emotional men, they readily become a vice; for when thrills are as easily obtained as cocktails, and as rapidly swallowed, if I may use the term, they form a tonic as difficult to resist as any digestive,[34] and are perhaps as harmful.

It would be useless for me to write down a typical drama, for the reader of to-day would not follow it, nor, for that matter, relish it more than he does those quaint old seventeenth-century plays where women dress up as men, and blood so freely flows. There were, indeed, a few from our own era, known as 'classics' and sometimes acted as curiosities in the neighbourhood of museums, but the earliest of these must date, I think, from at least 1940, and was a comedy with the strange title *The Psycho-Fans.*[35] A young man wishing to make a girl his bride, she puts him through a number of scientific, but comical tests, to prove his worth and his affection: she was afraid, I gathered, that he might turn out to be an 'introvert,' and not at all a suitable mate for an 'extravert' such as she was.[36] All I remember of the words is the opening of a sort of epilogue he spoke:

Oh had I wist
Before I kissed,
That you were a Behaviourist. . .

Normally, however, these dramas aimed at producing dread, and I naturally avoided one which was advertised as "Guaranteed to make your soul writhe." We are not yet made of such stern stuff as to derive courage to face the battle of life from art of this sort, though I have seen robust clerks stagger from these booths with white faces and a much increased zest for their humdrum labours. This being so, I had

34 Used in the sense of the French, *digestif*: an after-dinner drink to aid digestion, such as brandy, fortified wine or liqueurs.

35 Presumably punning on 'sychophants.'

36 C. G. Jung's terms distinguishing between inward-looking and outward-looking psychological types.

the temerity to suggest that one of these theatres might be installed in each government office for the use of civil servants, and am gratified to be able to say that my proposal was acted upon, only the Inland Revenue Department being excepted.

There was one theatre which several young people told me was the best, but as it did not meet with general favour, I did not go there until the end of my visit, for I have always felt an abhorrence of what is at all precious, and avoided the highbrow and snobbish. It was quite a mean place, not in the capital but in a small provincial town, and was regulated by a fairfusser who had never been able to make his way in a decent centre, owing to his poor skill in the art of advertisement.

The stage was much like that which we know, but though built in pleasant enough proportions was too simply decorated to be striking. The settings were so unobtrusive that at the end of a scene one could hardly say whether the framing had been good or bad, which I thought a pity, for one was in this manner robbed of a subject for conversation. The effects were obtained chiefly by the lighting, but unfortunately this was kept uniform throughout each scene, and thus one lost the pleasure of admiring the agility of the electrician, who nowadays, is, with his switches, as great a virtuoso as an organist with his stops. The most noticeable difference was the stage being only about two-thirds the size of ours, the reason for it that the actors were not people, but puppets, rather smaller than human beings.

I have always regarded these dolls as a mistake, for they must needs be anonymous, and how can one tell if the acting is good if one does not know the name of the performers? I think too, that if one is to have puppets at all, they should be either grotesque or fairy-like, and these were neither, resembling instead those early Egyptian or Indian sculptures we have so far outdistanced, or those Byzantine paintings, which, once thought beautiful, would look so oddly on the walls of Burlington House.[37] I must confess that their movements were graceful, their deftness above that of any human being, but not more so than one could imagine human beings capable of. They were actuated, not by strings, but by some invisible power, and everything they did seemed to be of such happy invention that one felt they had all nature at their command to use. But I detected a grave error in the way the fairfusser made the words issue from their lips, for they did not speak at all like actors, but simply and swiftly, as we all try to do in real life, and it is not for that we go to the theatre. Their speeches were so cadenced that they dwelt in the ear like a harmony in music, which is contrary to all experience, so that the characters did not seem like men ennobled, but, rather, fleshly embodiments of the thought or feeling it was their purpose to express. One enthusiast, eager to convert me, quoted to me the words of some foreign actress of our time:[38] "To save the Theatre, the Theatre must be destroyed, the actors and actresses must all die of the plague. They poison the air, they make art impossible," which is cruel and absurd, and in any case should not have come from an actress. For these semblances of mankind

37 The home of London's Royal Academy of Arts.

38 [Author's note] Eleonora Duse (Ed.) [Editor's note] One of the greatest actresses of her age, she was best known for acting in plays by Henrik Ibsen, and her lover Gabriele D'Annunzio. She was the first woman to appear (in 1923) on the cover of *Time* magazine.

by their remoteness banished all the accidental things which make a play realistic and warmly human, and all the personal emotion which makes us feel for an actor, and applaud him for the pain he has gone through. But we cannot feel for a puppet, or applaud him, even if he has played King Lear, for we know his sufferings were not real.

These plays were always made entirely by one man—for this fairfusser actually had one or two disciples—who directed each movement, whether of single persons or crowds, either tumultuous, or in the dances, which met with much applause, though they seemed to me even less comprehensible than some of the later Russian ballets which were lately in vogue for a short time. I was told he was always very careful about the 'rhythmic order' of the piece, whatever that may mean, and its groupings. All was done first on a little model, which in the end became a record repeated on the larger scale. For the voices he took human beings, going over and over each phrase until he got exactly the tone he wanted, and these he recorded, timing them afterwards with the movements, so that the whole play went, as it were, by clockwork. Thus there was nothing spontaneous about it, and this is a fault, since art, according to many serious philosophers, is a kind of game, and thus, surely, if any notion of being drilled creeps in, the pleasure evaporates.

The same man also, as a rule, wrote the words, which did not remain in my memory because I understood them so little, seeing that they dealt with thoughts and feelings which in our day we take small notice of. That this must be so is easy to see; for every age concerns itself with a different relation of man to what is outside him. We are now, to be sure, beginning to do what they were doing, in dealing with man's relation to man's idea of what he is; that is, so to say in the plain words of Albertus Magnus,[39] seeking "the causation of causes in the causes of things." But these plays bewildered me, as I own without shame, for would not Dr Johnson himself have been adrift at a play by Signor Pirandello? And since there was so much I could not fathom, I should only impart a false twist to the meaning however much I tried to give a true account.

But I fear these plays disordered me, for the unnatural is a sort of poison, and I have never since been able to feel real pleasure at any drama of to-day even in the best theatres in Paris, New York, or London. Indeed I have almost conceived an aversion from our stage; and it is only the importunities of my friends that make me go to a play once or twice in a year, so as not to seem unsociable. If the choice is left to me, we go to the English version of a French farce, for these are usually free from any meaning at all: and if one expects nothing, one cannot be disappointed.

39 Medieval German friar, scientist and philosopher to whom works on alchemy were falsely attributed.

The Future of Futurism[1]

John Rodker (1926)

Headnote

> The last twenty years have seen a revolution in the art of painting. This book is an attempt to see whether literature can support a similar revolution, and, if so, how can what is known as Futurism in painting be paralleled by a Futurism of literature, and will the result be a new vitality as in the case of painting? This essay answers these questions in the affirmative.

Thus the dust-jacket of the original Kegan Paul edition. Whoever wrote it evidently missed the note (n. 11 here) in which Rodker argues the term cannot be applied to literature. What he is concerned with, rather, is modernism. By 1926 Rodker (1894–1955) had already been involved with several groups of artists and writers now seen as modernist. He was one of the 'Whitechapel Boys' along with the painters Mark Gertler and David Bomberg, and the poet Isaac Rosenberg. He told Ford Madox Ford that his 1915 novel *The Good Soldier* was 'the finest French novel in the English Language.'[2] In 1919 he had set up the Ovid Press, publishing landmark works by Ezra Pound and T. S. Eliot, and portfolios of drawings by Henri Gaudier-Brzeska and the Vorticists Wyndham Lewis and Edward Wadsworth; and he was foreign editor of the American magazine, *The Little Review*, which serialized *Ulysses*. Yet the term 'modernism,' though it was used occasionally from at least 1908 to describe literature, was not used as it is now to describe a movement, a period or a style. These writers and artists were more often described as, or described themselves as, belonging to more closely defined movements such as Impressionism, Imagism, Futurism, or Vorticism, which we tend to see as contributors to a broader constellation of modernism. *The Future of Futurism* is a key document in the articulation of modernist aesthetics; and from its opening epigraph, is written in a recognizably modernist idiom, by a modernist poet.

******

1 [Author's note] From a definition of Cubism: "Mais il est incontestablement l'effort artistique qui en étant le plus important de notre époque, y a apporté le plus de confusion." ['It is incontestably the artistic movement which, being the most important of our era, has brought it most confusion'.]

2 Ford, *The Good Soldier*, ed. Max Saunders (Oxford: Oxford University Press, 2012), 5.

DOI: 10.4324/9781032625522-27

"... like the fortuitous encounter on a dissecting table of a sewing machine and an umbrella."

I: Introductory

Of Futurism so much has been said in the past fifteen years that, with a country very much divided on the subject of Mr Epstein's Rima,[3] some belated definition seems necessary.

But to call that work Futurist, as the Press does, rather than Archaic, Negro, or South American, traces of all of which influences it may be seen to bear, is to say nothing; for the misleading word does not in this case deal with the Future, but much with the Past. Indeed, primitive and savage motifs are not least important in what we know as Futurism. Thus a first confusion arises in the popular mind; the menace of continuity implied by the word is a second.

In our generation Marinetti, an Italian, first perceived the arts to be undergoing some similar change, and with certain exaggerations the poets, painters, and musicians he grouped round him exemplified such changes. What excitement was provoked by his manifestoes and exhibitions proved that there was in most places a reaction from the Romanticism of the preceding generation; that a new Romanticism of machinery, towns, violence was taking its place. War seemed its logical outcome, and, as was to be expected, became its extinguisher. But war could not entirely put back the clock, as the academics had hoped; while in some ways the revolutions accompanying war gave a wider tolerance. But Marinetti's first avatar left behind the generic term, and Futurism may now mean almost any unconventional activity in the contemporary arts, whether Vorticist, Cubist, Symbolist, Vers Librist, etc.

To-day, if writing is called Futurist, it is generally because words or ideas are found in free association. Cubism, too, whether good or bad, is called Futurism (though we hear of Braque in 1908), as is in some quarters a sort of Spirit-drawing; while in music Stravinsky has been claimed, which he cannot be if Pratella is.[4] So that for vagueness it would be difficult to find a better term. But generally the word will include all those artistic activities called revolutionary by the academies, incomprehensible by the man in the street.

That certain changes have taken place, changes seemingly peculiar to this generation, a quotation from Mr Goossens will show:

> ... The odd score of years... to the present day, have been more pregnant with revolutionary consequences to our art than any previous period in its history. Music... took on a new lease of life in the hands of...men...who... discarding

3 Stylized memorial in Hyde Park to the naturalist and writer W. H. Hudson by modernist sculptor Jacob Epstein (1880 –1959). Revealed in 1925, it was condemned as 'The Hyde Park Atrocity' by the *Daily Mail*.

4 Italian painter Attilio Pratella (1856–1949), known for impressionistic Neapolitan views.

> all the worn conventions . . . of the past, embarked on a series of experiments and innovations which . . . dealt the death-blow to old traditions...[5]

And Guillaume Apollinaire writing in 1912 said of painting:

> The secret aim of the young artists of the extreme schools is to produce pure painting. It is an entirely new plastic art. It is still in its first stages and is not yet as abstract as it would like to be. Most of the young painters work a great deal with mathematics without knowing it, but they have not yet abandoned nature, whom they patiently question so that she may teach them the way of life.... The painters have been led quite naturally and so to speak by intuition to pre-occupy themselves with possible new measures of space which in the language of modern studios has been designated briefly and altogether by the term the fourth dimension . . .

That mathematical pre-occupation has lately touched all the arts in some degree. In music Schoenberg, whom Mr Goossens fears is almost too mathematically precise to be of great creative significance, and in Literature Miss Stein, to whom we shall turn later.

But a so-called mathematical bias suddenly appearing in all the arts implies some inquiry into the properties of the units involved, and actually in the case of Futurism it was the material and abstract qualities of the constants or fundamentals of the arts, and their mutual repercussion, which, it may be assumed, some attempt was being made to record and systematize.

That was a very natural reaction, for the arts had so long wandered in the mazes of symbolism, that they had grown completely out of touch with that world which it was their function to interpret, a world which every day grew more mechanical, expressing a new fluidity in an infinity of geometric and rigid shapes. At the artist's side the scientist took seven-leagued strides. That turned him in emulation to something which would be analogous to the scientist's measuring sticks and instruments of precision, hoping to find thereby some fool-proof foundation for all his future structures.

He perceived also, and for the first time, that the centuries had accumulated about him—as though casually—a vast wealth of primitive forms and decoration. The first shock over, their rhythms were soon incorporated into his practice. But that was perhaps the last important decorative possibility to be tapped, and we imagine no such rich mine remains to be discovered. And it is possible that with the final exploiting of those influences and their final incorporation, the skeleton of the arts will be permanently fixed, though future elaborations grow never so fine.

I should therefore like now to dismiss the general problem which this title raises, and leave the arts of painting and music which are somewhat difficult to illustrate without

5 The English conductor and composer Eugene Goossens (1893–1962) had conducted the British concert premiere of Igor Stravinsky's modernist landmark *The Rite of Spring* in 1921.

examples, and confine myself to Literature about which everyone knows something, and the future of which should roughly illumine the other arts; and if we choose say A.D. 3000 as an approximate date, that will leave us plenty of room to turn round in.

II: Possible Changes in the Make-Up of Humanity

Mr Wells gives us a number of comprehensive but disturbing visions of posterity. Your libraries will turn on the speaking romances of that day, while other switches will fill the empty picture frame with any abstract arrangement of the units supplied, not omitting stage-noises, rustle of leaves, murmur of water, and battery of pastoral sounds.[6] These possibilities charm us, but if man has not by then overcome his instinct for nosing out overtones, volatilizing essences, he will find himself unhappily very much where now he finds himself, with his insuperable problems of soul, sex, etc., for we assume he will still yearn for an ultimate resolution of all his problems in some near future.

It is possible, however, that, before even so much happens, men may acquire a new sense or two, and with them new standards; but that seems unlikely, and with the best will in the world it is impossible to guess on what lines they would develop or if the arts would be thereby affected.

And, though the airman is said to be growing a new kind of mind,[7] as if like a bird, his spiritual skeleton grew hollow and could soar, and though life in submarines may cause profound organic changes, yet the expression of these activities has so far meant nothing more than a slight alteration in the hemoglobin index and additional bric-a-brac to the what-not[8] of the arts.

However, there has not been time enough for these vocations to modify deeply the minds of those engaged in them, but perhaps the mind itself is incapable of a transition commensurate with our taking to the air and the depths of the sea. Unless of course the change in atmospheric pressure stimulates some gland to renewed activity and a recrudescence of primitive impulses, in place of the highly refined ones to which we look forward.[9] So that in all this we have no ground for assuming any amelioration, and certainly not from the exactly opposite fact that, while

6 See, for example, Wells, *The Sleeper Awakes*, ed. Patrick Parrinder (London: Penguin, 2005), 56.

7 The earliest instance of the term 'air-minded' in the *OED* is from 1927; but as Rodker's remark indicates, it was current years earlier. See, for example, *New York Times* (23 October 1925), 11, in which Daniel Guggenheim is reported as wanting the nation to become more 'air-minded.'

8 'What not' can mean 'anything whatever'; but a 'what-not' is 'An article of furniture consisting of an open stand with shelves one above another, for keeping or displaying various objects, as ornaments, curiosities, books, papers, etc.'—in other words, 'bric-a-brac', or 'Miscellaneous small objects and ornaments which are typically old or antique but regarded as having little value' (*OED*).

9 Russell's *Icarus* had speculated that future societies might control individuals by regulating their hormones. See pp. 57–8. When T. S. Eliot reviewed *The Future of Futurism* (with another To-day and To-morrow book, Basil de Sélincourt's *Pomona, or the Future of English;* 1926), he wrote: 'Mr. Rodker is up-to-the-minute, if anyone is; we feel sure that he knows all about hormones [. . .]': 'Charleston, Hey! Hey!' *Nation and Athenaeum*, Vol. 40 (29 January 1927), 595. In *The Complete Prose of T. S. Eliot: The Critical Edition: Literature, Politics, Belief, 1927–1929,* ed. Frances Dickey, Jennifer Formichelli, and Ronald Schuchard (Baltimore: Johns Hopkins University Press, 2015), 25–29.

the possibilities of experience grow every day more common, the average use of words, far from widening with them, grows always more restricted.[10]

Indeed, if words had any vitality to-day, their reduction in number would be a source of strength (as we are accustomed to think all simplicity is), yet, instead, we seem to see about us the deliquescence of all non-specialized thinking, and arc left with the greater part of writing, as perhaps the most conservative of the arts, no better than the literature of the past taken at whatever point. That is not strange, for writing concerns itself with what it would have stable factors in mankind, the emotions, and where the reactions to these may be changed or modified, i.e. jealousy, most writers find it pays better and is more generally satisfactory to give them their full primitive value. In such works, therefore, the man of to-day is ready to be, and is perhaps even proud of being, on a par with his fellows of the stone age!

For the rest, it would seem that the best artistic expression has always been about equal to itself, that there has been no advance, no retrogression (there have of course been periods and places in which the arts were totally occluded), and if our problem were, is Futurism better or will it be better, this essay could now end, for we might be sure that the best work of any future time could not rank higher than the best work of to-day or of the past (though the whole world must be our province), and that what possibilities of permanence such work had, would lie in the possession of certain qualities mankind in the past has agreed to call classical; but from our modern use of the word we could not honestly exclude either savage art, cubist painting, the literature called subjective, and minor poetry.

In fact, the time has come to widen our definition of the classical, for futurism only exists as a state of flux, immediately it is accepted becoming the classical of its own generation and of posterity. Picasso for example in painting, Stravinsky in music, and in Literature let us say Joyce.[11]

We are not however concerned with the definition of what elements make the classical, or with those transitory elements which in A.D. 3000 will go to make some new school with the thirty-year life of a school, for a subsequent generation would be bound to react from it.

But the present condition of man's mind and its possible mutations does interest us, and the possibility of new inventions to keep his mind spongy and promise wider horizons. But in effect he takes, and may go on taking, most things for granted; and, like an aborigine surrounded by every marvellous contrivance, will still keep sacred his taboos and ceremonial gear. And like that world which is ransacked daily for those new tastes in which he finds so much medieval virtue, so everything which the mad scribbling child humanity has set down must find its billet in some mind. It is impossible to tell the truth so that it is understood and not

10 [Author's note] From this one need not deduce that the feelings are similarly impoverished. Indeed, man's present-day inarticulateness seems an added irritant to his sensibility.

11 [Author's note] But Joyce is not, never has been, a Futurist. And the term cannot be applied to Literature, for such experiments as have been made in the free association of words (the only formula comparable with an abstract painting) and which were mainly anonymous and pre-War, are of little importance.

be believed, said Blake, but to-day, in the general slackening of all standards, truth, however much understood, has less moral force than it ever had. So that, although some individual be affected to raving point or another utterly blasted, the mass has no use for it, and writing, searching for an absolute truth, goes on at the side of all other activities but influencing them less and less.

Yet though man is unconscious of the more obvious attacks of that civilization in which he lives, secretly and by various inappreciable influences his life does seem to be changing. Imperceptible in one generation, it is marked in two. Twenty generations and the possibilities are immense. It seems to me that the pettifogging clerk of St Petersburg in roughly the year 1870, the Slavonic contribution to the literature of the Universe,[12] is the most important indication of the direction Humanity must take. For it is his self-questioning, his doubts, his innocence, and his general incompetence, absent till then from literature and so presumably non-existent, which more and more will be translated into what we now call soul. When Dostoievsky said: "For humanity's daily needs, mere ordinary human sensibility ought to suffice, or about one-half or one-quarter of the sensibility which falls to the lot of the average educated man of this miserable 19th century", he had given humanity an enormous push along the road it was bound to travel. For, though Rousseau and then Restif de la Bretonne[13] had as if fearfully taken the first step or two in the path of extreme sensibility and the natural man, they could not suspect that the unnatural man was to be a thousand times more sensitive, more tortured, and that the conditions of modern life would encourage that morbidity. With them the path was closed until Dostoievsky pushed a way through.[14]

That this influence is not peculiar to literature, the following quotation from Mr Goossens will show:

> Finally we come to what is perhaps the most influential school and one more fraught with possibilities of expansion than any to-day, that of Russia; with which we immediately associate the names of two of the most remarkable composers of all time, Stravinsky and Scriabin...

So that the real hunt for the soul may perhaps be said now to have begun. Legions of writers between this and the year 3000 will earnestly till that field, and there can be no doubt that the complexities of Dostoievsky's characters, not to mention Tchekhov,

12 Rodker is presumably thinking of Dostoevsky's low-level bureaucrats in works like *The Double* (1846) and *Notes from Underground* (1864).

13 The Surrealists had rediscovered the French author Nicolas Restif de la Bretonne (1743–1806), autobiographer, pornographer, and early advocate of communism.

14 [Author's note] As to the factors that made this sensibility, they are too many to be capable of adequate definition; and, since it is now-a-days to be found among all classes, is not likely to be due to degeneracy resulting from the mixing of strains, as has been suggested. That, however, may be a contributing cause, and now that the races of Europe mix and intermarry on a large scale—if not actually—certainly as far as fashions, customs, and food go, we may expect sensibility to become everywhere exacerbated.

are about to descend upon these shores, indeed have descended; from whence they will spread and in that order to America, Australia, Africa, and Canada.[15]

Will man, then, following some super-Tchekhov always go one better, or imagine he is so doing, and shall we see him more nerve-racked, his physical difficulties almost solved but his emotional problems insuperable? How could he realize the necessity of combating them, in a world which has no use for his masculine reactions, the liens[16] which hold him conditioned by none but feminine standards. Our literature, our newspapers, are so filled with the feminine (naturally, when all writing is, has been, by men) and its reactions analysed in such detail, that insensibly these have now become our standards; and to-day it is that vision of the romantic, those yearnings, which dictate the outlook of whole nations. So, that, with to-day's preponderance of women, a vast revolution may be seen to be taking place, the gratification of the senses monopolizing all activity.

That finds its expression in the time-honoured chimpanzee game of poking some distracted insect (retired laundryman, miserly bookseller, virgins and dangerous ages) out of its hole with a straw. But with the perfecting of a mechanical world, mankind must assuredly react from so insulting a pre-occupation. That this is already happening is obvious from the advertisements, so prevalent in America and now beginning to appear here, which offer to free the subject from what is called Halitosis and other smells.

So that the time was ripe for Apollinaire and his:

> The modern school seems to me the most audacious that has ever been. It has put the question of what is beauty.[17] It wishes to visualize beauty disengaged from the pleasure which man causes man and since the dawn of historic times no European artist has dared to do that.

Unhappily it would seem that so far man cannot dissociate himself from the problems set him by that world from which he has abdicated; problems susceptible of infinite analysis, while ever remoter strata of consciousness remain to be tapped. Nor can we say there is a bedrock beyond which his sensibility cannot go, much as we should like to close this particular avenue of to-morrow's literature about the year 2500. Yet before then the pendulum must several times swing back to the innocence of the amœba.

As to the various schools I have already mentioned, Vorticist, Imagist, etc., it must not be forgotten that they do represent quite definite qualities of the mind, and that it would be easy to find equivalents for them in the past. Literature in the future therefore cannot but contain similar elements, so that our revolution does not lie in this direction.

15 [Author's note] It is to be regretted that Dostoievsky has been made the excuse for much of the temperament and lack of balance in various individuals in this country in the last twenty years. The simple and touching romanticism of these islands found rich aliment in the great Russian and revealed unsuspected possibilities to what was always the most romantic race in Europe.

16 A legal right to retain possession of something.

17 [Author's note] For beauty the reader must substitute "a work of art."

III: Literary Possibilities

But then we ask ourselves: What literary aliment will posterity seek? Will its literature be a free association of words or of ideas, or will words be built up in mechanical exercises like the fugue, or into geometric figures, the calligram;[18] or will they be matched with each other for colour and weight; Mallarmé and his: "Oui, la literature existe, et si l'on veut, a l'exclusion de tout",[19] or will the bias be towards a naturalistic writing; then will the refinements of a Tchekhov transcribe remoter vibrations of the individual and Dostoievsky's hysteria, his Rubens canvas, its welter of blood and tears, seem some schoolboy's exercise; or, if the bias be prophetic, will the symbolism of a super-Blake, wrapped in deeper silence, more distant and more dazzling, be equally obscure to posterity's subtler minds?

Or will humanity live in a purely documentary world, tolerating only the adventures of Nick Carter,[20] the Locus Solus of Roussel (the richest and most mechanical of worlds this),[21] *Faits Divers*[22] from the newspapers of 1925, mad-house lucubrations, advertizing matter, and the dreams of children, all standard units of the house of literature?

IV: Attempts to Revitalize the Language

But while words grow weaker, their outlines vaguer, their repercussion on the mind is of such a nature that from time to time some attempt is bound to be made to weight them and restore their primitive value. No one would accuse Guillaume Apollinaire of inventing his calligrams for the purpose of casting spells, nor have the mantrams of Miss Stein,[23] we imagine, any other motive than to grow and flower exactly as she wishes them to grow and flower. Her words are arranged not so much in free association, for their relation is obvious, but as though they were units and interdependent, and they are developed in recurring motifs as a fugue might be developed or a house built up of standard units. But it is an extravagant method and the amount of material involved quite astounding. Her latest book will

18 A caligram is 'A word or piece of text in which the form and layout of the letters creates a pictorial or ornamental design, esp. one related to the meaning of the words themselves' (*OED*). Apollinaire had published a volume called *Caligrammes* (1918).

19 Stéphane Mallarmé had written: 'la littérature existe et, si l'on veut, seule, à l'exclusion de tout": literature exists, and, if you like, alone, to the exclusion of everything', *La Musique et les lettres* (Paris: Perrin, 1895), 41.

20 A popular pulp fiction detective character.

21 See n. 54.

22 Miscellaneous news stories.

23 [Author's note] Mr. Wyndham Lewis says of her: "Miss Gertrude Stein is the best known exponent of a literary system that consists in a sort of gargantuan mental stutter. What she is exploiting in her method is the process of the demented. For any one less strong-minded than Miss Stein this might prove a dangerous operation... Her art is composed, first of repetition which lyricises her utterances on the same principle as that of Hebrew poetry. But the repetition is also in the nature of a photograph of the unorganized word dreaming of the mind when not concentrated on some logical functional purpose."

contain some thousand pages in the manner of the quotation I give later.[24] Such a method may prove some sort of discipline for a mind debauched (?) by too much compressed literature and the "mot juste".[25] (How far this last consideration removes us from the late strenuous generation is not to be measured in words.) As for asking that the method be refined, and a more essential oil extracted for posterity, it is possible someone may invent an even more repetitive manner. Then it must not be forgotten, that many people ask specially for 'long books', and this would seem to be an admission of the hypnotic value of repetition.

Miss Stein says:

> There are often two of them, both women. There were two of them, two women. There were two of them, both women. There were two of them. They were both women. There were two women and they were sisters. They both went on living. They were very often together then when they were living. They were very often not together when they were living. One was the elder and one was the younger.[26]

But at the other end of the scale we have Mr Cummings:

. . . "as usual I did not find him in cafés, the more dissolute atmosphere of a street superimposing a numbing imperfectness upon such peregrinations as twilight spontaneously by inevitable tiredness of flanging shop-girls impersonally affords furnished a soft first clue to his innumerable whereabouts"[27] . . . and this might more definitely help us in our search for the language of posterity though not its thoughts, for given the author's necessity to express this complicated relation it would be impossible to say it more clearly and compactly, yet keep all its atmosphere. And it is possible that as Mr Cummings[28] has grown out of the conventional manner in which his admirable novel, the "Enormous Room" is written, so perhaps posterity too may begin to think and speak in a similar way and may develop similar forms for the expression of the qualifying mazes of its adjectives, a problem which much preoccupied the later Henry James and to-day Mr Joyce. Thus it may entirely displace the simple grammatical constructions of to-day and make much of our present writing appear mere baby talk, as indeed Tennyson and Longfellow already seem; by whom thoughts and emotions of the very simplest degree of complexity were expressed in words of one syllable.

24 Probably *The Making of Americans*. Excerpts had appeared in Ford's *transatlantic review* in 1924. Contact Editions had published the book, which ran to 925 pages, in 1925, but Rodker may not have seen it when he wrote *The Future of Futurism*.

25 French: the just or exact word. The term is associated with Realist such as Flaubert and Maupassant, and was championed in English writing by Ford.

26 'Two Women', in *Contact Collection of Contemporary Writers* [ed. Robert McAlmon] (Paris: Contact Editions : Three Mountains Press, [1925]), 303–325, 303.

27 From 'Portraits', III, in *Tulips & Chimneys*, his first collection (New York: Thomas Seltzer, 1923), 63.

28 [Author's note] Mr Cummings is an American poet.

V: Help from the French and the Necessity for Abstractions

As for the quotation with which this essay begins: "above all like the fortuitous encounter on a dissecting table of a sewing machine and an umbrella", it is a very dry sentence but one which has given much pleasure to some minds.[29] In essence it is a development of the descriptions of scenery which began to appear in literature as a result of the agglomeration of men in towns (the countryman being blind to what we call the beauties of nature), but this particular point of view is we think, more intelligent and capable of much wider development. And if we look at the trend of certain latter aspects of writing, it may fairly safely be assumed that this 'still life' indicates an important development in the future of letters; though the image we hope will be more subtle, the leap more daring. The phrase however, a vague definition would be prophetic, is not particularly a French discovery though the passage quoted is from that language, for we have had, and may again have, writers with similar gifts and equally intense illumination.

But whereas that form of utterance has been spasmodic here, where for the moment it has no representative, in France to-day the many imitations of its devotees does not by much approach the admired masters, Baudelaire, Rimbaud, Lautréamont, Mallarmé. That does not condemn, nor suggest any impossibility of the further development of the method, which, undreamt of before the beginning of the last century, differs from other writing in that it is not concerned with the exhalations of humanity, a more or less complete catalogue of which the reader will find later, but is an arrangement of forms analogous to an abstract painting and dependent on qualities which do not trade on the emotions. The baser instincts play only a minor part.

In the past, inspired writers have occupied themselves with man, damning him, exalting him, in the finest most moving passages of literature, but they have never made the abstract activities of the mind and senses the field for their search. Yet there we have a possibility of more permanently satisfying ravings taken out of their all too human contexts. As we shall grow more refined, our nervous antennæ more delicately aware of new vibrations, all the invisible gamut of the spectrum should open to us, the image grow more daring and take a wider sweep.

But at the same time and within the same limits we shall have a directly contrary literature. It will occupy itself with the minutest variations; its descriptions will be the slow motion film of writing. The deliberate dropping stars of the tiny parachute[30] in the Naples aquarium will be taken in detail, caught at each momentary iridescent flash; or the detail of every least irregularity of some seemingly perfect surface analysed as far as human ingenuity will take it. Thus too some perfect picture of the Universe may be obtained.

29 'Comte de Lautréamont' (pseudonym of Isidore Lucien Ducasse; 1846–1870), *The Songs of Maldoror*. From Rodker's own translation of this work beloved by the Surrealists: *The Lay of Maldoror* (London: Casanova Society, 1924), Canto VI, p. 279. Man Ray made an assemblage, *L'Enigme d'Isidore Ducasse*, 1920, consisting of a sewing-machine wrapped in a blanket and string.

30 A bioluminescent jellyfish.

Then, we must have a literature, very large, very slow, like the development of one of Brancusi's[31] forms, which, under a mask of immobility, will epitomize all activity.

And then, since literature is inconceivable but as a mirror of ourselves, 5,000 years will have saturated us, so that we must turn elsewhere. But whichever way we turn, however cosmic our outlook, however limited, so far we have been brought hard up against ourselves. So we must hope that the phrase with which we started may grow wings, sweep wider, climb higher, though we cannot now follow it, and passing the peaks of Donne, Blake, Mallarmé, Roussel, drop boldly into that abyss on the brink of which these and other writers must have felt themselves hovering, to return with some earnest of that El Dorado.

It may be a stone, a gas, water of Styx, it matters not if only our descendants can somehow get away from themselves and the god in their own image.[32]

VI: The Russian Influence

By far the most important forces, however, which while not being inimical to each other, may possibly diverge, are the names I have mentioned on the one hand and on the other Dostoievsky and Tchekhov. In a way the immense calm of Tolstoi may be said to bridge both streams but I do not feel we shall hurriedly find another Tolstoi, though from time to time the continually diverging streams may be bridged by ever greater writers.

Yet the Russians are a puzzle with their contribution to the sensitivity of the world, for their own introspection it is said is due to white nights and long winter months spent on the stove; things undreamt of elsewhere. And as it is one of the latest European countries to be economically developed we cannot guess what flying, the railway and the telegraph may do to the race. So that here it must be the very instincts we deprecate which speak between them and us, progress passing through the spirit. For now we have been shown the way, we recognize ourselves as complicated and as hysterical as the best writing can make us.

For conditions will grow every day more stable, man will begin to think more of his soul and its place in the world, our engineers' world of the future should produce an immense hysterical self-consciousness.[33]

There we have the best of all reasons for constructive propaganda by the State and there is no doubt the way is being prepared for it, for introspection has always, and justly, been regarded as morbid and contrary to its best interests. And looking at America on the one hand and Russia on the other, it must now be obvious that the very near future will see the complete automatism of the individual in a benevolent state, manufacturing minds.

31 [Author's note] Brancusi. A Roumanian sculptor living in France.

32 [Author's note] Cf. Blake's definition of the most sublime poetry. "Allegory addressed to the intellectual powers while it is altogether hidden from the corporeal understanding."

33 [Author's note] Religion and the struggle for existence must have much simplified the general outlook of the majority in the past.

That must bring its own reaction. Writing, the arts, taking to the hills. So we shall have writers writing for each other in a tongue of which Mr Cummings is the vaguest forerunner. Then the earnest student will find himself crawling painfully among weighty exegeses, the whole business of the arts a priestly hierarchy of infinitely various degrees.[34]

VII: A Plea for Agility

I said earlier that the literature, and indeed all the arts of the future, might call for the exercise of greater agility than we use to-day, and indeed, no one looking at the writing of the younger Frenchmen, could at all doubt that the bias which the last seventy years has shown in that direction in France, above all in poetry, though I shall not attempt to define either poetry or prose, is not only a definite contribution to Literature, but is also applicable to all the arts; being the discovery of a path mankind is bound to take, though such is the astringent quality of the wind that sweeps the channel, that so soon as such manifestations of the spirit reach these shores they are known as Futurist.[35]

And here I must apologize for this continued reference to France, but it is the only country where the arts are practised in schools, and not as in England, in a sort of Thebaid[36] by the light of personal revelations. This reveals a certain weakness in the preceding plea, for the multiplicity of English words somewhat intersect each other, while French has fewer words and harder meanings, so that the languages may not be comparable; and then I am informed, our own language betrays an infantile ego-complex in those who use it as revealed by its unique use of the capital I, which may explain our Tennysons, Longfellows and racial ingenuousness.

Yet admitting that certain philosophic and even literary systems offer difficulties in languages other than their own, it seems to me that it would be an admirable thing if the writer here could develop a general gallic level of excitement about literature, and substitute hard general standards for the sentimental preferences that to-day mark his choice; for they are not necessarily an inevitable penalty of English to-day.

34 [Author's note] All this I shall be told has already happened in China, art has bloomed, and faded; its wisdom not wise enough to avoid the general scramble for the benefits of European civilization, and the consequent scrapping of its own institutions. But was it not rather that it had reached the final limit at which men may communicate and understand each other and that no further development in those directions was possible, apart from the inventing of the new language we plead for?

35 [Author's note] Stendhal says somewhere that genius loses 40 per cent. of its vitality in England. As to the commonalty Mr. Wyndham Lewis says: "And it is just the popularization and vulgarizing of art that is responsible for the innumerable swarms of dilettante competitors who make of every art a trifling pastime, so that it is impossible for the rest of the world to regard that occupation as a serious one, seeing into what hands it has fallen. So it would certainly be the redemption of art, and it would no doubt have the effect of protecting, and actually of encouraging science, if they were both removed from the wide superficial, 'democratic' (that is aristocratic) playground, and reinstated as a mystery or a craft."

36 Classical (or neo-classical) accounts of the sons of Oedipus, Eteocles and Polynices, who killed each other fighting over who would rule Thebes.

In any case the agility I am asking for is not completely that of Donne, where having landed the reader may stand awhile, meditate, survey the prospect, nor that of Mr Pound, where the difficulty is one of 'locale' and the reader's ignorance, which deficiency in the reader notes might remedy; but I ask for a poet's agility where these qualities and others would be compounded, and in which the inspiration though necessarily drawn from human sources would transcend (?) them and become part of some nonhuman world which is to be the delight of the inhabitants of the future.

It may be objected that the natural development of the artistic usages of to-day should fulfil these future needs, but the expression of nearly every mental activity is so circumscribed by traditional form, subject and accustomed reaction, and the accumulation of facts grown so vast in the last half century as to be unmanageable in this, that only some very complete break away seems to offer scope for an art which shall be adequate to the tremendous expansion which we feel the mechanical age has made flower within us. The artist's problem was simpler centuries ago when human activity was not so manifold, it was less simple at a later date with "War and Peace". It is perhaps still possible for Mr Pound, whose cantos[37] when completed may well prove to be the last epic between now and that mythical A.D. 3000; and no doubt some offspring of a still romantic and savage country, a Hamsun[38] or Conrad, may find himself leading an adventurous life familiar therefore with some new unsuspected milieux, resulting for us in a lesser Odyssey; but that will become increasingly difficult with easy communication, the general rise in standards of living, levelling mediocrity, the woman's page, society photographs and all the vast hotel-de-luxe display of travelled sight-seeing and polite anecdotage.

As to birth and death, and the whole butcher's shop of life which fills literature, the instinct by which a baby recognizes another baby at sight, or the deep interest of one human being in another of roughly the same size; the proper study of mankind being man and its setting apparently the nursery; mankind in any consistent world must assuredly violently react from a pre-occupation so strongly illumined by the psychology of Apes.

For we record it gladly: there has always been part of the community interested in the final elements of the emotions, their chemical formula if that were possible, and in ideas apart from their application; such an interest being not necessarily dependent on personal identifications.

Then, with the continual coming to consciousness of stratum upon stratum of each generation, Mr Cummings' ingenuity, and this was the Futurist process, and that of every artist whose method it is, or will be, to cut through his theme with the angles of curb-stones, street lamps, plate glass windows, all material objects and people, or bombard them with mathematical or astronomical dimensions; though superficially an innovation, is yet in essence identical with past practice—effects

37 Ezra Pound's long poem *The Cantos* appeared in installments (some later abandoned or revised) from 1915 to 1962.

38 The Norwegian writer Knut Hamsun (1859–1952) had been awarded the Nobel Prize in Literature in 1920. He is best-known for his 1890 novel *Hunger*.

of landscape, weather or natural objects on the hero—and so is assured of a life co-existent with the rest of the arts.

And one asks oneself why the suggestion of such experiment as shall induce mental agility need sound so particularly indecent in this country, as though such a quality were only appropriate to a frog-eating race; but without that there can be no progress and we shall continue to have our lonely Doughty's,[39] colossal and incomparable figures, belonging to no time, no school, and on whose shoulders no super-structure can be raised.

VIII: Language and Material

But if we want a new literature and new arts we must have a new language and that was Mallarmé's pre-occupation. To establish a language that should never be commonplace and solid enough to serve as foundation for the whole of succeeding poetry. And this necessity to get away from the banal, and to say a thing once and for all (and no "compte rendu"[40] of an emotional situation can do that) has always been and must continue to be the writer's preoccupation. It is impossible now-a-days to go on writing or even to put down once "At five o'clock the countess got back", or "sitting down she ordered a drink", there is no excuse for it. But as far as writing is concerned, it has been unable until now to dispense with these elementary bridges, and the fact that Dickens, Balzac, Tolstoi, our very great figures, did not bother about such trifles does not make the problem any less acute. For the problem of language and the significant, is essentially a modern hurdle the arts must clear before they can again get into their stride.

As for the writer's language we are seeking, and that others have sought, some shorthand of the emotions might possibly adumbrate not only the matter but the manner of the literature of the year 3000. For we no longer doubt that every set of associations has some detonator and the problem now is how to establish that code, that form of literary cablegram which should be true for the greatest number, and because "toute âme est un nœud rhythmique".[41] Thus our books would be much compressed, and simplified word endings and compound letters would take us back to the early Rolls. But again we should have to give up some of our economies, for the spacing of each word would become an important consideration. The waves of each explosion would be intersected by succeeding waves as a stone thrown into a pool makes a wave which succeeding stones cut at differing intervals with eccentric and rippling circles, the intervening silences having all the importance of sound.

Or perhaps it is time humanity again turned to some method of time notation and the pointing of its syllables.

39 English poet, writer, and explorer Charles Doughty (1843–1926), known for his *Travels in Arabia Deserta* (1888).

40 Report.

41 Every soul is a rhythmic knot. Mallarmé, *La Musique et les lettres*, 35.

With so much insistence on sound and tempo and the general problems of synchronization it cannot be long before the clavilux[42] is brought to perfection. The arts will begin to merge into each other, no distinguishing definition will remain valid for any of them, for a time all will sink into the welter of the countless millions of books, musical compositions and pictures the ages are busy piling about us.

And what indeed could be more disheartening than the adjoined catalogue of the junk of literature, and is it possible to contemplate with equanimity the likelihood of this sort of thing continuing for another 1,000 years?

"Perturbations, anxieties, depravities, death, exceptions to the physical or moral order, the spirit of negation, stupors, hallucination served by the will, torments, destruction, surprise, tears, insatiabilities, subjections, racking thoughts, novels, the unexpected, things one should not do, the chemical singularities of the mysterious vulture who spies on the carrion of some dead illusion, precocious and abortive experiences, obscurities, the terrible monomanias of pride, the inoculations of profound astonishments, funeral orations, longings, betrayals, tyrannies, impieties, irritations, acrimonies, aggressive outbursts, aberrations, melancholies, rationalized terrors, the strange disquiets which the reader will prefer not to experience, grimaces, phobias, the bloody drawplate through which logic at bay is made to pass, exaggerations, absence of sincerity, bores, platitudes, the gloomy, the lugubrious, births worse than murders, passions, the clan of assize court novelists, tragedies, odes, melodramas, extremes perpetually presented, reason hissed with impunity, cowardices, insipidities, frogs, octopuses, sharks, the desert simoom, what sleep-walks, is suspect, nocturnal, somniferous, night-walking, viscous, talking seal, equivocal, consumptive, spasmodic, aphrodisiac, anæmic, one-eyed, hermaphrodite, bastard, albino, perverted, aquarium phenomenon and whiskered woman; also, time drunk with silent discouragement, whims, tartnesses, monsters, demoralizing syllogisms, filth, what does not think like a child, desolation, the intellectual manchineel tree,[43] perfumed sores, the thighs of (la dame aux) camellias, the guilt of a writer who rolls down the slope of nothingness and despises himself with joyful cries, remorse, hypocrisies, diffuse perspectives which crush you in their unperceived cogs, grave spewings over sacred axioms, vermin and their insinuating ticklings, idiotic prefaces like those to Cromwell[,][44] Mlle. de Maupin[45] and of Dumas fils;[46] decrepitudes, impotences, blasphemies, asphyxiations, stiflings, fits of rage; before these filthy charnel houses which I blush to mention, it is time to react now against what shocks and utterly overwhelms us.["][47]

42 The Clavilux (Latin: 'Light [played by] key') was invented by the artist Thomas Wilfred (1889–1968) to create and perform light art. He produced different models through the 1920s and 30s.

43 Flowering plant from the Americas, every part of which is highly toxic.

44 Victor Hugo's vast play of 1827, unperformed till 1956. Its Preface is a manifesto for Romanticism.

45 Théophile Gautier's novel of 1835; its preface an early advocacy of 'art for art's sake.'

46 Alexandre Dumas the younger; author of *La Dame aux camélias* (1848).

47 Translated from Lautréamont, *Poésies I*. Slightly revised from the passage quoted in Remy de Gourmont's Introduction to Rodker's edition of The Lay of Maldoror, xv–xvi.

So we are left with Mallarm[é]'s "A savoir s'il y'a lieu d'écrire." [48]

And if this be applied to all the arts it is obvious that humanity must find some new universe of the spirit.

IX: A Helping Hand from the Sciences

But because the poet has not included scientific works in his catalogue, the reader is not to imagine that an oversight. Indeed he says somewhere: "O austere mathematics, I have not forgotten you since the days your learned lessons, sweeter than honey, trickled into my heart like a refreshing stream",[49] and it is clear that for lucidity and scope, the possibilities of a mathematical work, with its exact symbols and paucity of units, whose permutations are endless, must be unique. So we might hope that the impoverishment of the language which we see from day to day were a conscious process, ultimately to enrich us, but alas, words are too near their users and grow casual as they do, though indeed the repercussion may be mutual. For the error of that Sir Thomas Elyot who in 1533 (I quote from Ogden & Richards' *Meaning of Meaning*)[50] intended "to augment our English tongue whereby men should express more abundantly the thing they conceived in their hearts", is all too common. But if the writer's training gave him command of an exact language, either analytic or abstract, a suggestion of this kind would be unnecessary, but since as a rule he is profoundly ignorant of what he is doing and how he does it, any likelihood of his development of a richer, more adequate and significant language, would seem to lie in the remotest future.

And since the sciences do not concern themselves with the overtones, irrelevancies even (unless Newton's apple be so called) which are at once the accidents and beauties of writing (as indeed of all the arts) but also its confession of failure, the sciences alone can help us to the living and stable form we desire. For where the pre-occupation is not with limiting primitive impulses and reactions, the increase of workers can only carry further the sublime literature of the higher dimensions. As to writing one may safely say that no writer benefits by the discoveries of his predecessors (unless it be by the richer consciousness of his time) and so he must watch his scientific brother scaling the peaks of microns and parsecs[51] upon the shoulders of his fore-runners while he himself must linger with the sirens.[52]

48 'To know if there are grounds for writing': *La Musique et les lettres*, 37.

49 Lautréamont, *The Lay of Maldoror,* Canto II, p. 93.

50 The book Ogden wrote with literary critic I. A. Richards, *The Meaning of Meaning* (London: Kegan Paul, 1923), was published in Ogden's series, the International Library of Psychology, Philosophy and Scientific Method, and was a foundational text of Cambridge criticism.

51 A micron is one millionth of a meter. Parsecs are units to measure immense inter-stellar distances. 1 parsec = approx. 3.26 light-years or 30.9 trillion kilometers.

52 [Author's note] Tchekhov says in one of his stories: "As I yield up my last breath I shall still believe that science is the most important, the most splendid, the most essential thing in the life of man, that it always has been and will be the highest manifestation of love and that only by means of it will man conquer himself and nature."

Are the sirens then his sympathetic nervous system as Mr D. H. Lawrence would have us believe (fixed as far as we are concerned, but minutely changing); and his intuitions which in the end make the quality of his writing, functions of the same system. Certain writers, among them very consciously Mr Lawrence, do seem to reach out to their prey and store the answering vibrations as radium emanations are stored. And it may be we are wrong in asking our writer to think clearly, be mathematically conscious and able to turn on at will the elements he makes use of. But the fullest consciousness of himself is not incompatible with his own stable harmony, though this kind of scientific help cannot be of much use to him:

". . . To the student of the internal secretions, the lines, expressions, attitudes are important for they tell of the state of tensions and strains in the vegetative apparatus with which they are inseparably connected. It is when he comes to the consideration of the face as a complex of brows, eyes, nose, lips, and jaws that he becomes most interested. For in the modelling and tone of every one of the features each of the endocrine glands has something to say".[53]

But because we no longer need medicine-men or seers, and because our instruments and text-books will give us the same article with less fuss and more precision, can we not leave the intuitions for a bit and ask our writer or artist to be less bewildered by the forces which he uses, or prefers to think use him? Then he might endeavour to establish some set of constants on which to base his work and its possible developments, one advantage of schools being that before their theories are discarded, they have been so thoroughly worn as to be threadbare. In that way some consecutive effort might be born. Baudelaire's offer to teach poetry in thirty lessons is a perfectly sound and above board offer, for in this art as in others, apprenticeship plays a most important part and more can be done with properly directed average intelligences than the average intelligence imagines.

That might bring us to our priesthood, our families of artists under the very few deans who have tried to formulate an æsthetic of their art.

X: A New Story-Teller

In the meantime and while awaiting that series of concerted efforts which is to push back the limits of the universe and extend man's consciousness (though regrettably only on accustomed lines, but with wider scope if he will only consent to let go the life line of his fellows) we might make a very determined cast back to the past, particularly when it is as perfectly done as by Roussel.[54] Here is *The House that Jack Built*, and Hans Andersen, and *The Arabian Nights*, but the version is modern (let us call it that). The structure is infinitely rich and its elements, drawn from the remotest sources, documented to the last atom. This story of the scientist in his lonely

53 Louis Berman, *The Glands Regulating Personality: A Study of the Glands of Internal Secretion in Relation to the Types of Human Nature* (New York: Macmillan, 1921), 121.

54 Raymond Roussel (1877–1933); French poet, novelist, playwright, and musician, who influenced the Surrealists. His 1914 novel *Locus Solus* features an inventor who gives his house-guests a tour of his surreal inventions.

wood with his extraordinary inventions, takes us with every dramatic and curious detail through a host of incidents like the tanks of an aquarium one has watched for years, or like those other tanks of air where in the late Madam Tussaud the young man hurried down the road to ruin.

I cannot conceive that humanity, much as it may change and russianize, will move very far from this literature or be able to give up its story tellers. Let us hope that with Roussel, who now after the lapse of centuries seems again to combine in himself all the qualities of a tribal consciousness and the elements by which epics like *The Arabian Nights* grow or agglomerate, that the story is about to come into its own again and the marvellous, ousted by science and invention, become once more part of our lives. For what was capable of going into a formula seemed no longer marvellous to us.

XI: Conclusion

To recapitulate; posterity must choose from the various tendencies in writing which we have looked at, if it does not want them all; for we cannot postulate any real change in its make up or any ability to handle its material more consciously than we do, except in so far as it may drill into itself. This we do expect to reach a phenomenal degree of complication, with always however the saving clause as to the possible exploiting of the non-human and abstract image.

In any case, the general despair of our young men, apart from its being a post-war despair, plus the deceptions which attend upon the age of 30, seems to me a genuine russianization and no robust

> en l'an trentiesme de mon aage?
> que toutes mes hontes j'ay bues [55]

can be expected to-day.

As for the various forms already mentioned, the free association of words and ideas, first exploited seriously by Marinetti, has always existed as part of the properties of words and of the life they have in our minds and the habit they have of writing themselves when the mind is preoccupied and the hand holds a pen. This revelatory apparatus was applied by Freud. It is an important function of the mind and does not deserve the scorn it has met, since most mature minds will find a sentence more easily than the 'mot juste' or even a single word. And then it is a mistake to think of words as counters, they are rarely that, but generally the button which releases a sentence or an elaborate idea. But our russianization is making us more receptive, the mind already reacts more quickly to words and their content, and we shall turn from the ungrateful method of Miss Stein, who means words to be words only and to mean nothing but what she would have them mean.

55 The opening of François Villon's *Le Testament* (1461): 'In the thirtieth year of my life / When all my shames I have drunken'; a passage quoted by both T. S. Eliot and Ezra Pound in poems Rodker had published at the Ovid Press: 'A Cooking Egg' and *Hugh Selwyn Mauberley*.

And then as antithesis we have the complicated grammatical structure I quoted from Mr Cummings which seems infinitely more suitable a form for posterity, and of which Mr Joyce's[56] last jocular and Rabelaisian manner is an almost bewildering example, yet both allied by that 'Unanimisme'[57] which must go on through the ages. As for calligrams, abracadabra spells, charms of all kinds, we like them and no doubt posterity will like them too. But the elements in the composition are so simple that it must always be repetitive, and any variation unimportant, at whatever point of the future resuscitated.

With the prophetic we come to Dostoievsky, about whom we cannot help hoping that his particular hysteria finds its limit in him. That however is unlikely; his characters do exist, will go on existing, will become continually more complicated. Nor do we want to change them, nor could we. It is only by an important effort that we realize what hordes of people the world contains; how little the writer knows of any of them and how infinitely thin is the film under his microscope.

So let us have our Dostoievsky and with that everything; for through him we have the *Faits Divers* (Ivan and his newspaper cuttings) the dreams of children, the hypnogogic[58] visions of the night and of the madhouse.

Out of all this, for literature and the arts these possibilities of unlimited development emerge:

(1) The cerebrally prophetic, of which the opening quotation is a specimen and which at last seems to be becoming articulate and organized. It is susceptible of very rich development and it alone seems able to bear the application of a scientific method and to contain a promise of ever more grandiose development in keeping with the destinies of mankind.

And (2) we have the Russians as the supreme registering instrument and masters of the art of capturing remote vibrations. This is the great slide down which writers and humanity will go on slipping as long as there are towns, education, civilization.

56 . . . So much so that young Master Shemmy at the very dawn of history seeing himself such and such, when playing with words in his garden nursery, asked of all his little brothers and sweestureens the first riddle of the universe; when is a man not a man?; telling them take their time and wait till the tide stops (for whom from the first his day was a fortnight) and offering the prize of a bittersweet crab to the winner. Once said when the heavens are quakers, a second said when Bohemeand lips, a third said when he, no, when hold hard a jiffy, when he was a gnawstick and detarmined to, the next one said when the angel of death kicks the bucket of life, still another said when the wine is wife, and still another when lovely woman stoops to conk him, one of the littlest said me, me, Sem, when pappa papared the harbour, still one said when you are old I'm grey fall full of sleep, and still another when wee deader walkner, and another when he is just only after having being semisized, another when yea, he hath no mananas, and one when dose pigs they begin now that they will flies up intil the looft. All were wrong, so Shem himself took the cake, the correct solution being, all give it up?—When he is a Sham. [Excerpt from the early version of the 'Shem the Penman' episode first published as '"Extract from Work in Progress', *This Quarter,* Vol. 1, No. 2 (Autumn-Winter 1925–1926), 108–123, 108–109; *Finnegans Wake* was not published in book form till 1939 (where the extract corresponds to I.7; 169–170.]

57 Movement in French literature begun by Jules Romains with *La Vie unanime* (1904), seeking to understand collective consciousness, or the psychic life of groups of people.

58 Inducing or leading to sleep.

Feeling, feeling forever and everlasting, feeling about oneself, others, the whole creation, and no possibility of avoiding it if one would. But with it will go the large comforting blanket which understands, and soaks up all the violence and ebullition. But the State will perhaps think better to eradicate so demoralizing a preoccupation or a natural disgust will fling humanity into my first alternative.

So finally resuming, it seems to me we have on the one hand, Blake, Mallarmé, Roussel and the development of all those qualities we have called mental agility, the consciously prophetic, and so the consciously sublime, and then there is the sublimity of the bowels as in Tchekhov and Dostoievsky. The former will climb higher, its sweep grow more daring, the latter dig monstrously and fatally in before it finally disappears from the human stage or is merged into my first alternative, but both to become creatures whose splendour is to-day undreamed of.

As for what we called Futurism, that cannot exist, for if it is at all vital it must become part of the tradition of posterity despite its momentary envelope for to-morrow to-day is yesterday. For Futurism is a dress and not a quality of the soul, and what is good in it may be found scattered through the ages. But the skeleton of the arts had so long been buried under a mush of traditional and effete forms, that only some violent effort such as we have lately seen could disengage it, yet that effort must often be repeated in the time to come. What is new to humanity is the belated consciousness of its increasing sensibility. This may, nay must, reach some inconceivable degree, but only when, to our minds of to-day, it has grown utterly artificial.

Excerpts

From: *Euterpe; or, the Future of Art*

Lionel R. McColvin (1926)

Headnote

Lionel Roy McColvin (1896–1976) was an assistant at Croydon's central library when war broke out. He enlisted in 1915, and fought in the third battle of Ypres in October 1917. He returned to Croydon public libraries in 1919. After studying at University College, London, he was appointed chief assistant at Wigan Public Library in 1921, then chief librarian at Ipswich in 1924.

In 1931 McColvin took the role of chief librarian at Hampstead and curator of the Keats House Museum. He published a number of influential books on librarianship. In 1938 he became city librarian at Westminster, until he retired in 1961. His achievements at Westminster include opening a new branch library in Charing Cross Road and founding the central music library. He also worked for the British Council and UNESCO.

McColvin was an energetic amateur musician and playwright, and *Euterpe* draws on his varied experiences of different arts and their appreciation in different communities and classes. His sociological approach, and attention to how the varieties of mechanical reproduction affect the production and reception of art and literature, have striking affinities with Walter Benjamin's famous essay, written almost a decade later, 'The Work of Art in the Age of Mechanical Reproduction.'

******

I

At the outset it will be desirable to state that when I speak of the future of art I do not mean the "art of the future". Art can be considered from either an inside or an outside point of view; that is to say, we can deal either with its nature, problems, and performances—art itself, or with the amount and quality of the interest taken in art by men and women— the "art life" of the community. The latter subject is that dealt with here.

The "art life" of the civilized world is at present in a transition period, which is fraught with distinct, though maybe unrealized, dangers. Its problems are only indirectly related to the present and the future state of art-production: whether we

DOI: 10.4324/9781032625522-29

foresee development or retrogression in modem tendencies in literature, painting, music, and so on, these dangers will need to be faced, or they will, at least, minimize the value of the creative work of to-morrow. For we are concerned not with the production of art but with the enjoyment and appreciation of art. As the latter is the more important, since without it production would be sterile, it is an essential preliminary that the conditions necessary for the healthy growth of a more widespread, deeper-rooted love of the beautiful should exist. We are now viewing the situation as sociologists, as men, rather than as artists. The artist can be satisfied when he attains a certain level of performance; at least he can work with content and happiness while he is seeking to reach a may-be unattainable perfection. He is, naturally and rightly, concerned with absolute values; and the critic and the individual lover can maintain the same attitude. If a painting or a poem reaches perfection, he asks no more. But the sociologist must take a different attitude. To the artist and the critic the work is the end; to the sociologist it is the beginning. It is not enough for him to know that the painting is great, since to him it is only the means by which men attain artistic enjoyment; it has no significance until it has acted upon the minds of men. That being so he must ask other questions about it—firstly, *How many* men can see it? How many are able to appreciate its value intelligently, gaining the full æsthetic, spiritual, or intellectual stimulus from it? —in short, What is the aggregate of its human significance?

It does not follow, of course, that we can relate the quality of a work of art to the "quantity" of its appeal; it would, in fact, be absurd to suppose that it is necessarily better that 100,000 should know and appreciate the second-rate than that 100 should love the finest—neither, with certain reservations, need this necessarily be untrue. The point I would urge at present is simply that the value of art to humanity does depend very largely upon the desire and opportunity of men to take advantage of it. The poet whose works are ignored saving by the very few may be as impotent as a mute inglorious Milton.[1]

Therefore there are two factors—production and reproduction, or, shall we say, creation and distribution. A musician composes a symphony, a dramatist writes a play, a novelist a story—that is the first factor. If no one ever performed the symphony, produced the play, or published the novel, of what importance would this creation prove to the world?—Practically none. The art-product must be distributed before it can accomplish any part of its essential purpose. It necessarily follows, moreover, that the wider the distribution, the more adequately will it function. This is all very obvious, though often forgotten, and will disclose the next step in the argument, which is that, were it not for certain tendencies, increased means of reproduction and distribution would lead to a better developed, more valuable, and more active artistic life. That being so, the present, which is a period when mankind is enjoying the benefit of recent and important reproductive inventions, should be imbued with hopeful tendencies—Is it?

1 Thomas Gray, 'Elegy Written in a Country Churchyard' (1750), line 59: 'Some mute inglorious Milton here may rest.'

Yes and no. Let us take stock of our position. Reproduction is almost entirely a mechanical matter, depending upon non-artistic, purely material factors. Production is the business of the creative artists; reproduction that of the scientists. The latter have given us within recent years inventions which have revolutionized artistic conditions—the mechanical processes and innumerable secondary inventions such as stereotyping, and mechanical composition and binding,[2] which have facilitated the reproduction of printed matter, the three-colour and other photo-mechanical methods of reproducing pictorial matter, the gramophone, the piano-player, and wireless to aid the distribution of music, and so on, throughout the range of pure and applied art.

Until recent years the percentage of the population who were in direct contact with the fine arts had remained much the same in civilized countries from probably the earliest times. Art had almost invariably depended upon direct patronage of some kind or other, religious or secular, if not entirely at least to an important degree. I would not denounce this; one cannot, when one remembers that the system fostered art which has not been equalled under the new régime. But direct patronage by the few is rapidly declining and is to-day almost negligible. It has been replaced, simply as a result of the mechanical factor, by a more democratic economic basis. Some arts are still to some extent produced for the few, but others entirely for the many. The important fact is that wherever reproduction is easiest that art is the most democratic—books and music, for example; wherever least possible its range is narrower and its support less democratic, e.g., sculpture, household decoration, etc.

The character of our artistic experience has therefore to a very large extent been decided by purely non-artistic factors. That which can be reproduced has been reproduced, and opportunity has developed taste. This is a generalization, though not a fallacious one. We may assume that the artistic needs of men have been led into their different channels partly as a result of personal inclination, but very largely through the influence of opportunity. If a number of men were cast upon a desert island with only books to minister to their aesthetic needs, the majority would take what was to hand and be quite content. I am not saying that this is a good tendency but that it is a true hypothesis, applicable to modern life, and a contention which is tenable on historical grounds. The favourite pursuits of early civilizations were not those of to-day, and it is very unlikely that any one factor has done so much to change taste as the development of means of reproduction. The pursuit of once-popular arts need not die out; it need not even decline, since the numbers of those interested in all the arts is increasing; but the proportionate or relative interest alters. This being so, can we ignore the influence of the mechanical factor? It is operating in a striking manner to-day when *relatively* music is being appreciated

2 Stereotyping is the 'method or process of printing in which a solid plate of type-metal, cast from a papier-mâché or plaster mould taken from the surface of a forme of type, is used for printing from instead of the forme itself' (*OED*). A 'forme' is an arrangement of type made from individual letters by the compositor. 'Mechanical composition' is when that process has been automated, as in the hot metal presses then used to produce newspapers.

by more and literature by fewer people, when the theatre is attracting, relatively again, fewer every day than the cinema, when the graphic arts are becoming more significant than the plastic arts.

To ignore the mechanical factor is to put effect before cause. Certainly the character of taste has influenced the direction of invention to some extent, since the scientist would naturally turn first to fields where his work would be most effective. This aspect should not, however, be magnified. Sooner or later science has given all it was capable of giving to *every* form of art, regardless of its importance or popularity. And so we realize that the character of public taste—that is to say, the proportionate amount of interest in the various arts—has been dictated by the mechanical factor. We can go still further and assert that its quality has been largely determined by this same influence.

II

Before we can appreciate the truth of that assertion—that the quality of public taste has been influenced by mechanical methods of reproduction—we must be prepared to view the art life of the community as a whole. Too often we tend to regard only the better elements, the top layer, and to ignore the lower strata. We segregate a section of the populace—that which appreciates, or pretends to appreciate, Art (with a capital A)—and forget that the remainder, which indulges in jazz, 'the pictures', light fiction, Bovril pictures, and tin-chapel architecture,[3] is actuated by the same motives. The quality of their artistic experience and the standard of their taste and artistic education may be very different, yet they seek the same kind of experience as the others. It is entirely a matter of degree.

Therefore we must regard the art-life of a community, as we must and do regard its social, religious, or political life, as comprising a little good, much bad, and more that is indifferent. Once this is realized, and only then, the full significance of the mechanical factor is apparent.

Let us go back to the pre-mechanical era, when only a small number of people had any opportunity for contact with art and only a few had developed a love for and the ability to appreciate its higher manifestations. At the same time a similarly limited populace found satisfaction in the second, third—and fifth-rate. Probably then, as now, more enjoyed the second-best than the finest, and so on, though probably the contrast was not so great as it is now. However that may be, when a new reproductive process was introduced it was naturally applied to the lower types rather than to the better, for an obvious reason. It enabled *more people* to be brought into contact, and these newcomers must naturally be unaccustomed to and incapable of appreciating the best. The education of taste is a slow process, whereas the new invention was a sudden force, applied immediately in whatever direction offered

3 That is, commercial illustration used in advertising; tin churches, chapels, and mission halls (also called 'tin tabernacles' were built cheaply from corrugated iron.

it the greatest scope. And so we find at once an increase in the lower grades of appreciation which is out of proportion to the benefits bestowed upon the higher.

The trouble did not end there, however. Greater familiarity' tends to form taste, especially in these matters. Art serves most men chiefly as a luxury, a relaxation, a recreation; and in our quest for these we are apt to take that which is most easily obtained. The mechanical factor, by making the fourth-rate accessible, *generated a desire for the fourth-rate*: this desire stimulated further reproduction, and this, in turn, brought more into the artistic fold, at each step lowering the quality of the most accessible and the most desired.

The result is that to-day the average quality of the whole artistic consumption of the populace is considerably lower than it had ever been before in civilized times. Though every day more and more people are reading some kind of printed matter, witnessing plays—silent and audible—of a sort, looking at pictures, penny plain or twopence coloured, though the time is not far distant when every man will be interested to some extent in art in one or other of its forms, our art-life is developing not so much in quality as in quantity. [5–16]

[. . . .]

IX

We cannot close even a brief essay without some reference to the effect of some other mechanical devices, such as the gramophone, the piano-player, and wireless, and a note on that all important subject, commercial art.

The appreciation of no art shows such great possibilities of expansion in the near future as music. During the last few years it has been released from its most irksome bonds and is now just beginning to stretch its limbs. For technique has been the curse of music, and now it is becoming possible to gain enjoyment without exercising one's executive and interpretive powers.

Musicians are of two classes—executive and appreciative—those who perform and those who listen. True enjoyment of music belongs to the latter, just as true enjoyment of books, of pictures, of plays is the reward of the reader and the spectator—not of the writer, the painter, the actor, or the composer. *Their* joy is of another order—it is the joy of creation.

Without the assistance of modem mechanical aids the music-lover had either to listen to the music-making of his friends or of players at a concert, or he had to attempt to interpret for himself. The first was inconvenient and unsatisfactory. The selection of music was not his own but that of others; the time and place were not of his choosing. The alternative was even worse, since his appreciation was limited by his interpretive powers and marred by his deficiencies. The owner of a modern player-piano has the whole world of piano-music and a wealth of arrangements at his command. Even the lover of orchestral, instrumental, or vocal music has access, through the gramophone and the wireless, to a passable substitute for the real thing.

What effect will this have upon pianoforte-music? In the first place, we shall gradually rid ourselves of misplaced pride in the amateur's very limited technical powers. We shall no longer praise So and So for being able to play Chopin's *Studies* after a fashion, but shall consider him either a fool for wasting his time trying when he could much more easily enjoy Cortot's[4] performance of them, or sympathize with the poverty that prevents his purchasing this mechanical aid. Secondly, we shall not waste time and kill natural love of music by the dreary routine of "teaching the piano." Instead, we shall teach appreciation. If all the energy spent in acquiring a very inadequate technique were diverted to the real business of appreciation, we should be a more musical nation. Thirdly, we shall cease to tolerate the incompetent player now so often foisted upon us or even sought for want of any better, and the ostentatious "virtuoso" executant.

Before very long the piano-player will cost no more than an ordinary piano; in fact the ordinary instrument will no longer be manufactured. In our schools "piano-playing" will be erased from the curriculum and classes in appreciation substituted.

But what about non-pianoforte music? There is a big difference. While the piano-player produces exactly the same kind of musical tone as the hand-played instrument, the gramophone, or the wireless, does not reproduce at all exactly the timbre, quality or volume of the instruments recorded. It provides not the real thing but a substitute, which, though excellent, can never be entirely satisfactory. We do not care to assert dogmatically what science will or will not make possible in the future; at least, however, it is extremely doubtful that a mechanical violin as adequate as the mechanical piano will ever be invented. Wind instruments depend less upon human manipulation—the organ, for instance, is nothing but an imperfect essay in this direction. This is but idle speculation, however. As a practical proposition we may say that the perfect mechanical reproduction of music will be confined to the pianoforte.

So we are left with these problems. Shall we be tempted to seek the shadow and lose the substance—listen in often, but never attend an orchestral or chamber concert or a violin or vocal recital? The chances are that we shall, unless to enjoy the latter are greater than at present. Considerable loss would result. The ears of the next generation would become attuned to a diminished variety of tonal experiences, for one thing. For another, the psychological, even physical effects of large gradations in the volume of tone, such as can be experienced only in the concert-room, should not willingly be relinquished. And, again, it is not by any means the same thing to listen to music in the company of others, in the atmosphere of the concert-room, as it is to enjoy music in solitude. We may sometimes prefer the latter, but that fact does not remove the difference.

The second problem is that, though there is little physical or moral good to be found in solo instrumental playing, such good does result from singing and partaking in concerted music. There is no good reason why we should play the piano—rather than listen to it; but there are many reasons why we should sing or

4 The French piano virtuoso Alfred Cortot (1877–1962) made the first electrical recording of classical music in 1925.

play in chamber or orchestral music. By all means let us listen to more music of all kinds; increased facilities for listening should not, however, decrease our desire to perform when performance can benefit us.

Taking all these considerations together we may assume:

1 that pianoforte *playing* will decline though much more pianoforte music will be enjoyed.
2 that much of the practical energy now devoted to the pianoforte will be directed to the study of other instruments.
3 that, unless our musical life is to increase in volume but diminish in quality, more and not less concert-going and concerted instrumental playing and choral singing must be provided.

Books, music, pictures, sculpture, however, minister to only a small part of the artistic needs of the community. By far the most widespread, though not necessarily the most valuable, art-products are those which we may describe as commercial, or industrial, or, better, "applied" art. Only a minority, even in this age, concern themselves with the first-named, but we all wear clothes, use furniture, live, work, play, and worship in buildings, eat and drink out of vessels, and so on, through every one of our daily occupations. Into each of these art can, does, and must enter. We may wear clothes to keep as warm, but they must be either ugly or otherwise—their existence implies artistic properties, negative or positive. If they are we cannot avoid their ugliness, though it may dull our appreciative faculties. Of course this is true of all things. Every object, every occurrence almost, has its artistic aspect. With every manufactured article, every human production, however, this artistic quality is within our control. When we make a cup, a hat, or a church, we can make it as beautiful or as ugly as we like, subject to certain limitations, some of them real, some imaginary. But we must be sufficiently interested in its artistic value. It will seldom exist spontaneously, without conscious effort. [59–65]

From: *Scheherazade; or, the Future of the English Novel*

'John Carruthers' (pseud. of J. Y. T. Greig) (1927)

Headnote

John Carruthers was the pseudonym used for his novels by J. Y. T. Greig (1891–1963), otherwise a British literary scholar and biographer. He served in the First World War as an officer in the Northumberland Fusiliers. After the war, he studied at the University of Glasgow. He taught at Armstrong College in Newcastle-upon-Tyne, then in 1932 moved to the University of Witwatersrand in South Africa as chair of the English Department. His 1931 biography of the Scottish philosopher David Hume won the James Tait Black Award. He also wrote a biography of Thackeray, and a study called *The Psychology of Laughter and Comedy* (1923). He wrote four novels as Carruthers, all close to the date of *Scheherazade*: *The Virgin Wife* (1925); *Adam's Daughter* (1926); *A Man Beset* (1927); and *Lothian Cameron* (1928), also (under his own name) a second To-Day and To-Morrow volume, *Breaking Priscian's Head*; *or, English as she will be spoke and wrote* (1928).

Scheherazade begins with a survey of 'the oddly confusing literary and scientific influences that have been brought to bear on post-war novelists' (16): materialist science; literary giants such as Hardy, Tolstoy, and Dostoevsky; developments in psychology and psycho-analysis. Carruthers is an unusual case of a writer—a novelist as well as a critic and biographer—who has no doubt of the pre-eminence of the experiments by writers now regarded as canonical modernists, such as Proust, Joyce, Woolf, and Dorothy Richardson. *Ulysses* and *Mrs Dalloway* are his pinnacles of modern fiction. He writes with insight and admiration of *To the Lighthouse*, which had just appeared the same year. He hopes their achievements will be carried forward into the novel of the future. But the achievements come at a cost. He feels their aims are flawed; the *cul de sac* of a pre-war subjectivism unable to deliver the masterpieces of the post-war world. His view of what classic modernism lacks offers a provocative and chastening contrast to the hagiographic register of subsequent modernist studies.

******

DOI: 10.4324/9781032625522-30

From Section 6

[. . . .]

A summary statement like this of the chief tendencies that have helped to make our post-war fiction what it is, would be incomplete without references to Chehov, and to the reaction that has set in against the sociological methods of Mr Wells, Mr Galsworthy, and Mr Bennett. It might seem that Chehov had influenced only the short story. That, undoubtedly, he has done. Indeed, one might almost say that he has refashioned the short story for good. But certain post-war authors have adapted his method, or something very like it, to the writing of novels also; so that, instead of a continuous narrative of interconnected events, or a series of careful and detailed psychological analyses of mental processes, they give us a succession of vivid and revealing but seemingly disconnected episodes—such episodes as Baudelaire had in mind when he wrote:

> *Dans certains états de l'ame presque surnaturels, la profondeur de la vie toute entière se révèle dans le spectacle, si ordinaire qu'il soit, qu'on a sous les yeux. Il en devient le symbole.*[1]

Perhaps the best examples in English are to be found in Mrs Woolf's *Jacob's Room*, which is too discrete to be successful, *Mrs Dalloway*, which is a masterpiece of its kind, and *To the Lighthouse*, which is disappointing only because it does not mark a definite advance on *Mrs Dalloway* in the same or a different kind. For we have been led to expect a continuous advance in the work of Mrs Woolf. Her changes of method and the growth of her peculiar powers, from *The Voyage Out* and *Night and Day*, through *Monday and Tuesday*, to the three later novels, have been startling. It would seem that, having gone as far as she could in the psychological manner of Tolstoy and Dostoevsky, and been plainly dissatisfied with the results, she had then set herself to try out the manner of Chehov and Katherine Mansfield. And this manner, notwithstanding all appearances to the contrary, is poetical rather than psychological. It depends, not on detached, intellectual analysis of mental processes, but on the rigorous selection of details of high emotional potentiality. Thoughtful readers have puzzled and puzzled to explain how it is that Mrs Woolf can make quite ordinary life so thrilling, so enthralling; how it is that she can hold our interest through page after page in which hardly anything happens, and in which even the things that do happen are trivial in the extreme. Clarissa Dalloway goes to buy the flowers for her party, sews her green dress, says Thank you to Lucy her maid; Peter Walsh tramps up Whitehall, dawdles in Regent's Park, dines at his hotel; Mrs Ramsay walks into the village with the unpleasant young man Tansley: and it all seems important! It is important, not in itself, but in the moods and

1 [Author's note] I owe this quotation and its application to Chehov, to Mr Middleton Murry's *Discoveries*. See especially pp 141 ff. ['In certain almost supernatural states of soul, the profundity of life in its entirety reveals itself in the spectacle, be it ever so ordinary, before one's eyes. It becomes its symbol'; from *Fusées* in Baudelaire's *Intimate Journals*.]

interpenetrating emotions which Mrs Woolf uses it to evoke. She works embroideries on the commonplace, but it is the embroideries, and not their commonplace ground, that delight us.

It is Mrs Woolf, too, who has most openly expressed the reaction against the sociological methods of the Edwardians. In her *Mr Bennett and Mrs Brown* she begins by accepting a dictum of Mr Bennett's that 'the foundation of good fiction is character creating and nothing else.' Then she quarrels with Mr Bennett, Mr Wells, and Mr Galsworthy because they have failed to do this very thing that matters supremely. She says:

> Mrs Brown is human nature, Mrs Brown changes only on the surface . . . and not one of the Edwardian writers has so much as looked at her. They have looked very powerfully, searchingly, and sympathetically out of the window; at factories, at Utopias, even at the decoration and upholstery of the railway carriage; but never at her, never at life, never at human nature . . . They have laid an enormous stress upon the fabric of things. They have given us a house in the hope that we may be able to deduce the human beings who live there. To give them their due, they have made that house much better worth living in. But if you hold that novels are in the first place about people, and only in the second place about the houses they live in, that is the wrong way to set about it.[2]

An overstatement, certainly, as we should expect in a paper read before a university society and intended to provoke argument. And the implication that might be drawn from it, that Mrs Woolf herself works on the opposite method, looking directly at Mrs Brown and neglecting 'the fabric of things' is untrue. She is much too good a novelist to do anything of the kind. Indeed, the background of London on a summer's day is all-important in *Mrs Dalloway*. It is not the virtual hero of the book, as New York is the virtual hero of Mr John dos Passos's *Manhattan Transfer*; but the characters of Clarissa, Hugh Whitbread, Peter Walsh, Lucrezia Warren Smith, Doris Kilman, and the rest, emerge from the background and fade into it again, and between 'the fabric of things' and the fabric of human nature there is no hard and fast division.

The dissatisfaction with the methods of the Edwardians, however, which many of us feel, and which Mrs Woolf has only exaggerated a little in order to provoke discussion, is significant in this, that it proves but another indication of the trend of modern fiction from objective to subjective, from outer semblance to alleged inner reality.

7

Such being the principal tendencies that have gone to mould the English novel of to-day, what have we reason to expect from the English novel of to-morrow? Much—but only on one condition. The material lying ready to hand is rich and

2 The essay went through several versions—including being given as a paper to Ogden's Heretics Society on 18 May 1924—before being published as a pamphlet by the Woolfs' Hogarth Press as No. 1 of the Hogarth Essays series in 1924.

varied, and of technical ability to shape it there is enough and to spare. One is tempted to say that never before in the history of English or American literature has the material been so rich or the general level of craftsmanship so high. But one thing is lacking, and until that is supplied all of us who are writing novels to-day are only marking time. What we need and must have, if any of our work is to live for more than a publishing season, is a new philosophical synthesis, a new imaginative attitude towards the world. Nothing else will serve.

Fortunately, we need not just fold our hands and wait. The new philosophical synthesis has already been sketched for us, and all we have to do is to lay hold on it intellectually, and then transform it into terms of the creative imagination. The chief of our troubles is that most of us are hopelessly obsolete in our knowledge of science. We are still bogged fast in the dreary mechanical theories of the nineteenth century. The most advanced in appearance and their own estimation are in fact the most backward: the present philosophy of Mr Joyce could hardly have been farther out of date and more reactionary, had he remained a priest of the Roman Catholic Church; and between the philosophy of President Coolidge (supposing him to have one) and that of Mr John dos Passos, there is uncommonly little to choose. For the plain fact is that mechanistic materialism, the philosophical creed that served the turn of Mr Thomas Hardy because he reacted against it with passionate defiance on behalf of mankind, and which contemporary writers accept rather with boredom than with passion of any sort, is dead, stone dead. The leading thinkers of to-day, in metaphysics, psychology, biology, physiology, and mathematical physics, have just quietly killed it and passed on; and the only people old-fashioned enough to believe that it still lives are parsons (who got left behind in the seventeenth century and have never caught up since), chemists (always the worst educated and therefore the most bigoted of the scientists), surgeons (who are much too busy getting fortunes and titles to remember any science worth speaking of), 'advanced' poets anthropologically inclined like Mr T. S. Eliot, and 'advanced' prose-writers fast becoming incoherent like Mr James Joyce and Miss Gertrude Stein.

Let the novelists of to-day once realize this, and realize, besides, all the aesthetic significance that lies behind the alternative philosophy now being presented to us in one or another of its aspects by Professors Alexander, Lloyd Morgan, Whitehead, J. S. Haldane, McDougall, Köhler, Koffka, Eddington, and Rutherford—and English and American fiction will be transformed.[3] A veritable second Renaissance in thought is in progress, here and now, under our noses; and novelists, backward and complacent though they are, cannot remain unaware of it much longer.

3 Samuel Alexander (1859–1938), author of *Space, Time, and Deity* (1920), argued for the emergence of mind from body. Conwy Lloyd Morgan (1852–1936) was a British ethologist and psychologist who also theorized emergent evolution. John Scott Haldane (1860–1936) was a leading scientist and expert in respiration; but Carruthers almost certainly means his son, J. B. S. Haldane. *Scheherazade*, 59, mentions 'the newer *Gestalt* psychology of Köhler and Koffka. Ogden's International Library of Psychology, Philosophy and Scientific Method published *The Growth of the mind: An Introduction to Child Psychology* (1924) and *Principles of Gestalt Psychology* (1935) by Kurt Koffka; and *The Mentality of Apes* (1925) by Wolfgang Köhler.

Compared to it, all the work of Proust, Mr Joyce, Mr Eliot, Mr Conrad Aiken, Mr Sherwood Anderson, and Mr Carl Sandburg, is merest trifling with antiquated themes and methods. For the philosophy that is being propounded now—we may call it indifferently the philosophy of purpose, or of emergent evolution, or of organic realism, or, best of all perhaps, of form (*Gestalt*)—is overwhelming in its implications. Imaginatively apprehended, it changes our whole attitude towards the world; it displaces mechanism by purpose in vital affairs; it reaffirms values that nineteenth-century science did its best to annihilate: most important of all for literature, it reinstates the soul.

Form, shape, pattern, organic purpose—is not this precisely what the novel of to-day so manifestly lacks? Give the novelist, instead of a feeble, hesitant and rather shamefaced 'sense of form,' the burning conviction that form is the first principle of life, nay more, of the world as a whole, and what will he not be able to do? Henry James, Tolstoy, Do[s]toevsky, Professor Freud, Mr Joyce, Miss Richardson, and, to a smaller extent, Marcel Proust, have encouraged him to neglect inherent form, or plot, to abolish the story, to concentrate upon the inner life of relatively isolated characters. Analysis, analysis, analysis—that has been the wellnigh unvarying formula. The gain in subtlety and psychological understanding has been immense, and whatever happens, must be preserved for the future. We shall never go back to plain, untutored objectivity. But extreme subjectivity, aiming sincerely at truth, has missed it again and again, and that for two reasons. It has fostered the twin delusions, that a mind consists of its assembled parts, like a motor-car engine which can be taken to pieces and put together again; and that a single mind can be revealed in isolation from its fellows and unrelated to its inorganic surroundings. Stress the concept of organism instead of the concept of Cartesian matter, as the ultimate category of the world, and both these delusions are at once shown up. And that is exactly what is now being done. The concept of organism, which till recently was extended downwards in the scale of existence to cover nothing more elementary than plants, is now being used as the only possible concept to explain happenings in the inorganic world. The prophecy of Samuel Butler:

> Death is being defeated at all points. No sooner do we think that we have got a *bona fide* barrier than it breaks down. The divisions between varieties, species, genus, all gone; between instinct and reason, gone; between animals and plants, gone; between man and the lower animals, gone; so, ere long, the division between organic and inorganic will go and will take with it the division between mind and matter.[4]—

this prophecy, which at the time it was written must have sounded like another *boutade*[5] by the *enfant terrible* of literature and science, is now being literally fulfilled. A revolutionary change indeed! The world of matter, acting and reacting

4 [Author's note. Butler's] *Note[-]Books* [ed. Henry Festing Jones (London: A. C. Fifield, 1913)], p. 78.
5 French: outburst.

in space *and* time according to the mechanical laws of Newtonian physics, is discerned now to be no more than an abstract and temporary construction of the human intelligence, convenient for certain practical affairs, like engineering, but highly inconvenient or positively false for certain others, like biology or religion or economics or the comprehension of atomic structure.

Now, in the first place, an organism, how lowly soever, is more than the sum of its parts,[6] and its principle of coherence and endurance is its pattern. Without enduring pattern it would be the prey of environment, and its existence would be impossible; with enduring pattern it is able to deal uniformly with its environment, and by impressing itself upon this, attain, and retain, its identity. And in the second place, no organism is isolated. Its unity, its coherence, is not merely intrinsic, but extrinsic also; it is interfused with other organisms. Pattern is not private only, but dependent on the interaction between the thing patterned and other things similarly or differently patterned in space-time. 'Concrete fact', says Professor Whitehead, 'is process'.

> Its primary analysis is into underlying activity of prehension, [by which term Professor Whitehead means 'uncognitive apprehension' or, if you like, unconscious patterning], and into realized prehensive events. Each event is an individual matter of fact issuing from an individualisation of the substrate activity. But individualisation does not mean substantial independence. . . . We can be content with a provisional realism in which nature is conceived as a complex of prehensive unifications.... A prehension is a process of unifying. Accordingly, nature is a process of expansive development, necessarily transitional from prehension to prehension. What is achieved is thereby passed beyond, but it is also retained as having aspects of itself present to prehensions which lie beyond it.[7]

Apply this, not as cold metaphysics intellectually apprehended, but as hot conviction seized and impregnated by imagination, to the art of fiction, and at once the flat inadequacy of our contemporary novels is exposed. Contemporary novelists do not believe with all their heart in the whole, the comprehensive pattern which is made up of lesser patterns interlocking and interfusing, no one of which is sacrificed to any other but each one of which finds its reality and fulfilment in the others; they don't believe in it for fiction, because they don't believe in it for life. Even so delicate and beautiful a piece of work as *Mrs Dalloway* fails to stand the test. Mr Edwin Muir has said of Mrs Virginia Woolf that she is virtually alone among modern novelists in meeting her characters 'on the level,' on the same plane, without a barrier of irony, veiled hostility, or open disgust. 'She might walk into her novels and be at home in them.'[8] We may doubt that. Into *The Voyage Out* and *Night and Day*, yes, she might walk into these; but not into *Mrs Dalloway* or *To*

6 [Author's note] As General Smuts has pointed out (in 'Holism and Evolution') even water is something *more* than two parts of hydrogen and one of oxygen.

7 [Author's note] 'Science and the Modern World,' pp. 99, 101, 102.

8 [Author's note]: Edwin Muir, *Transition* [London: The Hogarth Press, 1926], p. 67.

the Lighthouse. For no mortal can walk into a flat canvas and be at home there. The beauty of *Mrs Dalloway* is the beauty of a two-dimensional picture, 'composed' according to the standards of pictorial art; it is fragile and delightful, as a beautiful picture is; but it lacks entirely just that four-dimensional solidity which human beings live in, that four-dimensional solidity reproduced in a novel by Mr Hardy, and, most wonderfully of all, in a novel by Tolstoy. Look only full-face at Clarissa and the other characters grouped about her, and you are charmed with the subtle beauty of the portraiture; yet you cannot walk round and view them from the varying angles of real life.

How could you, on the method Mrs Woolf has adopted? Faithful to contemporary tradition, she has not constructed a story. Several different stories, disconnected but juxtaposed, emerge fitfully in the course of the book, by means of a sensitive process of feeling backwards—the same sensitive and deliberate *recherche du temps perdu* that was undertaken so memorably by Proust. And though we may admire the brilliance and the delicacy of this searching for the past, with its sure and subtle manipulation of visual images and prose cadences to evoke the appropriate emotions; though we may declare, as I do, that there is nothing finer than this in postwar fiction; and though we may be sure that something of the same power will be carried forward into the novels of to-morrow; yet in the end we must admit that by this method not even Mrs Woolf can make life real and significant in the way it has been made real and significant in the past by novelists of lower gifts. None of the several stories in *Mrs Dalloway* is shaped into a concrete pattern, nor is any necessary, because organic, connection between them to be found. There is no orderly progression according to plan in time, or—to adopt the modern philosophical term that means so much more than it appears to—in space-time. And the reason seems plain. In this and other good contemporary novels the author has consciously or unconsciously accepted the belief, the product of incomplete and abstract scientific theory, that life does not progress according to plan, but just happens. As one of the characters in *To the Lighthouse* expresses it, a little mournfully

> What was it then? What did it mean? Could things thrust their hands up and grip one; could the blade cut; the fist grasp? Was there no safety? No learning by heart of the ways of the world? No guide, no shelter, but all was miracle, and leaping from the pinnacle of a tower into the air? Could it be, even for elderly people, that this was life?—startling, unexpected, unknown.

Well, the belief that life is *only* startling, unexpected, unknown, *only* a leaping from the pinnacle of a tower into the air, is a delusion. People in real life may be permitted to suffer from it, and so may characters in books; but novelists may not. Plan, shape, form, organic pattern, call it what you will—*that* is the ultimate fact about life, and not until novelists apprehend it, not until their apprehension of it possesses them like a mystical revelation of truth, will their work attain to classic rank.

> There might be lovers [we are still among the thoughts of Lily Briscoe in *To the Lighthouse*] whose gift it was to choose out the elements of things and

> place them together and so, giving them a wholeness not theirs in life, make of some scene, or meeting of people (all now gone and separate), one of those globed compacted things over which thought lingers, and love plays.

There might be? There must be. This, just this, must be the aim of the novelist of to-morrow. It will be, for he will have rid himself of the false beliefs oppressing the novelist of to-day, that the wholeness of things is 'not theirs in life,' and that all attempts to achieve wholeness in art must remain at best a sort of pretty-pretty deception. He will believe instead that in shaping his creation into an organic pattern he is working in the very spirit of life itself, and that the 'globed compacted things' he makes contain the same kind of reality, the same kind of truth, as he can find all round him in the world of hard, unadulterate fact. He will know that he selects only in order to reveal what *is*.

Does this mean a turning away again from the subjective to the objective in point of method? Not necessarily. But if it did, what then? Fine work has been done in the objective manner in the past; indeed, is being done now by some of our contemporaries, notably by Miss Willa Cather in America. Consider, for example, this passage from *A Lost Lady*—how effective it is in its stark objectivity, how little it would be improved by the intrusion of any psychological analysis:

> The sky was burning with the soft pink and silver of a cloudless summer dawn. The heavy, bowed grasses splashed him to the knees. All over the marsh, snow-on-the-mountain, globed with dew, made cool sheets of silver, and the swamp milkweed spread its flat, raspberry-coloured clusters. There was an almost religious purity about the fresh morning air, the tender sky, the grass and flowers with the sheen of early dew upon them. There was in all living things something limpid and joyous—like the wet, morning call of the birds, flying up through the unstained atmosphere. Out of the saffron east, a thin, yellow, wine-like sunshine began to gild the fragrant meadows and the glistening tops of the grove. . . .
>
> Under the bluffs that overhung the marsh he came upon thickets of wild roses, with flaming buds, just beginning to open. Where they had opened, their petals were stained with that burning rose-colour which is always gone by noon—a dye made of sunlight and morning and moisture, so intense that it cannot possibly last . . . must fade, like ecstasy. Niel took out his knife and began to cut the stiff stems, crowded with red thorns.
>
> He would make a bouquet for a lovely lady.... He would leave them just outside one of the French windows of her bedroom. When she opened her shutters to let in the light, she would find them—and they would perhaps give her a sudden distaste for coarse worldlings like Frank Ellinger.
>
> After tying his flowers with a twist of meadow grass, he went up the hill through the grove and softly round the still house to the north side of Mrs Forrester's own room, where the door-like green shutters were closed. As he bent to place the flowers on the sill, he heard from within a woman's soft laughter; impatient, indulgent, teasing, eager. Then another laugh, very different, a man's. And it was fat and lazy—ended in something like a yawn.

> Niel found himself at the foot of the hill on the wooden bridge, his face hot, his temples beating, his eyes blind with anger. In his hand he still carried the prickly bunch of wild roses. He threw them over the wire fence into a mudhole the cattle had trampled under the bank of the creek.

There is no need to suppose, however, that the novel of to-morrow will as a rule be starkly objective to this degree. On the contrary, it will almost certainly carry forward all the psychological subtlety and power of analysis that we find in the novel of to-day. But in it analysis will be checked and limited by the design of the whole, will be always controlled by the primary need for synthesis. William James used to call attention to the failure of English to distinguish between *scire* and *cognoscere*, *savoir* and *connaître*, *wissen* and *kennen*.[9] We *know* the transcendental deduction of the categories and the quantum theory, and we also *know* the housefly when we see it; we *know* our near relatives, and we also *know* the man we met yesterday for the first time and spoke but three words to. Obviously, knowledge ranges from the maximum of understanding to mere recognition, and other languages have felt a need for more than one word to cover all its degrees. And so James in his urbane way stressed the distinction between what he called 'knowledge-by-acquaintance' and 'knowledge-about,' a distinction which, admittedly, is only one of degree. Now, it is a strange fact of human experience that too much knowledge-about, if it is exclusively analytical, destroys knowledge-by-acquaintance, because it destroys the thing we were acquainted with. And thus has it happened in only too many psychological novels of to-day and yesterday. Novelists have been so eager to provide knowledge about their characters that they have dissected them out of existence. On a mechanical theory of mind this should not be possible, since the parts have only to be assembled again, and there you are! Yet it is not only possible, but common. And, as we know by experience, most of our reactions to real people and to people in books depend on our knowledge of them by acquaintance. And so they ought to, on any reasonable theory of mind. An organism—and mind is the highest form of organic development that we have experience of—is more than the sum of its parts, and each part being as it were representative of the whole, we need not indefinitely accumulate them in order to get to know it. We may know it, as we say, by intuition. Thus, in writing fiction, the novelist will no longer need to pile detail upon detail till the accumulated weight breaks down our patience. There is no advantage in enormously long books like, for instance, Mr Maurice Baring's *C.* and *Cat's Cradle* or Mr Brett Young's recent *Portrait of Clare*. So long as each part, be it of the book or of a character in the book, is made truly significant of the whole, a very few parts will suffice to give us the necessary knowledge-by-acquaintance, which is important, and too many will bore us without appreciably enlarging our knowledge-about, which in any case is much less important. Accordingly, we may guess that novels in future will tend to be shorter than they have been in the immediate past. Only writers indisputably of the first rank will be allowed to take their

9 Pairs of verbs in Italian, French, and German corresponding to the English verb 'to know,' but distinguishing between understanding, and being familiar with or recognizing.

time, and even they, obviously, will have to prove that they cannot say what they have to say more effectively in 80,000 words than in 300,000.

Knowledge-by-acquaintance is gained most rapidly, and knowledge-about is established most securely, by contrasts and comparisons. And this fact, too, many contemporary novelists seem to have forgotten. They tend to write all the time about the same kind of people, instead of throwing one into sharp and striking contrast with another. Hence their dreariness.

But above all, the organic view of life, applied to the art of fiction, will mean a fresh insistence on the story, on plot. Not on the old, arbitrary, artificial plot, glued to the surfaces of pasteboard figures; not on plot as understood by the average reader or writer of detective novels. Rightly understood, plot is what the characters do to one another; that is, plot is form *par excellence* in the novel. What the characters do to one another depends on three factors: on what they are, in and for themselves; on what they are as ingredients of their environment; and on what they are by reason of the environment's ingression into them. What they are, in and for themselves, will no longer serve alone as material to make novels out of, because, taken alone, it is incomplete, abstract, and to that extent untrue. Mrs Woolf may have done well to complain that the Edwardians missed the essential Mrs Brown by looking only at the house she lived in. But it is even easier to miss the essential Mrs Brown by not looking at her house at all. Her house is an ingredient in the enduring pattern that we call Mrs Brown; and so are Mr Brown her husband, and Johnny Brown her son, and the milkboy who delivers her morning milk, and the lawyer who makes her will, and the undertaker who finally buries her; and so too are the war memorial that she passes (fortunately without looking at it) every day, and the Sunday newspapers that she reads, and the ugly tombstone with which her sorrowing husband disfigures her grave. Mrs Woolf is much too good a novelist to ignore all this in practice. She is always hinting at the nearly invisible threads that bind one character to another and each to the inorganic environment: that is one reason why she is the most important of our post-war novelists. But nearly invisible threads are not enough. The pattern inherent in life, depending on the interrelations of the individual patterns that are individual human beings, must be made visible and significant by creative selection. That, first and last, is the novelist's job.

And the point of view, if it means, as with Henry James and his followers, the point of view of a single character in the book—that, too, must be abandoned, and for the same reason as before. The pattern in only one aspect is false; the true pattern combines different aspects, different prehensions, no one of which must be sacrificed to any other. Whichever way we turn, to this we come back in the end: the imperative need for organic pattern. The lack of it in our post-war culture, and accordingly in our post-war fiction, makes all our technical skill in matters of detail a mere whistling of jigs to a milestone. It will be manifest in the English and American fiction of to-morrow. We can be certain of that. And we can be reasonably certain, too, of the kind of agents it will need, and find—reincarnations of Scheherazade, the Teller of STORIES. [66–95]

From: *Lars Porsena; or, the Future of Swearing and Improper Language*

Robert Graves (1927)

Headnote

Robert von Ranke Graves (1895–1985) was a prolific English poet, historical novelist, and critic. He is particularly known for his war memoir, *Good-Bye to All That* (1929); and his novels about the Roman Emperor, *I, Claudius* (1934) and *Claudius the God* (1935).

He won a classical exhibition to St John's College, Oxford, but enlisted in the Royal Welsh Fusiliers when the war broke out, and fought in the Battle of the Somme. The report of his death turned out to have been exaggerated. After the war he took up his place at Oxford, but switched to English Literature.

In 1926, he took a job as professor of English Literature at Cairo University, accompanied by his wife and their children, but also his lover, the poet Laura Riding. He and Riding soon went to live in Deià, Majorca.

Graves also wrote *Mrs. Fisher; or, the Future of Humour* (1928) for To-day and To-morrow, which, like *Lars Porsena*, turns on the difficulty for future historians of reconstructing topics which depend upon performance, and are frequently unrecorded in writing. It is not hard to see how these projects got him thinking about how to write about earlier periods, such as Ancient Greece and its mythology, or Imperial Rome.

As with Vera Brittain, the high-spirited whimsy of his writing for To-day and To-morrow may surprise readers familiar with the dead-pan trauma of his more famous war reminiscences published soon afterwards.

******

[. . . .]

Might not a useful addition be made to this *To-day and To-morrow* series, by some worthier, more energetic, and more scholarly hand than mine? To be called *Lars Porsena; or The Future of Swearing*. Lars Porsena, if we may trust Lord Macaulay, was more fortunate than ourselves: he had no less than nine gods to swear by, and every one of them in Tarquin's time was taken absolutely seriously. How would the argument run? On the lines perhaps of the following synopsis:

DOI: 10.4324/9781032625522-31

The imaginative decline of popular swearing under industrial standardization and since the popular Education Acts of fifty years ago; the possibility that swearing under an anti-democratic régime will recover its lost prestige as a fine art; following the failure of the Saints and Prophets, and the breakdown of orthodox Heaven and Hell as supreme swearing-stocks, the rich compensation offered by newer semi-religious institutions, such as the "League of Nations" and "International Socialism", and by superstitious objects such as pipes, primroses, black-shirts, and blood-stained banners; the chances of the eventual disappearance of the sex-taboo and of the slur on bastardy, but in the near future the intentional use of Freudian symbols as objurgatory material; the effect on swearing of the gradual spread of spiritistic belief, of new popular diseases such as botulism and sleepy-sickness; of new forms of chemical warfare, of the sanction which the Anglican Church is openly giving to contraception, thereby legitimizing the dissociation of the erotic and progenitive principles and of feminism challenging the view that hard swearing is a proof of virility. Research would be suggested on the variations of taboo in different English-speaking lands[1] on the alliterative emphasis and rhythm of swearing, on the maximum nervous reaction that can be got from a normal subject by combinations and permutations of the oath, the results to be recorded on a highly sensitive kymograph.[2] Finally this valuable and carefully documented work might treat of the prospects of Pure Swearing; by which is not meant sterilized swearing or "Cliff Clawsonism",[3] but Swearing without a practical element, with only a musical relation between the images it employs. Swearing of universal application and eternal beauty, following the recent sentimental cult for Pure Poetry.

"But how is this?" the reader asks. "Isn't what I'm reading called *Lars Porsena, or the Future of Swearing*". I apologize for a little joke, somewhat resembling those advertisements in *Snappy Bits*, which promise erotic delights to any schoolboy who will send five shillings and a statement that he is not a minor: only to job him off with badly printed photographs of classical paintings and statuary—for to send indecent matter by post is illegal. No doubt the Chic-Art Publishing Company would not object to dealing more faithfully with its clients if it could, and perhaps the delight of expectation is worth the ensuing disappointment of only getting the Venus of Milo and a Rubens or two to gloat over. But though a joke is a joke, this volume goes as far as it decently can in containing

1 [Author's note] A man charged recently at Hoxton with using language calculated to make a breach of the peace complained that at Bethnal Green, where he lived, he could have said all that and more with impunity. He suggested a swearing-directory for the London district, which should indicate what you might say where.

2 An antecedent of computer plotters, which drew a graph of information against time, on a drum or roll of paper.

3 On pp. 16–17 Graves explains: 'there are no forms of humour more boring than guaranteed non-alcoholic substitutes for the true wine of swearing "Great Jumping Beans!," "Ye little fishes!," "Snakes and ladders!," and "Mind your step, you irregular old Pentagon!" If Sinclair Lewis has done nothing else in *Martin Arrowsmith*, he has at least nailed up as an abominable type Cliff Clawson, the medical student who indulged perpetually in this form of heartiness.'

at least a few classically draped forecasts and an honest inquiry into the taboos which prevent publication of the real *Lars Porsena*. And, anyhow this is nearest to a *Lars Porsena* that will ever be published. For as soon as there is sufficient weakening of the taboos to permit an accurate and detailed account of swearing and obscenity, then, by that very token, swearing and obscenity can have no future worth prophesying about, but only a past more or less conjectural because undocumented.

Though Samuel Butler's definition of "Nice People" as "people with dirty minds" can be misunderstood by critics who refuse to differentiate between the humorously obscene and the obscenely obscene, I like it. No nice person is uncritical; and yet we are all hedged round with an intricate system of taboos against "obscenity". To consent uncritically to the taboos, which are often grotesque, is as foolish as to reject them uncritically. The nice person is one who good-humouredly criticizes the absurdities of the taboo in good-humoured conversation with intimates; but does not find it necessary to celebrate any black masses as a proof of his emancipation from it. This book is written for the Nice People. Then, though it is in its first intention a detached treatise on swearing and obscenity, it cannot claim a complete innocence of obscenity, while consenting to the publishers' limitations of what is printable and what is not. Observe with what delicacy I have avoided and still avoid writing the words x——and y——, and dance round a great many others of equally wide popular distribution. I have yielded to the society in which I move, which is an obscene society: that is, is acquiesces emotionally in the validity of the taboo, while intellectually objecting to it. I have let a learned counsel go through these pages with a blue pencil and strike through paragraph after paragraph of perfectly clean writing. My only self-justification is that the original manuscript is to be kept safe for a more enlightened posterity in the strong-room of one of our greater libraries. [48–54]

[. . . .]

When a future historian comes to treat of the social-taboos of the nineteenth and twentieth centuries in a fourteen-volume life-work, his theories of the existence of an enormous secret-language of bawdry and an immense oral literature of obscene stories and rhymes known, in various degrees of initiation, to every man and woman in the country, yet never consigned to writing or openly admitted as existing, will be treated as a chimerical notion by the enlightened age in which he writes. As Sir James Frazer took, as the text for his inquiries, the Golden Bough legend of Aricia and the primitive ceremonies there surviving until Imperial times,[4] so this new Sir James may take *The Bottom Legend* recorded by a contemporary historian Roberts as his text. As follows:

'Shortly before the "Great War for Civilization" (the indecisive conflict, 1914–1918, between rival European confederations to decide which was to have the right of defining Civilization) there was a student at Oxford University famous for his

4 Frazer's *The Golden Bough: A Study in Comparative Religion* was first published in two volumes in 1890; It was retitled *The Golden Bough: A Study in Magic and Religion* in its 3-volume second edition (1900), and grew to 12 volumes for its third edition (1906–1915).

"practical joking."[5] He is said to have been one of the rare persons of the day to whom a peculiar licence was given for such "practical joking" and for deriding the most sacred taboos of the time. It was he who first defiled a local altar, "The Martyr's Memorial", by climbing to the very summit at night-time and planting a chamber-pot—a stringently taboo'd vessel—on the cross which crowned it. The civic authorities had great difficulty in removing this scandalous object, because climbing the Memorial was no easy feat, and the chamber-pot, being made of enamelware and not, as was first thought, of porcelain, could not be dislodged by rifle fire. On another occasion, this same student is said to have impersonated an African potentate and, with a suite of disguised companions, to have been officially welcomed with a Royal Salute aboard a battleship of the English Navy, and to have aggravated this quasi-blasphemous performance (for the Fleet was a religious institution of greater dignity and efficiency than the Church itself) by the bestowal of medals on the ship's officers.

'But the most interesting breach of taboo with which he is credited was a dinner-party which he gave at a Cathedral town in the Midlands. He spent over a year, and a great deal of money, in scraping acquaintance under an assumed name with every person in the town whose surname contained the syllable "bottom": Ramsbottom, Longbottom, Sidebottom, Winterbottom, Higginbottom, Whethambottom, Bottomwhetham, Bottomwallop, Bottomley, and plain Bottom; he insinuated himself into the friendship of every one of these families, but separately, without allowing them to meet in his presence, until finally he was able to invite them all together to a huge dinner-party at his hotel. When each name in turn had been announced by a particularly loud-voiced hotel-servant, he withdrew, promising to return in a few minutes, and begging them to begin dinner without him. The meal consisted merely of rump-steak, and the host was already in a railway train, riding swiftly towards London, and leaving no address.

'This story is regarded by Roberts and others as a most amusing one, though the point of the joke will need explaining to readers of this thirtieth century.

'Apparently "bottom" was the common equivalent, in the secret language which I postulate, of the word "buttocks" Now, among primitive peoples *no man will utter common words which coincide with or merely resemble in sound taboo[e]d names*, and, though the twentieth century refused to admit itself primitive, we cannot now understand on what grounds this refusal could have been plausibly justified. The principle I have italicized is a direct quotation from a contemporary treatise on taboo.[6] The author, whose name has been lost with the title-page of the unique copy in the Jerusalem Library, was only able to state this principle in the case of

5 Horace De Vere Cole (1881–1936). See Richard Davenport-Hines' unparalleled entry in the *Oxford DNB*: 'Cole's most famous hoax (10 February 1910) was perpetrated in concert with Duncan Grant, Adrian Stephen and his sister Virginia Woolf, and others. Together they impersonated the emperor of Abyssinia and his suite on a stately visit to HMS Dreadnought commanded by Admiral Sir William May at Weymouth.'

6 In fact it is quoted (with omission) from *The Golden Bough*; the volume *Taboo: and the Perils of the Soul*, from the third edition (London: Macmillan, 1911), Chapter 6, 'Tabooed Words', 376.

the South African Zulus and other savage tribes; but there is little doubt in my mind that the point of the joke lay in the sensitivity of the Bottom families to the obscene connotations of their name. That the buttocks should have been taboo'd is a surprising idea, but apparently a morbid prolongation of the lavatory-taboo accounts for it: or so Mannheim holds. The Bottom names either had no original connexion with the buttocks as in Bottomwallop, which is a geographical name, or, as in Longbottom, they were inherited from an age when the taboo had not yet hardened. Be that as it may, the unfortunates who were born at this period to a name containing the taboo'd syllable were in a quandary. If they changed their names by Deed Poll, the expense and embarrassment would be considerable. Yet not to change meant that they would continue to be aware of repressed snickering wherever they went beyond the immediate circle of their friends. Most of them, therefore, changed the spelling merely from "Bottom" to "Botham", and thus thought to circumvent the taboo. Indeed, as Roberts tells the story, the Bottom-guests were all disguised as Bothams or Bottomes. One family, the Sidebottoms or Sidebothams, went so far as to pronounce their name "Siddybotaam" and in Bigland's *Life and Times of H. Bottomley* (1954) there is mention of one of these "Siddybotaams" to whom Bottomley (a famous practical joker)[7] is said to have introduced himself as "H. Bumley, Esq.", "bum" being a common, but strongly taboo'd, shortening of "bottom".

'Now, the secret language, which was generally known as "smut"—possibly the idea of defilement is latent in this word, since another synonym was "The Dirty Talk" or "The Foul Language" was so rich in its vocabulary, and drew so copiously on the legitimate language for secret obscene usages of common words, that the greatest ingenuity was needed in legitimate speech to avoid the appearance of obscenity. Thus so common a word as "bottom" meaning a *base*, a *bed*, a *fundament*, a *cause*, owing to its use in smut as an equivalent for "buttocks", could never be used in the legitimate language in any context where a *double entendre* might be understood. The word "parts" becoming a synonym in Smut of the organs of generation had to be used with great care, and these are merely two isolated instances of a principle so strong that when two persons who had been initiated into the third or fourth degree of the secret language began a conversation, practically not a single phrase could be used by them without this *double entendre*, causing hysterical laughter. *And not merely the names themselves but any words that sound like them are scrupulously avoided, and other words used in their place. A custom of this sort, it is plain, may easily be a potent agent of change in language, for, where it prevails to any considerable extent, many words must constantly become obsolete and new ones spring up.*[8]

7 A jab at Horatio Bottomley (1860–1933), a financier, journalist, editor of *John Bull* magazine, newspaper proprietor, and M.P. When his business associate Reuben Bigland accused him in print of swindling, Bottomley sued for libel; the case leading to his own trial for fraud. In 1922 he was imprisoned and expelled from Parliament. On his release in 1927 he made a living lecturing and appearing in music halls.

8 Frazer, *Taboo: and the Perils of the Soul*, conflating sentences from pp. 342, 358.

'This is a quotation from the same anonymous ethnologist, who is here discussing the taboos in Melanesia and Australia on the mention of the names of certain relatives, whether dead or alive, but it also explains many linguistic changes in the vocabulary of the nineteenth, twentieth, and twenty-first centuries: for instance, the constant out-of-dating of popular equivalents to the words "whore" and "harlot" which being Biblical alone remained in constant use as pure descriptive terms; and the disappearance from common use of the phrase "a man of parts", meaning "a man of great attainments", and the phrase "he (or she) has no bottom", meaning that the person referred to has no stability of character. It will be seen that this furtive language must have had a great influence on the legitimate language. [55–63]

[. . . .]

To conclude, swearing as an art is at present in low water. National passion seldom runs high, invention is numbed, and there is no appeal of a politico-religious nature which will meet everywhere with the same respect. The only taboo strong enough to be worth breaking is the sexual one, and swearing shows every sign of continuing standardized on that basis for some time. It may be that "bastard", and similar words, may gradually creep into legitimate speech, but only because obscener equivalents have been found.

The only really effective form of swearing that I know is this: Suppose you quarrel violently with a fellow-traveller in a crowded railway-carriage, perhaps about opening windows or the disposition of luggage. You get worsted. "Very well", you say, with a sigh, "have it your own way." "By the way", you add, with a peculiar intensity, "I happen to know that in three weeks' time you will have a dangerous illness." If the quarrel has been very violent, you may even sentence your adversary to death.

You have not used obscene or threatening language, or expressed a wish that your adversary should suffer. You have not used God's name. If you had done any of these things you would not only be putting yourself in danger of prosecution and alienating the sympathy of the other travellers, but you would further be weakening the effect of your curse. "God damn you", says Jones to Brown. Brown says to himself: "Good; Jones is thoroughly annoyed with me, and afraid to do anything but curse." And Brown considers himself on good terms with God, and cannot imagine the latter being influenced by any angry petitions of Jones. But "You will have a dangerous illness in three weeks' time" is a different matter. For all the traveller knows, you may be a specialist, giving a free diagnosis of his condition. Pride will keep him from asking you on what grounds you said what you did. If he does ask, he cannot force a reply from you without assault. Keep silence for the rest of the journey, and watch his nerves gradually go. He is pinned in that corner-seat with you opposite him: he has no refuge from your curse because he does not understand it. The more often he tells himself that he should pay no attention to you, the more irritating will be the superstitious reactions. When eventually you part, he takes the curse home with him—not your curse, but his own. For this is an individualistic age: the community has little power over the individual, and, if you would curse effectively, it must not be done in the name of the community or the formula of the community. You must put it into your adversary's mind to curse himself with his own fears. "Injuries only come from the heart" quoth my Uncle Toby.[9]

A final word and a most important one. No critic of this essay will be satisfied unless fuller mention is made of James Joyce's *Ulysses* than has here been given. But they must remain unsatisfied. Though *Ulysses* could be studied as a complete manual of contemporary obscenity, such a study will get no encouragement here. It is true that *Ulysses* is forbidden publication in England as indecent and that it contains more words classified by law as indecent than any other work published this century; but on the other hand it also contains more obscure poetic and religious references than any other work published this century and the choice of language in the blameless passages is as scholarly as Mr. Sainsbury's[10] and as English as Charles Doughty's. So far from being a work of merely pornographic intention or even a serious work given the pornographic sugar-coating that Rabelais gave his politico-philosophic pills, it is a deadly serious work in which obscenity is anatomized as it has never been anatomized before. To call Joyce obscene is like calling the Shakespeare of the *Sonnets* lustful: true, both have had the intimate experiences that their writing implies, but Joyce has brought himself as far beyond obscenity as Shakespeare got beyond the lust of which he makes frank confession. Bloom, gross obscenity incarnate, is presented in *Ulysses* directly without the prejudice of tenderness or harshness. Stephen Daedalus[11] whose early history had been given (semi-autobiographically) in "A Portrait of the Artist as a Young Man" is presented as a type of the oversophisticated intellectual, a poet who has failed as a poet because he is unable to find any strong enough reality to make foundation for his poetry. In the contemporary religious and literary scene, though a man of strong natural religious feelings and great literary capacity, he finds only emptiness. Irish nationalist politics are no better. The only life that has any appearance of reality to him is the obscene life as lived by Bloom, the middle-aged married commercial traveller and by Mulligan, a forceful young medical student who lodges with Daedalus. Daedalus, who makes his living by schoolmastering in an old-fashioned school, is philosophically inclined to the obscene life because Bloom and Mulligan, who live it seriously, are in this respect at least superior to the priests, the schoolmasters, and the little Celtic-Twilight poets (Joyce himself began as one) whose lives have no such absorption in a ruling idea. Yet as a sensitive person Daedalus is utterly repelled by the badness and rankness which obscenity exudes; and in the spiritual conflict between an artist's love of reality and an artist's hatred of obscenity the plot of the book lies. The only character in the book with whom Daedalus has a strong natural sympathy is his father, the only one man who is able to harmonize reli-

9 Laurence Sterne, *The Life and Opinions of Tristram Shandy, Gentleman* (London: Dent, 1912), 121.

10 Presumably the conservative academic George Saintsbury (1845–1933).

11 Graves gives the name its mythological spelling, rather than Joyce's version, 'Dedalus.'

gion, politics, and obscenity into something like an artistic reality. Old Daedalus swears admirably. Though most of his oaths are on the censored list there is no disgust stirred by them:

> A tall black-bearded figure, bent of a stick, stumping round the corner of Elvery's elephant-house showed them a curved hand on his spine.
>
> "In all his pristine beauty," Mr. Power said. Mr. Daedalus looked after the stumping figure and said mildly: "The devil break the hasp of your back!"

But Stephen has a bitter quarrel with his father since his mother's death, and anyhow finds no sympathy in him for the intellectual sophistication which is one of the chief causes of unrest. The book rises to a scream of dreadful pain when we come on Stephen drunk in Mabbot Street in company with Bloom, a bawd-mistress, and several harlots, two English private soldiers, and a whole fantastic crowd of the imaginary characters of Stephen's brain: dying away in a monstrously droned account of the trivialities of lust and obscenity to which early middle-age has brought Bloom and his wife.

It is quite right that *Ulysses* should be censored, since its chief public in England could at the best of times be only an obscene one: and it is not an obscene book, but on the contrary perhaps the least obscene book ever published: that is why it is censored. And there is every reason why Shakespeare's sonnets should be censored at the same time, and more strictly, because the public even blinds its eyes to the painful history that the sequence gives and makes it 'extravagant flattery of a patron' or 'an academic exercise.' Joyce is read as obscene instead of successfully past obscenity: Shakespeare instead of being read as past lust is not even read as lusting. [85–94]

Index

advertising 28, 50, 57
Aeolus (Oliver Stewart) 8
aeroplane 163–7, 283
aesthetics 5, 36, 42
Africa 14, 56, 230, 307, 412
AI 6, 11, 13, 14, 150, 151–2
anthropology 47–8, 55–8, 173, 411–12
Aphrodite (Ralph de Pomerai) 4, 6, 7, 11, 12, 14, 307–29
Apollinaire, Guillaume 373, 377–8
Archimedes (L. L. Whyte) 6, 14, 24, 25, 79–105
Aristotle 34, 124, 141, 363, 366
art 5–6, 7, 27, 33, 87, 100–1, 105, 116, 128, 132, 280, 304, 318, 341, 345–6, 358, 360, 365–6, 368, 371–90, 392–8
artificial human 92–7, 123–4
astronomy 21, 88–92, 108–9, 173–81, 215, 216, 218–19, 321
atom 14, 80, 91, 101, 216
Augé, Marc 10
Automaton (H. S. Hatfield) 11, 13, 149–72
automobile 283, 323, 403

BASIC English 1, 160
behaviorism 98, 360, 368
Bergson, Henri 84–9
Bernal, J. D. 3, 6–7, 10, 42, 77, 106–37
'Bernal Sphere' 112–13
biochemistry 28
biology 23, 31–45, 47, 80–1, 86, 102–3, 128, 139–48, 404
birth control 32, 48, 55–6, 75, 188, 190, 200, 206, 225–7, 236, 240, 243–4, 251–2, 258, 315, 321–2
body 37–9, 109, 117–24, 130, 135, 151–2, 195–206, 243, 279, 285–6
Bohr, Niels 81, 216
brain 9, 13, 68, 73, 81, 91–2, 98–9, 103, 108, 119–24, 149–52, 171, 181, 216–17, 280, 285
Brave New World 12, 353
Brittain, Vera 2–3, 8, 13, 14, 237–75, 409
Butler, Samuel 21, 28, 342, 403, 411

Callinicus (J. B. S. Haldane) 7, 20
censorship 243, 245, 311, 367, 415–16
Chekhov, Anton 376–8, 381, 386, 390, 400
chemistry 23, 28–9, 47, 81, 109, 110
Chesterton, G. K. 22, 201
China 294
Chiron (M. Chaning Pearce) 11, 14, 296–306
Churchill, Winston 7
cinema 12, 21, 51–2, 65–6, 73, 259, 278, 283, 288, 323, 380, 395
civilization 5, 22, 26, 38, 54, 59, 69–70, 171–2, 190, 199, 203, 205, 265, 270, 274, 282–95, 299, 301, 323, 328, 389, 412
class 49, 54, 134–5, 147, 168, 187–9, 193–4, 196, 206–8, 226, 233, 244, 261, 264, 298–301, 334, 376
classics 28, 193, 211, 298, 339, 367–8, 375, 405
communication 25–6, 70, 162–3, 259, 278, 280, 349
communism 20
complexity 94, 98–9, 108–9, 119, 279, 313, 316, 364, 379, 381, 388
computer 6–7, 11, 13, 14, 163
consciousness 92, 94, 96–101, 105, 124, 279–80, 287, 291, 299, 360, 377, 387; *see also* hive mind
creativity 126, 318, 323, 347
crime 127, 208, 220, 226, 228, 231, 248–50, 311, 325

culture 13, 14, 298, 328
Cummings, E. E. 379, 382–3, 389
cyborg 6, 13, 118–24

Daedalus 32–3, 45, 46
Daedalus (J. B. S. Haldane) 1, 7, 12, 13, 14, 19–45, 46, 135, 261, 267–8, 335, 343
Darwin, Charles 34, 38, 41, 47, 80, 94
de Pomerai, Ralph 4, 7, 13, 307–29
death 35, 37, 39, 43, 55, 117–119, 219, 279, 316
desire 107–8, 127–9, 222, 262, 273
divorce 227, 242, 252–3, 311
DNA 11
Dobrée, Bonamy 9, 353–70
Dostoevsky, Fyodor 376–8, 381, 389, 390, 399–400, 403

economics 47–9, 54
ectogenesis 37–8, 120, 190, 268–9
Eddington, Arthur 84–9, 402
education 4, 5, 50, 136–7, 140–1, 193, 270–1, 286, 288, 296–306, 389, 410
Einstein, Albert 7, 23–4, 84–9, 90, 95
electron 90–1, 101–2, 216, 218
Eliot, T. S. 3, 7, 8, 10, 12, 13, 374, 388, 402–3
Ellis, Havelock 244, 316, 365
Empire 127
environment 36
Eos (James Jeans) 12, 173–81
eugenics 4, 14, 31, 35, 48, 56–7, 117, 145–8, 273, 350
Euterpe (Lionel McColvin) 6, 13, 15, 392–8
evolution 4, 20, 80, 96, 130, 133, 291, 321

fascism 334, 339, 410
federalism 301–3
feminism 13, 185–214, 247, 249, 251, 263, 268
flight 9, 23, 32, 33, 75, 77, 260–2, 374, 381
food 30, 110–11, 125, 130
form (*Gestalt*) 403, 405, 408
Forster, E. M. 12
Fournier d'Albe, E. E. 156, 158, 277–81
Freud, Sigmund 78, 128, 314–15, 328, 388, 403, 410
Fukuyama, Francis 10
The Future of Futurism (John Rodker) 13, 14, 371–90

Gallio (J. W. N. Sullivan) 6
gender 343, 355
genetics 14, 20, 33, 36, 103, 139–48
Gibbon, Lewis Grassic 9
glands 38, 57, 59, 374, 387
The Golden Bough (J. G. Frazer) 342, 411–13
gramophone 66, 152, 323, 358, 363–4, 394, 396–7
Graves, Robert 3, 8, 13, 14, 366, 409–16

Haire, Norman 185, 246, 250, 261, 267
Halcyon (Vera Brittain) 2, 11, 13, 237–75
Haldane, J. B. S. 1, 4, 7, 14, 19–45, 46, 48, 117, 120, 135, 261, 267–8, 359, 402
Hall, Radclyffe 245
Hanno (J. Leslie Mitchell) 9
happiness 46, 48, 231–6
Hatfield, H. Stafford 149–72
Heard, Gerald 8, 318
Heretics, Discussion Society 2, 19
history 14, 35, 44, 108
hive mind 122–3, 132–3
homosexuality 248, 263, 270
hormones 39, 267
Huxley, Aldous 2, 7, 129, 325, 353
Huxley, Julian 310
Huxley, T. H. 44, 341
Hypatia (Dora Russell) 2, 4, 7, 12, 13, 14, 185–214, 242, 296
hypnotism 40

Icarus 33
Icarus (Bertrand Russell) 2, 7, 10, 12, 19, 46–59, 135, 147, 215, 335, 374
idealism 24
image recognition 157–8
Imagism 371, 377
India 21, 35, 229, 294
intelligence 2, 3, 4, 20, 56–8, 118, 125, 132, 134, 187, 191, 198–9, 201, 206–7, 211, 229–30, 234, 241, 264, 266, 273–4, 333–52, 363, 387
International Library of Psychology, Philosophy and Scientific Method (ed. Ogden) 1, 128, 133, 239, 386, 402
internet 13
invention 2, 7, 10–11, 22–3, 28, 31–3, 38, 43, 50–2, 60–1, 70, 75, 147, 149, 150, 154, 159, 160, 164, 169, 170–1, 258, 268, 278, 284, 319, 369, 388, 394–5

James, Henry 379, 403
Jeans, Sir James 3, 116–17, 173–81
Jennings, H. S. 13, 139–48, 202

Joyce, James 7, 13, 375, 379, 389, 399, 402–3, 415–16

Kalki (S. Radhakrishnan) 12, 14, 282–95
Kellogg Pact 294–5, 300

Lady Chatterley's Lover (D. H. Lawrence) 12
language 1, 12, 25, 126, 160, 278–9, 304, 338, 371, 378–9, 382, 384–6, 407, 409–16
Lars Porsena (Robert Graves) 13, 409–16
Lautréamont, Comte de 380, 385
Lawrence, D. H. 12, 128–9, 387
League of Nations 42, 53, 94, 232, 265, 294–5, 299, 302, 304, 410
Lee, Vernon 3, 13–14, 333–52
Lewis, Percy Wyndham 378, 382
liberalism 53
life 79–105, 178–81
life force (élan vital; vitalism) 118–19, 281, 317, 337, 354
light 25, 69–71, 73, 77, 80, 86, 91, 105, 111, 124, 156–8, 162, 176, 216, 248
literature 27, 41, 333–416
Low, A. M. 13, 60–78, 119
Ludovici, Anthony 189, 206, 263–4
Lysistrata (Anthony Ludovici) 7, 189–90, 193, 196, 206, 263–4

Macdonald, Ramsay 6
machine (mechanical) 129, 130, 135, 323, 373, 377, 394–7, 402, 407
malevolence 2, 58, 231–4
Mallarmé, Stéphane 378, 380–1, 384, 386, 390
Marinetti, Filippo Tommaso 372, 388
marriage 197–200, 237–75, 307–29
Marx, Karl 25, 40, 43
maternity leave 209, 258
Maurois, André 3
Maxwell, James Clerk 95, 111
McColvin, Lionel 6, 13, 392–8
media 13, 50–2, 187–8, 240, 278, 377
medicine 34–5, 39, 48, 55, 81, 103, 142, 144, 219, 220, 244, 279, 414
Mendel, Gregor 33–4, 47, 103, 139, 140–1, 340
modernism 5, 8
money 188, 201, 208–9, 288, 309, 336
morality 43, 191, 197, 213, 220–9, 231, 238, 241, 243, 245, 247, 259, 286–7, 292, 307, 313, 322, 327–9, 341–5
Murry, John Middleton 6, 400
Mussolini 9

Narcissus (Gerald Heard) 8
nationalism 47, 49–50, 52–3, 293–4, 339
Newton, Isaac 24, 82–3, 86, 90, 95, 97, 102, 404
Nietzsche, Friedrich 100, 335, 341, 349
novel 288, 399–408

Ogden, C. K. 1, 3–4, 5, 7, 25, 160, 386, 401
Orwell, George 3
Otherness 337–8, 340, 345

philosophy 41, 96–101, 219, 284, 299, 334, 402, 404–5
photo-electric effect 4, 71, 156, 158–9, 170
photography 71, 73
physics 4, 23, 25, 47, 48–50, 71, 79–105, 110, 126, 216, 219
Picasso, Pablo 375
Planck, Max 25, 91
Plato 31, 228–9, 346
poetry 13, 27–8, 42, 45, 47, 120, 128, 147–8, 198, 213, 230, 243, 288, 308, 324, 338, 345, 348, 358, 363, 365–6, 371–2, 375, 378, 381–2, 386–7, 393, 402, 409–10, 415
politics 13, 22, 25, 35, 41, 50–52, 68, 76, 115, 127, 136–7, 187, 212–13, 233, 248–51, 284, 294, 297–8, 305–6, 307, 359
Pound, Ezra 13, 383, 388
power 25–7, 43–4, 48, 57–9, 168, 280, 288, 298
prediction 107–10, 128, 215, 336; *see also* prophecy
Prometheus 32, 202
Prometheus (H. S. Jennings) 13, 139–48, 202
prophecy 3, 10, 14, 23, 25, 41, 42, 61, 80, 105, 107–10, 129, 130, 157, 277, 351, 403, 410; *see also* prediction
prostitution 231, 239, 248, 256, 260–1, 263, 272, 322, 325
Proteus (Vernon Lee) 13, 14, 333–52
Proust, Marcel 399, 403
psychical research 40, 217–18
psychology 4, 20, 39–40, 47, 57–9, 80–1, 87, 96–101, 103–5, 107–37, 161, 199, 206, 234–5, 307, 313–15, 342, 358, 360, 364, 366, 368, 374, 381, 383, 397, 399, 406

quantum mechanics 4, 6, 25, 87, 99, 101, 110
Quo Vadimus? (E. E. Fournier d'Albe) 12, 13, 277–81

race 142–3, 231, 279, 284, 287, 298, 306, 350, 376
radiation 4, 21, 27, 87–8, 91, 96–7, 101, 124, 178
radio 23, 50, 60–78, 94, 115, 119, 122, 157, 164, 174, 176, 258, 278, 293, 323, 394, 396–7
reason 40–1, 48, 51, 58, 127–8, 132, 214, 288–9, 340, 349–50, 352, 403
recording 151, 160–1, 170, 278, 362–3, 370
rejuvenation 103, 266–7
relativity 4, 87–8, 99, 176, 354
religion 5, 20, 32–3, 40, 42, 44, 50, 87, 103, 105, 107, 128, 197, 199, 200, 204, 207–8, 212–14, 218–19, 224–5, 228–30, 242, 253, 280, 284, 287–92, 299, 305, 309–10, 315, 320–1, 327–9, 339, 342, 344, 346, 404, 410
reproduction 1, 66, 308, 316–17, 393–7
Richards, I. A. 1, 6, 359, 367, 386
Richardson, Dorothy 399, 403
Rivers, W. H. R. 47, 315
Rodker, John 371–90
Roussel, Raymond 378, 380–1, 387–8, 390
Russell, Bertrand 1–2, 4, 10, 11, 46–59, 126, 129, 135, 147, 215–36, 242, 258, 318
Russell, Dora 2, 7, 11, 13, 185–214, 242, 296
Russia 22
Rutherford, Ernest 81, 402

Scheherazade ('John Carruthers') 12, 14, 399–408
Schrödinger, Erwin 319
science 1, 2, 4–5, 6, 7, 13, 20–181, 189–90, 193, 195, 199, 201, 203–6, 216–18, 220, 225, 231–6, 272, 278, 280, 282, 288, 298–9, 301, 304–5, 321, 334, 336, 339–40, 342, 347–9, 351–2, 360, 386–8, 395, 397, 399, 403, 405
senses 68–9, 74, 78, 88, 91, 121, 123, 150, 278, 285, 290–1, 311, 324–6, 374
sex 1, 4, 6, 13, 14, 31, 37, 128, 132–3, 185–214, 225, 239–75, 285, 307–29, 374, 414
Shakespeare 192, 362–3, 415–16
Shaw, George Bernard 119, 128, 326, 349, 350
Shelley, Percy Bysshe 28, 230, 320
social sciences 5, 7, 13, 47, 51, 298, 301, 393
socialism 7, 24–5, 339, 410
solar power 112–13, 170
Solnit, Rebecca 10
sound 156, 158, 159, 357
space 6, 9, 13, 23–4, 27, 31, 41, 70, 77, 88–9, 99–100, 103, 111–17, 130, 135, 172, 173–81, 216, 218, 259, 261, 272, 281, 404–5
Stein, Gertrude 13, 373, 378–9, 388, 402
Stewart, Oliver 8
Stopes, Marie 245, 316, 320
Stravinsky, Igor 372–3, 375–6
Sullivan, J. W. N. 6, 12, 87, 319

taboo 310, 312–13, 321, 323–4, 375, 410–14
telegraph 28, 50, 278, 362, 381
telephone 13, 50, 62–3, 66–7, 70, 169–70, 238, 260, 278, 287, 293, 362, 367–8
television 69–75, 122, 259
theatre 353–70, 395
thermodynamics 116, 170
time 4, 23–4, 27, 42, 77, 82, 86–9, 92–7, 99–100, 104, 111, 123, 130, 168–9, 172, 173–5, 178–9, 216, 259, 404–5
The Time Machine (H. G. Wells) 12, 353–4
Timotheus (Bonamy Dobrée) 9, 12–15, 353–70
travel 25, 50, 159, 165, 167–8, 254, 259–60, 278–80, 306, 349, 381, 414

Ulysses (James Joyce) 12, 13

voice recognition 160
vorticism 371–2, 377

war 2, 8, 9, 20–1, 24, 42–3, 48–50, 52, 55–6, 67, 74–6, 94, 103, 105, 125, 127, 143, 186–91, 200, 201, 206–8, 211, 231–3, 240–3, 249, 262, 265–6, 270, 278–9, 286–8, 293–4, 299, 303, 307, 323, 345, 349–50, 357, 359, 372, 409, 411
Waugh, Evelyn 7, 11
Wells, H. G. 3, 8, 9, 23, 42, 77, 242, 261, 317, 333, 351, 353–4, 374, 400–1
West, Rebecca 11
Westermarck, Edward 238–9, 262–3, 271, 318
What I Believe (Bertrand Russell) 215–36, 258
Whitehead, Alfred North 80, 84–9, 91, 320, 404

Whyte, L. L. 6, 79–105, 123
Wilson, Woodrow 53
wind-power 37
wireless *see* radio
Wireless Possibilities (A. M. Low) 13, 14, 60–78, 160
Wittgenstein, Ludwig 1
Woolf, Virginia 7, 12, 399–401, 404
The World, the Flesh and the Devil (J. D. Bernal) 6–7, 11, 13, 14, 106–37
world state 42, 54, 56, 59, 279, 284
Wyndham Lewis, Percy 7, 12

X-rays 4, 70, 110, 119, 279

Yeats, W. B. 30, 243